Lakhmi C. Jain, Mika Sato-Ilic, Maria Virvou, George A. Tsihrintzis,
Valentina Emilia Balas, and Canicious Abeynayake (Eds.)

Computational Intelligence Paradigms

Studies in Computational Intelligence, Volume 137

Editor-in-Chief
Prof. Janusz Kacprzyk
Systems Research Institute
Polish Academy of Sciences
ul. Newelska 6
01-447 Warsaw
Poland
E-mail: kacprzyk@ibspan.waw.pl

Further volumes of this series can be found on our homepage: springer.com

Vol. 116. Ying Liu, Aixin Sun, Han Tong Loh, Wen Feng Lu and Ee-Peng Lim (Eds.)
Advances of Computational Intelligence in Industrial Systems, 2008
ISBN 978-3-540-78296-4

Vol. 117. Da Ruan, Frank Hardeman and Klaas van der Meer (Eds.)
Intelligent Decision and Policy Making Support Systems, 2008
ISBN 978-3-540-78306-0

Vol. 118. Tsau Young Lin, Ying Xie, Anita Wasilewska and Churn-Jung Liau (Eds.)
Data Mining: Foundations and Practice, 2008
ISBN 978-3-540-78487-6

Vol. 119. Slawomir Wiak, Andrzej Krawczyk and Ivo Dolezel (Eds.)
Intelligent Computer Techniques in Applied Electromagnetics, 2008
ISBN 978-3-540-78489-0

Vol. 120. George A. Tsihrintzis and Lakhmi C. Jain (Eds.)
Multimedia Interactive Services in Intelligent Environments, 2008
ISBN 978-3-540-78491-3

Vol. 121. Nadia Nedjah, Leandro dos Santos Coelho and Luiza de Macedo Mourelle (Eds.)
Quantum Inspired Intelligent Systems, 2008
ISBN 978-3-540-78531-6

Vol. 122. Tomasz G. Smolinski, Mariofanna G. Milanova and Aboul-Ella Hassanien (Eds.)
Applications of Computational Intelligence in Biology, 2008
ISBN 978-3-540-78533-0

Vol. 123. Shuichi Iwata, Yukio Ohsawa, Shusaku Tsumoto, Ning Zhong, Yong Shi and Lorenzo Magnani (Eds.)
Communications and Discoveries from Multidisciplinary Data, 2008
ISBN 978-3-540-78732-7

Vol. 124. Ricardo Zavala Yoe
Modelling and Control of Dynamical Systems: Numerical Implementation in a Behavioral Framework, 2008
ISBN 978-3-540-78734-1

Vol. 125. Larry Bull, Bernadó-Mansilla Ester and John Holmes (Eds.)
Learning Classifier Systems in Data Mining, 2008
ISBN 978-3-540-78978-9

Vol. 126. Oleg Okun and Giorgio Valentini (Eds.)
Supervised and Unsupervised Ensemble Methods and their Applications, 2008
ISBN 978-3-540-78980-2

Vol. 127. Régie Gras, Einoshin Suzuki, Fabrice Guillet and Filippo Spagnolo (Eds.)
Statistical Implicative Analysis, 2008
ISBN 978-3-540-78982-6

Vol. 128. Fatos Xhafa and Ajith Abraham (Eds.)
Metaheuristics for Scheduling in Industrial and Manufacturing Applications, 2008
ISBN 978-3-540-78984-0

Vol. 129. Natalio Krasnogor, Giuseppe Nicosia, Mario Pavone and David Pelta (Eds.)
Nature Inspired Cooperative Strategies for Optimization (NICSO 2007), 2008
ISBN 978-3-540-78986-4

Vol. 130. Richi Nayak, Nikhil Ichalkaranje and Lakhmi C. Jain (Eds.)
Evolution of the Web in Artificial Intelligence Environments, 2008
ISBN 978-3-540-79139-3

Vol. 131. Roger Lee and Haeng-Kon Kim (Eds.)
Computer and Information Science, 2008
ISBN 978-3-540-79186-7

Vol. 132. Danil Prokhorov (Ed.)
Computational Intelligence in Automotive Applications, 2008
ISBN 978-3-540-79256-7

Vol. 133. Manuel Graña and Richard J. Duro (Eds.)
Computational Intelligence for Remote Sensing, 2008
ISBN 978-3-540-79352-6

Vol. 134. Ngoc Thanh Nguyen and Radoslaw Katarzyniak (Eds.)
New Challenges in Applied Intelligence Technologies, 2008
ISBN 978-3-540-79354-0

Vol. 135. Hsinchun Chen and Christopher C. Yang (Eds.)
Intelligence and Security Informatics, 2008
ISBN 978-3-540-69207-2

Vol. 136. Carlos Cotta, Marc Sevaux and Kenneth Sörensen (Eds.)
Adaptive and Multilevel Metaheuristics, 2008
ISBN 978-3-540-79437-0

Vol. 137. Lakhmi C. Jain, Mika Sato-Ilic, Maria Virvou, George A. Tsihrintzis, Valentina Emilia Balas and Canicious Abeynayake (Eds.)
Computational Intelligence Paradigms, 2008
ISBN 978-3-540-79473-8

Lakhmi C. Jain
Mika Sato-Ilic
Maria Virvou
George A. Tsihrintzis
Valentina Emilia Balas
Canicious Abeynayake
(Eds.)

Computational Intelligence Paradigms

Innovative Applications

Springer

Professor Dr. Lakhmi C. Jain
University of South Australia
South Australia, Australia
Lakhmi.jain@unisa.edu.au

Professor Dr. Mika Sato-Ilic
University of Tsukuba
Tsukuba, Ibaraki, Japan

Professor Dr. Maria Virvou
University of Piraeus,
Greece

Professor Dr. George A. Tsihrintzis
University Piraeus,
Greece

Professor Dr. Valentina Emilia Balas
University of Arad, Romania

Dr. Canicious Abeynayake
Defence Science and
Technology Organisation
Edinburgh, Australia

ISBN 978-3-642-09839-0 e-ISBN 978-3-540-79474-5

DOI 10.1007/978-3-540-79474-5

Studies in Computational Intelligence ISSN 1860949X

Softcover reprint of the hardcover 1st edition 2008

Printed in acid-free paper

9 8 7 6 5 4 3 2 1

springer.com

Preface

System designers are faced with a large set of data which has to be analysed and processed efficiently. Advanced computational intelligence paradigms present tremendous advantages by offering capabilities such as learning, generalisation and robustness. These capabilities help in designing complex systems which are intelligent and robust.

The book includes a sample of research on the innovative applications of advanced computational intelligence paradigms. The characteristics of computational intelligence paradigms such as learning, generalization based on learned knowledge, knowledge extraction from imprecise and incomplete data are extremely important for the implementation of intelligent machines. The chapters include architectures of computational intelligence paradigms, knowledge discovery, pattern classification, clusters, support vector machines and gene linkage analysis. We believe that the research on computational intelligence will simulate great interest among designers and researchers of complex systems. It is important to use the fusion of various constituents of computational intelligence to offset the demerits of one paradigm by the merits of another.

We are grateful to the authors and the reviewers for their great contribution. We appreciate the editorial support of Scientific Publishing Services Pvt. Ltd.

Lakhmi C. Jain
Mika Sato
Maria Virvou
George A. Tsihrintzis
Valentina Emilia Balas
Canicious Abeynayake

Contents

1 An Introduction to Computational Intelligence Paradigms
Lakhmi C. Jain, Shing Chiang Tan, Chee Peng Lim 1

2 A Quest for Adaptable and Interpretable Architectures of Computational Intelligence
Witold Pedrycz 25

3 MembershipMap: A Data Transformation for Knowledge Discovery Based on Granulation and Fuzzy Membership Aggregation
Hichem Frigui 51

4 Advanced Developments and Applications of the Fuzzy ARTMAP Neural Network in Pattern Classification
Boaz Lerner, Hugo Guterman 77

5 Large Margin Methods for Structured Output Prediction
Elisa Ricci, Renzo Perfetti 109

6 Ensemble MLP Classifier Design
Terry Windeatt 133

7 Functional Principal Points and Functional Cluster Analysis
Nobuo Shimizu, Masahiro Mizuta 149

8 Clustering with Size Constraints
Frank Höppner, Frank Klawonn 167

9 Cluster Validating Techniques in the Presence of Duplicates
Ravi Jain, Andy Koronios 181

10 Fuzzy Blocking Regression Models
Mika Sato-Ilic 195

11 Support Vector Machines and Features for Environment Perception in Mobile Robotics
Rui Araújo, Urbano Nunes, Luciano Oliveira, Pedro Sousa, Paulo Peixoto 219

12 Linkage Analysis in Genetic Algorithms
Miwako Tsuji, Masaharu Munetomo 251

Author Index 281

An Introduction to Computational Intelligence Paradigms

Lakhmi C. Jain[1], Shing Chiang Tan[2], and Chee Peng Lim[3]

[1] School of Electrical & Information Engineering
University of South Australia, Australia
[2] Faculty of Information Science and Technology
Multimedia University, Malaysia
[3] School of Electrical & Electronic Engineering
University of Science Malaysia, Malaysia

Abstract. This chapter presents an introduction to computational intelligence (CI) paradigms. A number of CI definitions are first presented to provide a general concept of this new, innovative computing field. The main constituents of CI, which include artificial neural networks, fuzzy systems, and evolutionary algorithms, are explained. In addition, different hybrid CI models arisen from synergy of neural, fuzzy, and evolutionary computational paradigms are discussed.

1 Introduction

Computational Intelligence (CI) is an emerging field that comprises a highly interdisciplinary framework useful for supporting the design and development of intelligent systems. Defining CI is a challenging task as it is a new, emerging computing field. Indeed, CI involves innovative models that often come with a high level of machine learning quotient. Nevertheless, the main computing paradigms that are normally categorized under the umbrella of CI are Artificial Neural Networks (ANNs), Fuzzy Systems (FSs), and Evolutionary Algorithms (EAs). Some excerpts of CI definitions, as highlighted in Gorzalczany (2002), are as follows.

Perhaps one of the earliest definitions of CI is provided by James C. Bezdek. According to Bezdek (1994), "… a system is computationally intelligent when it: deals only with numerical (low-level) data, has a pattern recognition component, and does not use knowledge in the AI sense; and additionally when it (begins to) exhibit (i) computational adaptivity; (ii) computational fault tolerance; (iii) speed approaching human-like turnaround, and (iv) error rates that approximate human performance …".

Other definitions of CI are also suggested by Fogel (1995), Marks (1993), and Pedrycz (1999), which identify ANNs, FSs, and EAs as the constituents of CI paradigms. Fogel (1995) argued that "… these technologies of neural, fuzzy, and evolutionary systems were brought together under the rubric of computational intelligence, a relatively new trend offered to generally describe methods of computation that can be used to adapt solutions to new problems and do not rely on explicit human knowledge …".

Marks (1993) explained that "… neural networks, genetic algorithms, fuzzy systems, evolutionary programming, and artificial life are the building blocks of CI …".

L.C. Jain et al. (Eds.): Comp. Intel. Para.: Innov. Applications, SCI 137, pp. 1–23, 2008.
springerlink.com

On the other hand, Pedrycz (1999) pointed out that "... it is clear that the already identified components of CI encompass neural networks, a technology of granular information and granular computing as well as evolutionary computation. In this synergistic combination, each of them plays an important, well-defined, and unique role ..."

In this chapter, we attempt to present how the CI constituents could provide a platform for designing innovative "intelligent" systems, which can exhibit the following characteristics: the ability to learn from examples; the ability to generalize based on learned knowledge; the ability to process imprecise, uncertain information; the ability to extract knowledge from data; the ability to give explanation for decisions made; and, the ability to perform adaptation through parallel search and optimization.

The organization of this chapter is as follows. In Section 2, an overview of each CI constituents, i.e., ANNs, FSs, and EAs, as well as their hybrid models, is presented. In Section 3, a summary of the work presented in each chapter of this book is given. Some concluding remarks are included in Section 4. Finally, a list of CI resources is included in Section 5.

2 Computational Intelligence Models

2.1 Neural Computational Models

Interests in human and animal performance have triggered the motivation of understanding and modeling various functions of the brain. The brain is a hub for numerous biological processing systems of which their structures are networks of nerve cells. These systems are complex, yet highly organized. They are robust and self-adaptive biological devices responsible for monitoring and controlling activities of the body. The brain has the ability to perform activities such as pattern recognition, perception and motor control in perhaps just a few milliseconds. In addition, the brain is able to capture, store, and generalize information, i.e., the so-called learning ability.

ANNs are learning systems that are inspired by the operation of nervous systems in the brain. McCulloch and Pitts (McCulloch & Pitts, 1943; Pitts and McCulloch, 1947) were the pioneers who proposed mathematical modeling of a neuron. However, it was the work by Hebb (1949) that introduced the concept of adapting the connections between nodes, and suggested the idea of incorporating learning in ANNs. Ever since the publication of Hebb's seminal study, a variety of different ANN architectures and learning paradigms have been introduced in the literature. Examples of some classic models include the "Perceptron" (Rosenblatt, 1958), the "Adaline" (Widrow and Hoff, 1960), the Hopfield network (Hopfield, 1982, 1984). Other popular ANN examples include the feedforward networks (especially the Multilayered Perceptron (MLP) (Rumelhart *et al.*, 1986) and Radial Basis Function (RBF) (Moody & Darken, 1989) based models), the Adaptive Resonance Theory family of networks (Carpenter and Grossberg, 1987a; Carpenter and Grossberg, 1987b; Carpenter *et al.*, 1992), the self-organizing feature maps (Kohonen, 1982, 1988). To date, there is a proliferation of ANN architectures and learning algorithms in the literature. Some recent examples include a weighted probabilistic neural network (Song *et al.*, 2007), a self-organizing RBF network (Lian *et al.*, 2008), a delayed Hopfield neural network (Mou, *et al*, 2008), the Lyapunov-function-based feedforward networks (Behera *et al.*, 2006),

Polytype ARTMAP (Amorin *et al.*, 2007), and the Density-Driven GRNN (Goulermas *et al.*, 2007).

In general, research activities in ANNs can be categorized into two main streams: one to model mathematically biological nervous systems at the microscopic level of neurons and synapses, and another to devise machine learning algorithms that mimic certain functions of the brain at the macroscopic level such as capturing and processing information.

From the machine learning point of view, ANNs are capable of making non-linear mapping from a set of inputs to a set of outputs. They can be employed as universal functional approximators, which can offer accurate approximation of an unknown system on provision of data samples. The functionality of ANNs can be enhanced if they are equipped with a capability of absorbing information continually and autonomously without forgetting previously learned information. Such ability is favorable for ANNs to operate in non-stationery environments with greater autonomy and less dependency on humans.

2.2 Fuzzy Computational Models

The theory of fuzzy logic was developed by Zadeh (1965). The fuzzy set theory provides a methodology for representing and computing data and information that are uncertain and imprecise. Fuzzy logic provides an inference mechanism on a set of *if-then* rules for reasoning. The rules are defined with fuzzy sets, in which the fuzzy sets generalize the concept of the conventional set by extending membership degree to be any value between 0 and 1. Such "fuzziness" feature occurs in many real-world situations, where it is ambiguous to decide if something can be categorized exactly into a specific class or not.

A typical example of uncertainty and vagueness associated with human activities is the linguistic uncertainty of a natural language (Zadeh, 1973). FSs, which assimilate the concepts from fuzzy set theory and fuzzy logic, provide a framework for handling commonsense knowledge represented in a linguistic or an uncertain numerical form. There are two main characteristics of FSs that give rise to their popularity: they are suitable for uncertain or approximate reasoning, especially for systems where a mathematical model is difficult to derive; fuzzy logic allows decision to be inferred using incomplete or uncertain information with linguistic variables that are easily comprehensible by humans.

In general, an FS normally comprises three components: a knowledge base that is formed by a group of fuzzy *if-then* rules; a data base that include all membership functions used in the fuzzy rules; and a reasoning mechanism that conducts an inference procedure by using rules and known facts to reach a conclusion. Each fuzzy rule explicitly describes the relationship between its inputs (i.e., antecedents) and output (i.e., consequence). Such a relationship indicates the behavioral link between the antecedents and the consequence locally. Therefore, FSs, which aggregate the hypothesis of rules through the inference procedure for an outcome can be viewed as a non-linear mapping from the fuzzy region in the input space to the fuzzy region in the output space.

While conventional FSs use the conventional fuzzy sets, or type-1 fuzzy sets, another variant of fuzzy sets, i.e., type-2 fuzzy sets (Zadeh, 1975; John and Coupland,

2007; and Mendel 2007), is available for constituting FSs that exhibit a higher degree of approximation. Generally speaking, type-2 fuzzy sets can be perceived as a generalization of those of type-1 by taking into consideration uncertainty that exists in linguistic variables as well as in membership function definitions. In this case, type-1 FSs only deal with uncertainty from the linguistic variables whereas type-2 FSs perform mapping by considering the uncertainty from both the linguistic variables and membership functions.

2.3 Evolutionary Computational Models

EAs are a collection of algorithms based on the principles of evolution, heredity, and natural selection in living population. EAs are adaptive, robust algorithms, are particularly useful for applications that require search and optimization. They are population-based algorithms for which any communication and interaction are carried out within the population. Therefore, they possess a high degree of implicit parallelism.

In general, there are five major types of EAs, *viz.*, genetic algorithms (GA) (Holland, 1962), evolutionary programming (EP) (Fogel *et al.*, 1966), evolution strategies (ES) (Schwefel, 1981; Rechenberg, 1994), genetic programming (GP) (Koza, 1992), and learning classifier systems (LCS) (Holland, 1986). While the detail dynamics of each EA variants are different, the fundamental idea behind them is similar. On availability of a population of individuals, natural selection, which is based on the principle of survival of the fittest following the existence of environmental pressures, is exercised to choose individuals that could better fit the environment. The operations involved are as follows.

Given a fitness function to be optimized, a set of candidate solutions (offspring) is randomly generated. The candidate solutions are assessed by the fitness function that indicates an abstract of the fitness measure. Based on the fitness values, fitter candidate solutions have a better chance to be selected to seed the next generation through the application of recombination (crossover) and/or mutation. On one hand, the recombination operator is applied to two (or more) parent candidate solutions, and one (or more) new offspring is produced. On the other hand, the mutation operator is applied to one parent candidate solution, and one new candidate solution is produced. The new candidate solutions compete among themselves, based on the fitness values, for a place in the next generation. Even though candidate solutions with higher fitness values are favorable, in some selection schemes, solution candidates with relatively low fitness values are included in the next generation. The idea is to maintain diversity of the population. The processes of selection, recombination, and mutation are repeated from one generation of population to another until a terminating criterion is satisfied, i.e., either a candidate solution with sufficient quality has been obtained, or the algorithm has been executed for a predefined number of generations.

2.4 Hybrid Computational Models

Despite the main constituents of CI–ANNs, FSs, and EAs–can be applied independently to solve real-world problems, more effective solutions can be obtained if they are used in combination. Examples of hybrid CI paradigms include neural-fuzzy, neural-genetic, fuzzy-genetic, and neural-fuzzy-genetic models. Each combination brings synergy to the resulting system in such a way that the hybrid system exploits the advantages of the constituent techniques and, at the same time, avoids their shortcomings.

A significant disadvantage of ANNs is that they do not have interpretability. ANNs are generally criticized as the black-box models since they are unable to provide an explanatory facility on decisions that they have made. On the other hand, FSs rely heavily on human experts to construct fuzzy rules and to identify the associated membership functions. They do not possess a learning capability for dealing with the changes in the external environment. Two main research directions have been endeavored for combining ANNs and FSs, which subsequently lead to the synthesis of neural-fuzzy and fuzzy-neural systems. Neural-fuzzy systems attempt to incorporate the ANN learning capability into the FS framework that is intrinsically endowed with the interpretability property. In this regard, ANNs are employed to perform numerical processing of fuzzy sets (e.g., to identify the shape of membership functions through learning from data) and to enforce network-like implementations of fuzzy rules. ANFIS (Jang, 1993) is a typical example of network-like FS that performs rule learning. It uses the error gradient information in the feedforward pass to adjust fuzzy partitions in the network hidden layer in the feedback pass. Even though the structure of ANFIS is network-like, its functionality is essentially equivalent to a fuzzy inference system. Other neural-fuzzy models include NEFCLASS (Nauck and Kruse, 1997), EFuNN (Kasabov, 2001) and Inverted HNFB (Gonçalves *et al.*, 2006) for pattern classification; NEFPROX (Nauck and Kruse, 1999) and ADFNFN (Su and Yang, 2002) for function approximation; as well as DENFIS (Kasabov and Song, 2002) and NFI (Song and Kasabov, 2005) for time-series prediction.

On the other hand, fuzzy-neural systems inherit the basic properties and structure of conventional ANNs. Besides, fuzzy set methods are integrated within the ANN framework for improving the performance of the network. Attempts of devising fuzzy-neural systems are exemplified by the introduction of fuzzy neurons (e.g., Gupta and Qi (1992); Yamakawa *et al.* (1992)); the proposal of a fuzzy backpropagation learning algorithm (e.g., Hayashi *et al.* (1993); Liu and Li (2004)); and the use of fuzzy rules for modifying the learning coefficients of ANNs. Fuzzy-neural systems can also be regarded as a generalization of conventional ANNs, which are capable of performing learning and generalization using quantitative numerical data as well as qualitative linguistic information (given in terms of fuzzy sets). Examples of such fuzzy-neural systems include fuzzy MLP (Mitra *et al.*, 1997), granular neural networks (Zhang *et al.*, 2000), SuPFuNIS (subsethood-product fuzzy neural inference system) (Paul and Kumar, 2002), and ASuPFuNIS (asymmetric subsethood-product fuzzy neural inference system) (Velayutham and Kumar, 2005).

The gradient methods are not suitable for tuning the membership functions of fuzzy rules in FSs when they are represented as non-differentiable fuzzy numbers such as trapezoids and triangles. EAs, however, provide a means of global optimization that is independent of gradient information. In the development of fuzzy-genetic systems, EAs can be used to optimize not only the shapes and number of membership functions (e.g., Fuzzy CoCo (Peña-Reyes and Sipper, 2001)) but also to generate fuzzy rules (e.g., SOGARG (Pal and Pal, 2003); and the models introduced by Casillas *et al.* (2007a) and Casillas *et al.* (2007b)).

The main purpose of training an ANN is to obtain a set of parameters that could minimize some error measure. These parameters include the network weights, thresholds, transfer functions, and structure. As such, the motivation of combining ANNs and EAs lies in reserving the well-defined characteristics of both entities, for which ANNs provide an architecture for learning and accurate computation whereas EAs provide a robust and efficient search and optimization procedure to determine the ANN parameter set for achieving improved solutions. The combination between ANNs and EAs leads to a special class of ANNs known as evolutionary ANNs (EANNs). EANNs capitalize evolution as another fundamental form of adaptation in addition to learning (Yao, 1999). These two forms of adaptation allow EANN models to perform efficiently and effectively in a dynamic environment. Some examples of EANN models include GNARL (Angeline *et al.*, 1994), EPNet (Yao and Liu, 1997), COVNET (García-Pedrajas *et al.*, 2002a, 2003), MOBNET (Garcia-Pedrajas *et al.*, 2002b), MGNN (Palmes *et al.*, 2005), and GARBFN (Manrique *et al.*, 2006).

Viewing from another perspective, EAs can also be integrated with neural-fuzzy systems to provide an alternative learning facility for constructing a rule base from data samples. In this regard, EAs are used for learning in case traditional optimization techniques defy attempts to give good results. Examples include the use of genetic algorithms to train a recurrent fuzzy network (Juang, 2002), an Interval Type-2 Fuzzy Neural Network (Wang, 2004), and a recurrent wavelet-based neural-fuzzy system (Lin and Hsu, 2007).

3 Chapters Included in This Book

The book includes a sample of most recent theoretical foundation and practical applications of computational intelligence paradigms. This chapter presents introduction to computational intelligence paradigms. Chapter two presents the adaptable and interpretable architectures of computational intelligence paradigms. Chapter three is on knowledge discovery based on granulation and fuzzy membership aggregation. Chapters four presents a fuzzy ARTMAP neural network for pattern classification. Chapter five is on clustering with size constraints.

Chapter six presents an ensemble of MLP classifiers. Chapter seven presents large margin methods for structured output prediction. Chapter eight is on functional principal points and functional cluster analysis. Chapter nine presents a comparative study of three cluster validation techniques. Chapter ten presents fuzzy clustering based blocking regression models. Chapter eleven is on support vector machines for environment perception in mobile robotics. The final chapter is on gene linkage analysis for genetic algorithms. This technique is useful for solving complex problems.

4 Summary

In this chapter, an overview of different CI paradigms is presented. A number of CI definitions are first extracted from the literature to give a general concept of this emerging, innovative computing field. The main constituents of CI, i.e., ANNs, FSs, and EAs, are then explained. In addition, hybrid CI models, which are formulated by exploiting the advantages of different CI paradigms and, at the same time, avoiding their shortcomings, are discussed. Finally, an overview of the work presented in each chapter in this book is highlighted and a list of CI resources is included in the next section. It is envisaged that innovative CI paradigms will ultimately play an important role in solving many real-world problems that occur in our daily life.

5 Resources

Following is a sample of resources on computational intelligence.

5.1 Journals

- International Journal of Knowledge-Based intelligent Engineering systems, IOS Press, The Netherlands.
  http://www.kesinternational.org/journal/
- International Journal of Hybrid Intelligent Systems, IOS Press, The Netherlands.
- Intelligent Decision Technologies: An International Journal, IOS Press, The Netherlands.
- IEEE Intelligent Systems, IEEE Press, USA.
  www.computer.org/intelligent/
- IEEE Transactions on Neural Networks.
- IEEE Transactions on Evolutionary Computing.
- IEEE Transactions on Fuzzy Systems.
- IEEE Computational Intelligence Magazine.
- Neural Computing and applications, Springer.
- Neurocomputing, Elsevier.
- International Journal of Intelligent and Fuzzy Systems, IOS Press, The Netherlands.
- Fuzzy Optimization and Decision Making, Kluwer.
- AI Magazine, USA
  www.aaai.org/

5.2 Special Issue of Journals

- Ghosh, A., Seiffert, U. and **Jain, L.C.**, Evolutionary Computation in Bioinformatics, Journal of Intelligent and Fuzzy Systems, IOS Press, The Netherlands, Volume 18, Number 6, 2007.
- Abraham, A., Smith, K., Jain, R. and **Jain, L.C.**, Network and Information Security: A Computational Intelligence Approach, Journal of Network and Computer Applications, Elsevier Publishers, Volume 30, Issue 1, 2007.
- Palade, V. and Jain, L.C., Practical Applications of Neural Networks, Journal of Neural Computing and Applications, Springer, Germany, Volume 14, No. 2, 2005.

- Abraham, A. and Jain, L.C., Computational Intelligence on the Internet, Journal of Network and Computer Applications, Elsevier Publishers, Volume 28, Number 2, 2005.
- Abraham, A., Thomas, J., Sanyal, S. and Jain, L.C., Information Assurance and Security, Journal of Universal Computer Science, Volume 11, Issue 1, 2005.
- Abraham, A. and Jain, L.C., Optimal Knowledge Mining, Journal of Fuzzy Optimization and Decision Making, Kluwer Academic Publishers, Volume 3, Number 2, 2004.
- Palade, V. and Jain, L.C., Engineering Applications of Computational Intelligence, Journal of Intelligent and Fuzzy systems, IOS Press, Volume 15, Number 3, 2004.
- Alahakoon, D., Abraham, A. and Jain, L.C., Neural Networks for Enhanced Intelligence, Neural Computing and Applications, Springer, UK, Volume 13, No. 2, June 2004.
- Abraham, A., Jonkar, I., Barakova, E., Jain, R. and Jain, L.C., Special issue on Hybrid Neurocomputing, Neurocomputing, Elsevier, The Netherlands, Volume 13, No. 2, June 2004.
- Abraham, A. and Jain, L.C., Knowledge Engineering in an Intelligent Environment, Journal of Intelligent and Fuzzy Systems, IOS Press, The Netherlands, Volume 14, Number 3, 2003.
- Jain, L.C., Fusion of Neural Nets, Fuzzy Systems and Genetic Algorithms in Industrial Applications, IEEE Transactions on Industrial Electronics, USA, Volume 46, Number 6, December 1999.
- De Silva, C. and Jain, L.C., Intelligent Electronic Systems, Engineering Applications of Artificial Intelligence, Pergamon Press, USA, Volume 11, Number 1, January 1998.
- Jain, L.C., Intelligent Systems: Design and Applications - 2, Journal of Network and Computer Applications, Elsevier, **Vol. 2**, April 1996.
- Jain, L.C., Intelligent Systems: Design and Applications - 1, Journal of Network and Computer Applications, Elsevier, **Vol.1**, January, 1996.

5.3 Conferences

- KES International Conference Series
  www.kesinternational.org/
- AAAI Conference on Artificial Intelligence
  www.aaai.org/aaai08.php
- European Conferences on Artificial Intelligence (ECAI)

5.4 Conference Proceedings

- Apolloni, B., Howlett, R.J. and Jain, L.C. (Editors), Knowledge-Based Intelligent Information and Engineering Systems, Lecture Notes in Artificial Intelligence, **Volume 1, LNAI 4692**, KES 2007, Springer-Verlag, Germany, 2007.

- Apolloni, B., Howlett, R.J.and Jain, L.C. (Editors), Knowledge-Based Intelligent Information and Engineering Systems, Lecture Notes in Artificial Intelligence, **Volume 2, LNAI 4693,** , KES 2007, Springer-Verlag, Germany, 2007.
- Apolloni, B., Howlett, R.J. and Jain, L.C. (Editors), Knowledge-Based Intelligent Information and Engineering Systems, Lecture Notes in Artificial Intelligence, **Volume 3, LNAI 4694,** KES 2007, Springer-Verlag, Germany, 2007.
- Nguyen, N.T., Grzech, A., Howlett, R.J. and Jain, L.C., Agents and Multi-Agents Systems: Technologies and Applications, Lecture Notes in artificial Intelligence, **LNAI 4696,** Springer-Verlag, Germany, 2007.
- Howlett, R.P., Gabrys, B. and Jain, L.C. (Editors), Knowledge-Based Intelligent Information and Engineering Systems, Lecture Notes in Artificial Intelligence, KES 2006, Springer-Verlag, Germany, Vol. **4251**, 2006.
- Howlett, R.P., Gabrys, B. and Jain, L.C. (Editors), Knowledge-Based Intelligent Information and Engineering Systems, Lecture Notes in Artificial Intelligence, KES 2006, Springer-Verlag, Germany, Vol. **4252**, 2006.
- Howlett, R.P., Gabrys, B. and Jain, L.C. (Editors), Knowledge-Based Intelligent Information and Engineering Systems, Lecture Notes in Artificial Intelligence, KES 2006, Springer-Verlag, Germany, Vol. **4253**, 2006.
- Liao, B.-H., Pan, J.-S., Jain, L.C., Liao, M., Noda, H. and Ho, A.T.S., Intelligent Information Hiding and Multimedia Signal Processing, IEEE Computer Society Press, USA, 2007. ISBN: 0-7695-2994-1.
- Khosla, R., Howlett, R.P., and Jain, L.C. (Editors), Knowledge-Based Intelligent Information and Engineering Systems, Lecture Notes in Artificial Intelligence, KES 2005, Springer-Verlag, Germany, Vol. **3682**, 2005.
- Skowron, A., Barthes, P., Jain, L.C., Sun, R., Mahoudeaux, P., Liu, J. and Zhong, N.(Editors), Proceedings of the 2005 IEEE/WIC/ACM International Conference on Intelligent Agent Technology, Compiegne, France, IEEE Computer Society Press, USA, 2005.
- Khosla, R., Howlett, R.P., and Jain, L.C. (Editors), Knowledge-Based Intelligent Information and Engineering Systems, Lecture Notes in Artificial Intelligence, KES 2005, Springer-Verlag, Germany, Vol. **3683**, 2005.
- Khosla, R., Howlett, R.P., and Jain, L.C. (Editors), Knowledge-Based Intelligent Information and Engineering Systems, Lecture Notes in Artificial Intelligence, KES 2005, Springer-Verlag, Germany, Vol. **3684**, 2005.
- Khosla, R., Howlett, R.P., and Jain, L.C. (Editors), Knowledge-Based Intelligent Information and Engineering Systems, Lecture Notes in Artificial Intelligence, KES 2005, Springer-Verlag, Germany, Vol. **3685**, 2005.
- Negoita, M., Howlett, R.P., and Jain, L.C. (Editors), Knowledge-Based Intelligent Engineering Systems, KES 2004, Lecture Notes in Artificial Intelligence, Vol. **3213**, Springer, 2004.
- Negoita, M., Howlett, R.P., and Jain, L.C. (Editors), Knowledge-Based Intelligent Engineering Systems, KES 2004, Lecture Notes in Artificial Intelligence, Vol. **3214**, Springer, 2004.
- Negoita, M., Howlett, R.P., and Jain, L.C. (Editors), Knowledge-Based Intelligent Engineering Systems, KES 2004, Lecture Notes in Artificial Intelligence, Vol. **3215**, Springer, 2004.

- Palade, V., Howlett, R.P., and Jain, L.C. (Editors), Knowledge-Based Intelligent Engineering Systems, Lecture Notes in Artificial Intelligence, Vol. **2773**, Springer, 2003.
- Palade, V., Howlett, R.P., and Jain, L.C. (Editors), Knowledge-Based Intelligent Engineering Systems, Lecture Notes in Artificial Intelligence, Vol. **2774**, Springer, 2003.
- Damiani, E., Howlett, R.P., Jain, L.C. and Ichalkaranje, N. (Editors), Proceedings of the Fifth International Conference on Knowledge-Based Intelligent Engineering Systems, **Volume 1**, IOS Press, The Netherlands, 2002.
- Damiani, E., Howlett, R.P., Jain, L.C. and Ichalkaranje, N. (Editors), Proceedings of the Fifth International Conference on Knowledge-Based Intelligent Engineering Systems, **Volume 2**, IOS Press, The Netherlands, 2002.
- Baba, N., Jain, L.C. and Howlett, R.P. (Editors), Proceedings of the Fifth International Conference on Knowledge-Based Intelligent Engineering Systems (KES'2001), **Volume 1,** IOS Press, The Netherlands, 2001.
- Baba, N., Jain, L.C. and Howlett, R.P. (Editors), Proceedings of the Fifth International Conference on Knowledge-Based Intelligent Engineering Systems (KES'2001), **Volume 2,** IOS Press, The Netherlands, 2001.
- Howlett, R.P. and Jain, L.C.(Editors), Proceedings of the Fourth International Conference on Knowledge-Based Intelligent Engineering Systems, IEEE Press, USA, 2000. **Volume 1**.
- Howlett, R.P. and Jain, L.C.(Editors), Proceedings of the Fourth International Conference on Knowledge-Based Intelligent Engineering Systems, IEEE Press, USA, 2000. **Volume 2**.
- Jain, L.C.(Editor), Proceedings of the Third International Conference on Knowledge-Based Intelligent Engineering Systems, IEEE Press, USA, 1999.
- Jain, L.C. and Jain, R.K. (Editors), Proceedings of the Second International Conference on Knowledge-Based Intelligent Engineering Systems, **Volume 1**, IEEE Press, USA, 1998.
- Jain, L.C. and Jain, R.K. (Editors), Proceedings of the Second International Conference on Knowledge-Based Intelligent Engineering Systems, **Volume 2**, IEEE Press, USA, 1998.
- Jain, L.C. and Jain, R.K. (Editors), Proceedings of the Second International Conference on Knowledge-Based Intelligent Engineering Systems, **Volume 3**, IEEE Press, USA, 1998.
- Jain, L.C. (Editor), Proceedings of the First International Conference on Knowledge-Based Intelligent Engineering Systems, **Volume 1**, IEEE Press, USA, 1997.
- Jain, L.C. (Editor), Proceedings of the First International Conference on Knowledge-Based Intelligent Engineering Systems, **Volume 2**, IEEE Press, USA, 1997.
- Narasimhan, V.L., and Jain, L.C. (Editors), The Proceedings of the Australian and New Zealand Conference on Intelligent Information Systems, IEEE Press, USA, 1996.

5.5 Book Series

5.5.1 Advanced Intelligence and Knowledge Processing, Springer-Verlag, Germany: www.springer.com/series/4738

- Nguyen, N.T., Advanced Methods for Inconsistent Knowledge Management, Springer-Verlag, London, 2008.
- Meisels, A., Distributed Search by Constrained Agents, Springer-Verlag, London, 2008.
- Camastra, F. and Vinciarelli, A., Machine Learning for audio, Image, and Video Analysis, Springer-Verlag, London, 2008.
- Kornai, A., Mathematical Linguistics, Springer-Verlag, London, 2008.
- Prokopenko, M. (Editor), Advances in Applied Self-Organising Systems, Springer-Verlag, London, 2008.
- Scharl, A., Environmental Online Communication, Springer-Verlag, London, 2007.
- Pierre, S. (Editor), E-Learning Networked Environments and Architectures, Springer-Verlag, London, 2007.
- Karny, M. (Editor), Optimized Bayesian Dynamic Advising, Springer-Verlag, London, 2006.
- Liu, S. and Lin, Y., Grey Information: Theory and Practical Applications, Springer-Verlag, London, 2006.
- Maloof, M.A. (Editor), Machine Learning and Data Mining for Computer Security, Springer-Verlag, London, 2006.
- Wang, J.T.L., et al. (Editors), Data Mining in Bioinformatics, Springer-Verlag, London, 2005.
- Grana, M., et al. (Editors), Information Processing with Evolutionary Algorithms, Springer-Verlag, London, 2005.
- Fyfe, C., Hebbian Learning and Negative Feedback Networks,, Springer-Verlag, London, 2005.
- Chen-Burger, Y. and Robertson, D., Automatic Business Modelling,, Springer-Verlag, London, 2005.
- Husmeier, D., et.al. (Editors), Probabilistic Modelling in Bioinformatics and Medical Informatics, Springer-Verlag, London, 2005.
- Tan, K.C., et al., Multiobjective Evolutionary Algorithms and Applications, Springer-Verlag, London, 2005.
- Bandyopadhyay, S., et. al. (Editors), Advanced Methods for Knowledge Discovery from Complex Data, Springer-Verlag, London, 2005.
- Stuckenschmidt, H. and Harmelen, F.V., Information Sharing on the Semantic Web, Springer-Verlag, London, 2005.
- Abraham, A., Jain, L.C. and Goldberg, R., Evolutionary Multiobjective Optimization, Springer-Verlag, London, 2005.
- Gomez-Perez, et al., Ontological Engineering, Springer-Verlag, London, 2004.
- Zhang, S., et. al., Knowledge Discovery in Multiple Databases, Springer-Verlag, London, 2004.
- Ko, C.C., Creating Web-based Laboratories, Springer-Verlag, London, 2004.

- Mentzas, G., et al., Knowledge Asset Management, Springer-Verlag, London, 2003.
- Vazirgiannis, M., et al., Uncertainty Handling and Quality Assessment in Data Mining, Springer-Verlag, London, 2003.

5.5.2 Advanced Information Processing, Springer-Verlag, Germany

- Harris, C., Hong, X. and Gan, Q., Adaptive Modelling, Estimation and Fusion from Data, Springer-Verlag, Germany, 2002.
- Ohsawa, Y. and McBurney, P. (Editors), Chance Discovery, Springer-Verlag, Germany, 2003.
- Deen, S.M. (Editor), Agent-Based Manufacturing, Springer-Verlag, Germany, 2003.
- Gasós J. and Hirsch B., e-Business Applications, Springer-Verlag, Germany, 2003.
- Chen, S.H. and Wang, P.P. (Editors), Computational Intelligence in Economics and Finance, Springer-Verlag, Germany, 2004.
- Liu, J. and Daneshmend, L., Spatial reasoning and Planning, Springer-Verlag, Germany, 2004.
- Wang, L. and Fu, X., Data Mining with Computational Intelligence, Springer-Verlag, 2005.
- Ishibuchi, H., Nakashima, T. and Nii, M., Classification and Modeling with Linguistic Information Granules, Springer-Verlag, Germany, 2005.

5.5.3 Computational Intelligence and Its Applications Series, IGI Publishing, USA

- Chen, S.-H., Jain, L.C. and Tai, C.-C. (Editors), Computational Economics: A Perspective from Computational Intelligence, IGI Publishing, 2006.
- Begg, R. and Palaniswami, M., Computational Intelligence for Movement Sciences, IGI Publishing, 2006.
- Fulcher, J. (Editor), Advances in Applied Intelligences, IGI Publishing, 2006.
- Zhang, D., et al., Biometric Image Discrimination technologies, IGI Publishing, 2006.

5.5.4 Knowledge-Based Intelligent Engineering Systems Series, IOS Press, The Netherlands

- Phillips-Wren, G. and Jain, L.C. (Editors), Intelligent Decision Support Systems in Agent-Mediated Environments, IOS Press, The Netherlands, Volume 115, 2005.
- Nakamatsu, K. and Abe, J.M., Advances in Logic Based Intelligent Systems, IOS Press, The Netherlands, Volume 132, 2005.
- Abraham, A., Koppen, M. and Franke, K. (Editors), Design and Applications of Hybrid Intelligent Systems, IOS Press, The Netherlands, Volume 104, 2003.
- Turchetti, C., Stochastic Models of Neural Networks, IOS Press, The Netherlands, Volume 102, 2004.

- Wang, K., Intelligent Condition Monitoring and Diagnosis Systems, IOS Press, The Netherlands, Volume 93, 2003.
- Abraham, A., et al. (Editors), Soft Computing Systems, IOS Press, The Netherlands, Volume 87, 2002.
- Lee, R.S.T. and Liu, J.H.K., Invariant Object Recognition based on Elastic Graph Matching, IOS Press, The Netherlands, Volume 86, 2003.
- Loia, V. (Editor), Soft Computing Agents, IOS Press, The Netherlands, Volume 83, 2002.
- Motoda, H., Active Mining, IOS Press, The Netherlands, Volume 79. 2002.
- Namatame, A., et al. (Editors), Agent-Based Approaches in Economic and Social Complex Systems, IOS Press, The Netherlands, Volume 72, 2002.

5.5.5 The CRC Press International Series on Computational Intelligence, The CRC Press, USA

- Teodorescu, H.N. Kandel, A. and Jain, L.C. (Editors), Intelligent systems and Technologies in Rehabilitation Engineering, CRC Press USA, 2001.
- Jain, L.C. and Fanelli, A.M. (Editors), Recent Advances in Artificial Neural Networks: Design and Applications, CRC Press, USA, 2000.
- Medsker, L., and Jain, L.C. (Editors) Recurrent Neural Networks: Design and Applications, CRC Press, USA, 2000.
- Jain, L.C., Halici, U., Hayashi, I., Lee, S.B. and Tsutsui, S. (Editors) Intelligent Biometric Techniques in Fingerprint and Face Recognition, CRC Press, USA, 2000.
- Jain, L.C. (Editor), Evolution of Engineering and Information Systems, CRC Press USA, 2000.
- Dumitrescu, D., Lazzerini, B., Jain, L.C. and Dumitrescu, A., Evolutionary Computation, CRC Press USA, 2000.
- Dumitrescu, D., Lazzerini, B., Jain, L.C., Fuzzy Sets and their Applications to Clustering and Training, CRC Press USA, 2000.
- Jain, L.C. and De Silva, C.W. (Editors), Intelligent Adaptive Control, CRC Press USA, 1999.
- Jain, L.C. and Martin, N.M. (Editors), Fusion of Neural Networks, Fuzzy Logic and Evolutionary Computing and their Applications, CRC Press USA, 1999.
- Jain, L.C. and Lazzerini, B., (Editors), Knowledge-Based Intelligent Techniques in Character Recognition, CRC Press USA, 1999.
- Teodorescu, H.N. Kandel, A. and Jain, L.C. (Editors), Soft Computing Techniques in Human Related Science, CRC Press USA, 1999.
- Jain, L.C. and Vemuri, R. (Editors), Industrial Applications of Neural Networks, CRC Press USA, 1998.
- Jain, L.C., Johnson, R.P., Takefuji, Y. and Zadeh, L.A. (Editors), Knowledge-Based Intelligent Techniques in Industry, CRC Press USA, 1998.
- Teodorescu, H.N., Kandel, A. and Jain, L.C. (Editors), Fuzzy and Neuro-fuzzy Systems in Medicine, CRC Press USA, 1998.

5.5.6 International Series on Natural and Artificial Intelligence, AKI http://www.innoknowledge.com

- Apolloni, B., et al, Algorithmic Inference in Machine Learning, Advanced Knowledge International, Australia, 2006.
- Lee, R.S.T., Advanced Paradigms in Artificial Intelligence, Advanced Knowledge International, Australia, 2005.
- Katarzyniak, R, Ontologies and Soft Methods in Knowledge Management, Advanced Knowledge International, Australia, 2005.
- Abe, A. and Ohsawa, Y. (Editors), Readings in Chance Discovery, Advanced Knowledge International, Australia, 2005.
- Kesheng Wang, K., Applied Computational Intelligence in Intelligent Manufacturing Systems, Advanced Knowledge International, Australia, 2005.
- Murase, K., Jain, L.C., Sekiyama, K. and Asakura, T. (Editors), Proceedings of the Fourth International Symposium on Human and Artificial Intelligence Systems, University of Fukui, Japan, Advanced Knowledge International, Australia, 2004.
- Nguyen N.T., (Editor) Intelligent Technologies for inconsistent Knowledge Processing, Advanced Knowledge International, Australia, 2004.
- Andrysek, J., et al., (Editors), Multiple Participant Decision Making, Advanced Knowledge International, Australia, 2004.
- Matsuda, K., Personal Agent-Oriented Virtual Society, Advanced Knowledge International, Australia, 2004.
- Ichimura, T. and Yoshida, K. (Editors), Knowledge-Based Intelligent Systems for Healthcare, Advanced Knowledge International, Australia, 2004.
- Murase, K., and Asakura, T. (Editors), Dynamic Systems Approach for Embodiment and Sociality: From Ecological Psychology to Robotics, Advanced Knowledge International, Australia, 2003.
- Jain, R., et al. (Editors), Innovations in Knowledge Engineering, Advanced Knowledge International, Australia, 2003.
- Graziella, T., Jain, L.C., Innovations in Decision Support Systems, Advanced Knowledge International, Australia, 2003.
- Galitsky, B., Natural Language Question Answering System Technique of Semantic Headers, Advanced Knowledge International, Australia, 2003.
- Guiasu, S., Relative Logic for Intelligence-Based Systems, Advanced Knowledge International, Australia, 2003.

5.5.7 Series on Innovative Intelligence, World Scientific: http://www.worldscientific.com/

- Jain, L.C., Howlett, R.J., Ichalkaranje, N., and Tonfoni, G. (Editors), Virtual Environments for Teaching and Learning, World Scientific Publishing Company Singapore, Volume 1, 2002.
- Jain, L.C., Ichalkaranje, N. and Tonfoni, G. (Editors), Advances in Intelligent Systems for Defence, World Scientific Publishing Company Singapore, Volume 2, 2002.

- Howlett, R., Ichalkaranje, N., Jain, L.C. and Tonfoni, G. (Editors), Internet-Based Intelligent Information Processing, World Scientific Publishing Company Singapore, Volume 3, 2002.
- Zaknich, A., Neural Nets for Intelligent Signal Processing, World Scientific Publishing Company Singapore, Volume 4, 2003.
- Hirose, A., Complex Valued Neural Networks, World Scientific Publishing Company Singapore, Volume 5, 2003.
- Shapiro, A.F. and Jain, L.C. (Editors), Intelligent and Other Computational Intelligence Techniques in Insurance, World Scientific Publishing Company Singapore, Volume 6, 2003.
- Pan, J-S., Huang, H.-C. and Jain, L.C. (Editors), Intelligent Watermarking Techniques, World Scientific Publishing Company Singapore, Volume 7, 2004.
- Hasebrook, J. and Maurer, H.A., Learning Support Systems for Organisational Learning, World Scientific Publishing Company Singapore, Volume 8, 2004.

5.6 Books

- Jain, L.C., Sato, M., Virvou, M., Tsihrintzis, G., Balas, V. and Abeynayake, C. (Editors), Computational Intelligence Paradigms: Volume 1 – Innovative Applications, Springer-Verlag, 2008.
- Phillips-Wren, G., Ichalkaranje, N. and Jain, L.C.(Editors), Intelligent Decision Making-An AI-Based Approach, Springer-Verlag, 2008.
- Fulcher, J. and Jain, L.C., Computational Intelligence: A Compendium, Springer-Verlag, 2008.
- Sordo, M., Vaidya, S. and Jain, L.C.(Editors), Advanced Computational Intelligence Paradigms in **Healthcare 3**, Springer-Verlag, 2008.
- Virvou, M. And Jain, L.C.(Editors) , Intelligent Interactive Systems in Knowledge-Based Environments, Springer-Verlag, 2008.
- Tsihrintzis, G. and Jain, L.C., Multimedia Services in Intelligent Environments, Springer-Verlag, 2008.
- Jarvis, J., Ronnquist, R, Jarvis, D. and Jain, L.C., Holonic Execution: A BDI Approach, Springer-Verlag, 2008.
- Jain, L.C., Palade, V. and Srinivasan, D. (Editors), Advances in Evolutionary Computing for System Design, Springer-Verlag, 2007.
- Baba, N., Handa, H. and Jain, L.C. (Editors), Advanced Intelligent Paradigms in Computer Games, Springer-Verlag, 2007.
- Chahl, J.S., Jain, L.C., Mizutani, A. and Sato-Ilic, M. (Editors), Innovations in Intelligent Machines 1, Springer-Verlag, 2007.
- Zharkova, V. and Jain, L.C. (Editors), Artificial Intelligence in Recognition and Classification of Astrophysical and Medical Images, Springer-Verlag, 2007.
- Pan, J-S., Huang, H.-C., Jain, L.C. and Fang, W.-C. (Editors), Intelligent Multimedia Data Hiding, Springer-Verlag, 2007.

- Yoshida, H., Jain, A., Ichalkaranje, A., Jain, L.C. and Ichalkaranje, N. (Editors), Advanced Computational Intelligence Paradigms in **Healthcare 1**, Springer-Verlag, 2007.
- Vaidya, S., Jain, L.C. and Yoshida, H. (Editors), Advanced Computational Intelligence Paradigms in **Healthcare 2**, Springer-Verlag, 2007.
- Jain, L.C, Tedman, R. and Tedman, D. (Editors), Evolution of Teaching and Learning in Intelligent Environment, Springer-Verlag, 2007.
- Sato, M. and Jain, L.C., Innovations in Fuzzy Clustering, Springer-Verlag, 2006.
- Patnaik, S., Jain, L.C., Tzafestas, S.G., Resconi, G. and Konar, A. (Editors), Innovations in Robot Mobility and Control, Springer-Verlag, 2006.
- Apolloni, B., Ghosh, A., Alpaslan, F., Jain, L.C. and Patnaik, S. (Editors), Machine Learning and Robot Perception, Springer-Verlag, 2006.
- Palade, V., Bocaniala, C.D. and Jain, L.C. (Editors), Computational Intelligence in Fault Diagnosis, Springer-Verlag, 2006.
- Holmes, D. and Jain, L.C. (Editors), Innovations in Machine Learning, Springer-Verlag, 2006.
- Ichalkaranje, N., Ichalkaranje, A. and Jain, L.C. (Editors), Intelligent Paradigms for Assistive and Preventive Healthcare, Springer-Verlag, 2006.
- Seiffert, U., Jain, L.C. and Schweizer, P. (Editors), Bioinformatics Using Computational Intelligence Paradigms, Springer-Verlag, 2005.
- Ghosh, A. and Jain, L.C. (Editors), Evolutionary Computation in Data Mining, Springer-Verlag, Germany, 2005.
- Phillips-Wren, G. and Jain, L.C.(Editors), Intelligent Decision Support Systems in Agent-Mediated Environments, IOS Press, The Netherlands, 2005.
- Silverman, B., Jain, A., Ichalkaranje, A. and Jain, L.C. (Editors), Intelligent Paradigms in Healthcare Enterprises, Springer-Verlag, Germany, 2005.
- Ghaoui, C., Jain, M., Bannore, V., and Jain, L.C. (Editors), Knowledge-Based Virtual Education, Springer-Verlag, Germany, 2005.
- Pal, N. and Jain, L.C.(Editors), Advanced Techniques in Knowledge Discovery and Data Mining, Springer-Verlag, London, 2005.
- Khosla, R., Ichalkaranje, N. and Jain, L.C.(Editors), Design of Intelligent Multi-Agent Systems, Springer-Verlag, Germany, 2005.
- Abraham, A., Jain, L.C. and van der Zwaag, B.(Editors), Innovations in Intelligent Systems, Springer-Verlag, Germany, 2004.
- Tonfoni, G. and Jain, L.C., Visualizing Document Processing, Mouton De Gruyter, Germany, 2004.
- Fulcher, J. and Jain, L.C.(Editors), Applied Intelligent Systems, Springer-Verlag, Germany, 2004.
- Damiani, E., Jain, L.C. and Madravio, M. (Editors), Soft Computing in Software Engineering, Springer-Verlag, Germany, 2004.
- Resconi, G. and Jain, L.C., Intelligent Agents: Theory and Applications, Springer-Verlag, Germany, 2004.
- Abraham, A., Jain, L.C. and Kacprzyk, J. (Editors), Recent Advances in Intelligent Paradigms and Applications, Springer-Verlag, Germany, 2003.

- Tonfoni, G. and Jain, L.C., The Art and Science of Documentation Management, Intellect, UK, 2003.
- Seiffert, U. and Jain, L.C. (Editors), Self-Organising Neural Networks, Springer-Verlag, Germany, 2002.
- Jain, L.C., Howlett, R.J., Ichalkaranje, N., and Tonfoni, G. (Editors), Virtual Environments for Teaching and Learning, World Scientific Publishing Company Singapore, 2002.
- Schmitt, M. Teodorescu, H.-N., Jain, A., Jain, A., Jain, S. and Jain, L.C. (Editors), Computational Intelligence Processing in Medical Diagnosis, Springer- Verlag, 2002.
- Jain, L.C. and Kacprzyk, J. (Editors), New Learning Paradigms in Soft Computing, Springer-Verlag, Germany, 2002.
- Jain, L.C., Chen, Z. and Ichalkaranje, N. (Editors), Intelligent Agents and Their Applications, Springer-Verlag, Germany, 2002.
- Jain, L.C. and De Wilde, P. (Editors), Practical Applications of Computational Intelligence Techniques, Kluwer Academic Publishers, USA, 2001.
- Howlett, R.J. and Jain, L.C. (Editors), Radial Basis Function Networks 1, Springer-Verlag, Germany, 2001.
- Howlett, R.J. and Jain, L.C. (Editors), Radial Basis Function Networks 2, Springer-Verlag, Germany, 2001.
- Teodorescu, H.N., Jain, L.C. and Kandel, A. (Editors), Hardware Implementation of Intelligent Systems, Springer-Verlag, Germany, 2001.
- Baba, N. and Jain, L.C., Computational Intelligence in Games, Springer-Verlag, 2001.
- Jain, L.C., Lazzerini, B. and Halici, U. (Editors), Innovations in ART Neural Networks, Springer-Verlag, Germany, 2000.
- Jain, A., Jain, A., Jain, S. and Jain, L.C. (Editors), Artificial Intelligence Techniques in Breast Cancer Diagnosis and Prognosis, World Scientific Publishing Company, Singapore, 2000.
- Jain, L.C. (Editor), Innovative Teaching and Learning in Intelligent Environment, Springer-Verlag, 2000.
- Jain, L.C. and Fukuda, T. (Editors), Soft Computing for Intelligent Robotic Systems, Springer-Verlag, Germany, 1998.
- Jain, L.C. (Editor), Soft Computing Techniques in Knowledge-Based Intelligent Engineering Systems, Springer-Verlag, Germany, 1997.
- Sato, M., Sato, S. and Jain, L.C., Fuzzy Clustering Models and Applications, Springer-Verlag, Germany, 1997.
- Jain, L.C. and Jain, R.K. (Editors), Hybrid Intelligent Engineering Systems, World Scientific Publishing Company, Singapore, 1997.
- Vonk, E., Jain, L.C. and Johnson, R.P., Automatic Generation of Neural Networks Architecture Using Evolutionary Computing, World Scientific Publishing Company, Singapore, 1997.
- Van Rooij, A.J.F., Jain, L.C. and Jain, L.C., Neural Network Training Using Genetic Algorithms, World Scientific Publishing Company, Singapore, December 1996.

6.7 Book Chapters

- Pedrycz, W., Ichalkaranje, N., Phillips-Wren, G., and Jain, L.C., Introduction to Computational Intelligence for Decision Making, Springer-Verlag, 2008, pp. 75-93, Chapter 3.
- Tweedale, J., Ichalkaranje, N., Sioutis, C., Urlings, P. and Jain, L.C., Future Directions: Building a Decision Making Framework using Agent Teams, Springer-Verlag, 2008, pp. 381-402, Chapter 14.
- Virvou, M. and Jain, L.C., Intelligent Interactive Systems in Knowledge-Based Environments: An Introduction, Springer-Verlag, 2008, pp. 1-8, Chapter 1.
- Tsihrintzis, G. and Jain, L.C., An Introduction to Multimedia Services in Intelligent Environments, Springer-Verlag, pp. 1-8, 2008, Chapter 1.
- Zharkova, V.V. and Jain, L.C., Introduction to Recognition and Classification in Medical and Astrophysical Images, Springer-Verlag, 2007, pp. 1-18, Chapter 1.
- Yoshida, H., Vaidya, S. and Jain, L.C., Introduction to Computational Intelligence in Healthcare, Springer-Verlag, 2007, pp. 1-4, Chapter 1.
- Huang, H.C., Pan, J.S., Fang, W.C. and Jain, L.C., An Introduction to Intelligent Multimedia Data Hiding, Springer-Verlag, 2007, pp. 1-10, Chapter 1.
- Jain, L.C., et al., Intelligent Machines :An Introduction, Springer-Verlag, 2007, pp. 1-9, Chapter 1.
- Jain, L.C., et al., Introduction to Evolutionary Computing in System Design, Springer-Verlag, 2007, pp. 1-9, Chapter 1.
- Jain, L.C., et al., Evolutionary Neuro-Fuzzy Systems and Applications, Springer-Verlag, 2007, pp. 11-45, Chapter 1.
- Do, Q.V, Lozo, P. and Jain, L.C., Vision-Based Autonomous Robot Navigation, in Innovations in Robot Mobility and Control, Springer-Verlag, 2006, pp. 65-103, Chapter 2.
- Tran, C., Abraham, A. and Jain, L., Soft Computing Paradigms and Regression Trees in Decision Support Systems, in Advances in Applied Artificial Intelligence, Idea Group Publishing, 2006, pp. 1-28, Chapter 1.
- Jarvis, B., Jarvis, D. and Jain, L., Teams in Multi-Agent Systems, in IFIP International Federation for Information Processing, Vol. **228**, Intelligent Information Processing III, Springer, 2006, pp. 1-10, Chapter 1.
- Abraham, A. and Jain, L.C., Evolutionary Multiobjective Optimization, Springer-Verlag, 2005, pp. 1-6, Chapter 1.
- Sisman-Yilmaz, N.A., Alpaslan, F. and Jain, L.C., Fuzzy Multivariate Auto-Regression Method and its Application, in Applied Intelligent Systems, Springer, 2004, pp. 281-300.
- Lozo, P., Westmacott, J., Do, Q., Jain, L.C. and Wu, L., Selective Attention ART and Object Recognition, in Applied Intelligent Systems, Springer, 2004, pp. 301-320.
- Wang, F., Jain, L.C. and Pan, J., Genetic Watermarking on Spatial Domain, in Intelligent Watermarking Techniques, World Scientific, 2004, pp. 481-514, Chapter 17.

- Wang, F., Jain, L.C. and Pan, J., Watermark Embedding System based on Visual Cryptography, in Intelligent Watermarking Techniques, World Scientific, 2004, pp. 377-394, Chapter 13.
- Sioutis, C., Urlings, P., Tweedale, J., Ichalkaranje, N., Forming Human-Agent Teams within Hostile Environments, in Applied Intelligent Systems, Springer-Verlag, 2004, pp. 255-279.
- Jain, L.C. and Chen, Z., Industry, Artificial Intelligence In, in Encyclopedia of Information Systems, Elsevier Science, USA, 2003, pp. 583-597.
- Jain, L.C. and Konar, A., An Introduction to Computational Intelligence Paradigms, in Practical Applications of Computational Intelligence Techniques, Springer, 2001, pp. 1-38.
- Tedman, D. and Jain, L.C., An Introduction to Innovative Teaching and Learning, in Teaching and Learning, Springer, 2000, pp. 1-30, Chapter 1.
- Filippidis, A., Russo, M. and Jain, L.C., Novel Extension of ART2 in Surface Landmine Detection, Springer-Verlag, 2000, pp.1-25, Chapter 1.
- Jain, L.C. and Lazzerini, B., An Introduction to Handwritten Character and Word Recognition, in Knowledge-Based Intelligent Techniques in Character Recognition, CRC Press, 1999, 3-16.
- Filippidis, A., Jain, L.C. and Martin, N.N., "Computational Intelligence Techniques in Landmine Detection," in Computing with Words in Information/Intelligent Systems 2, Edited by Zadeh, L. and Kacprzyk, J., Springer-Verlag, Germany, 1999, pp. 586-609.
- Halici, U., Jain, L.C. and Erol, A., Introduction to Fingerprint Recognition, in Intelligent Biometric Techniques in Fingerprint and Face Recognition, CRC Press, 1999, pp.3-34.
- Teodorescu, H.N., Kandel, A. and Jain, L., Fuzzy Logic and Neuro-Fuzzy Systems in Medicine: A historical Perspective, in Fuzzy and Neuro-Fuzzy Systems in Medicine, CRC Press, 1999, pp. 3-16.
- Jain, L.C. and Vemuri, R., An Introduction to Intelligent Systems, in Hybrid Intelligent Engineering Systems, World Scientific, 1997, pp. 1-10, Chapter 1.
- Karr, C. and Jain, L.C., Genetic Learning in Fuzzy Control, in Hybrid Intelligent Engineering Systems, World Scientific, 1997, pp. 69-101, Chapter 4.
- Karr, C. and Jain, L.C., Cases in Geno- Fuzzy Control, in Hybrid Intelligent Engineering Systems, World Scientific, 1997, pp. 103-132, Chapter 5.
- Katayama, R., Kuwata, K. and Jain, L.C., Fusion Technology of Neuro, Fuzzy, GA and Chaos Theory and Applications, in Hybrid Intelligent Engineering Systems, World Scientific, 1997, pp. 167-186, Chapter 7.
- Jain, L.C., Medsker, L.R. and Carr, C., Knowledge-Based Intelligent Systems, in Soft Computing Techniques in Knowledge-Based Intelligent Systems, Springer-Verlag, 1997, pp. 3-14, Chapter 1.
- Babri, H., Chen, L., Saratchandran, P., Mital, D.P., Jain, R.K., Johnson, R.P. and Jain, L.C., Neural Networks Paradigms, in Soft Computing Techniques in Knowledge-Based Intelligent Systems, Springer-Verlag, 1997, pp. 15-43, Chapter 2.

- Jain, L.C., Tikk, D. and Koczy, L.T., Fuzzy Logic in Engineering, in Soft Computing Techniques in Knowledge-Based Intelligent Systems, Springer-Verlag, 1997, pp. 44-70, Chapter 3.
- Tanaka, T. and Jain, L.C., Analogue/Digital Circuit Representation for Design and Trouble Shooting in Intelligent Environment, in Soft Computing Techniques in Knowledge-Based Intelligent Systems, Springer-Verlag, 1997, pp. 227-258, Chapter 7.
- Jain, L.C., Hybrid Intelligent System Design Using Neural Network, Fuzzy Logic and Genetic Algorithms - Part I, Cognizant Communication Corporation USA, 1996, pp. 200-220, Chapter 9.
- Jain, L.C., Hybrid Intelligent System Applications in Engineering using Neural Network and Fuzzy Logic - Part II, Cognizant communication Corporation USA,1996, pp. 221-245, Chapter 10.
- Jain, L.C., Introduction to Knowledge-Based Systems, Electronic Technology Directions to the Year 2000, IEEE Computer Society Press USA, 1995, pp. 17-27, Chapter 1.
- Jain, L.C. and Allen, G.N., Introduction to Artificial Neural Networks, Electronic Technology Directions to the Year 2000, IEEE Computer Society Press USA, 1995, pp. 36-62, Chapter 2.
- Jain, L.C. and Karr, C.L., Introduction to Fuzzy Systems, Electronic Technology Directions to the Year 2000, IEEE Computer Society Press USA, 1995, pp. 93-103, Chapter 3.
- Jain, L.C. and Karr, C.L., Introduction to Evolutionary Computing Techniques, Electronic Technology Directions to the Year 2000, IEEE Computer Society Press USA, 1995, pp. 121-127, Chapter 4.
- Sato, M., Jain, L.C. and Takagi, H., Electronic Design and Automation, Electronic Technology Directions to the Year 2000, IEEE Computer Society Press USA, 1995, pp. 402-430, Chapter 9.
- Furuhashi, T., Takagi, H. and Jain, L.C., Intelligent Systems using Artificial Neural Networks, fuzzy Logic and Genetic Algorithms in Industry, Electronic Technology Directions to the Year 2000, IEEE Computer Society Press USA, 1995, pp. 485-495, Chapter 10.

References

Amorim, D.G., Delgado, M.F., Ameneiro, S.B.: Polytope ARTMAP: Pattern classification without vigilance based on general geometry categories. IEEE Transactions on Neural Networks 18, 1306–1325 (2007)

Angeline, P.J., Saunders, G.M., Pollack, J.B.: An evolutionary algorithm that constructs recurrent neural networks. IEEE Transactions on Neural Networks 5, 54–65 (1994)

Behera, L., Kumar, S., Patnaik, A.: On adaptive learning rate that guarantees convergence in feedforward networks. IEEE Transactions on Neural Networks 17, 1116–1125 (2006)

Bezdek, J.C.: What is a computational intelligence? In: Zurada, J.M., Marks II, R.J., Robinson, C.J. (eds.) Computational Intelligence: Imitating Life, pp. 1–12. IEEE Press, Los Alamitos (1994)

Carpenter, G.A., Grossberg, S.: A massively parallel architecture for a self-organising neural pattern recognition machine. Computer Vision, Graphics and Image Processing 37, 54–115 (1987a)

Carpenter, G.A., Grossberg, S.: ART 2: Stable self-organisation of pattern recognition codes for analogue input patterns. Applied Optics 26, 4919–4930 (1987b)

Carpenter, G.A., Grossberg, S., Markuzon, N., Reynolds, J.H., Rosen, D.B.: Fuzzy ARTMAP: A neural network architecture for incremental supervised learning of analogue multidimensional maps. IEEE Transactions on Neural Networks 3, 698–712 (1992)

Casillas, J., Carse, B., Bull, L.: Fuzzy-XCS: A michigan genetic fuzzy system. IEEE Transactions on Fuzzy Systems 15, 536–550 (2007b)

Casillas, J., Cordón, O., del Jesus, M.J., Herrera, F.: Genetic tuning of fuzzy rule deep structures preserving interpretability and its interaction with fuzzy rule set reduction. IEEE Transactions on Fuzzy Systems 13, 13–29 (2007a)

Fogel, D.: Review of Computational Intelligence: Imitating Life. IEEE Trans. on Neural Networks 6, 1562–1565 (1995)

Fogel, L.J., Owens, A.J., Walsh, M.J.: Artificial Intelligence through Simulated Evolution. John Wiley, New York (1966)

García-Pedrajas, N., Hervás-Martinez, C., Muñoz-Perez, J.: Symbiont: A cooperative evolutionary model for evolving artificial neural networks for classification. In: Bouchon-Meunier, B., Gutierrez-Rios, J., Magdalena, L., Yager, R.R. (eds.) Technologies for constructing intelligent systems: tools, vol. 2, pp. 341–354. Springer, Berlin (2002a)

García-Pedrajas, N., Hervás-Martinez, C., Muñoz-Perez, J.: Multi-objective cooperative coevolution of artificial neural networks (multi-objective cooperative networks). Neural Networks 15, 1259–1278 (2002b)

García-Pedrajas, N., Hervás-Martinez, C., Muñoz-Perez, J.: Covnet: A cooperative coevolutionary model for evolving artificial neural networks. IEEE Transactions on Neural Networks 14, 820–834 (2003)

Gonçalves, L.B., Vellasco, M.M.B.R., Pacheco, M.A.C., de Souza, F.J.: Inverted hierarchical neuro-fuzzy bsp system: a novel neuro-fuzzy model for pattern classification and rule extraction in databases. IEEE Transactions on Systems, Man, and Cybernectics–C 36, 236–248 (2006)

Gorzalczany, M.B.: Computational Intelligence Systems and Applications: Neuro-Fuzzy and Fuzzy Neural Synergisms. Physica-Verlag, Heidelberg (2002)

Goulermas, J.Y., Liatsis, P., Zeng, X.-J., Cook, P.: Density-driven generalized regression neural networks (DD-GRNN) for function approximation. IEEE Transactions on Neural Networks 18, 1683–1696 (2007)

Gupta, M.M., Qi, J.: On fuzzy neuron models. In: Zadeh, L.A., Kacprzyk, J. (eds.) Fuzzy Logic for the Management of Uncertainty, pp. 479–491. J. Wiley & Sons, New York (1992)

Hayashi, Y., Buckley, J.J., Czogala, E.: Fuzzy neural network with fuzzy signals and weights. International Journal of Intelligent Systems 8, 527–537 (1993)

Hebb, D.O.: The Organisation of Behaviour. John Wiley and Sons, New York (1949)

Holland, J.H.: Outline for a logical theory of adaptive systems. J. ACM 3, 297–314 (1962)

Holland, J.H.: Escaping brittleness: The possibility of general-purpose learning algorithms applied to parallel rule-based systems. In: Michalski, R.S., Carbonell, J.G., Mitchell, T.M. (eds.) Mach. Learn., pp. 593–624. Morgan Kaufmann, Los Altos (1986)

Hopfield, J.J.: Neural networks and physical systems with emergent collective computational abilities. Proceedings of the National Academy of Science 79, 2554–2558 (1982)

Hopfield, J.J.: Neurons with graded response have collective computational properties like those of two-state neurons. Proceedings of the National Academy of Science 81, 3088–3092 (1984)

Jang, J.S.R.: ANFIS: Adaptive-network-based-fuzzy-inference-system. IEEE Transactions on Systems, Man, and Cybernectics 23, 665–685 (1993)

John, R., Coupland, S.: Type-2 fuzzy logic: A historical view. IEEE Computational Intelligence Magazine 2, 57–62 (2007)

Juang, C.-F.: A TSK-Type Recurrent Fuzzy Network for Dynamic Systems Processing by Neural Network and Genetic Algorithms. IEEE Transactions on Fuzzy Systems 10, 155–170 (2002)

Kasabov, N.K.: Evolving fuzzy neural networks for supervised/unsupervised online knowledge-based learning. IEEE Transactions on Systems, Man, and Cybernectics–B 31, 902–918 (2001)

Kasabov, N.K., Song, Q.: DENFIS: Dynamic evolving neural-fuzzy inference system and its application for time-series prediction. IEEE Transactions on Fuzzy Systems 10, 144–154 (2002)

Kohonen, T.: Self-organising formation of topologically correct feature maps. Biological Cybernetics 43, 59–69 (1982)

Kohonen, T.: Self-Organization and Associative Memory. Springer, Heidelberg (1988)

Koza, J.R.: Genetic Programming. The MIT Press, Cambridge (1992)

Lian, J., Lee, Y., Sudhoff, S.D., Żak, S.H.: Self-organizing radial basis function network for real-time approximation of continuous-time dynamical systems. IEEE Transactions on Neural Networks 19, 460–474 (2008)

Lin, C.-J., Hsu, Y.-C.: Reinforcement hybrid evolutionary learning for recurrent wavelet-based neurofuzzy systems. IEEE Transactions on Fuzzy Systems 15, 729–745 (2007)

Liu, P., Li, H.: Efficient learning algorithms for three-layer regular feedforward fuzzy neural networks. IEEE Transactions on Neural Networks 15, 545–558 (2004)

Manrique, D., Rios, J., Rodriguez-Paton, A.: Evolutionary system for automatically constructing and adapting radial basis function networks. Neurocomputing 69, 2268–2283 (2006)

Marks, R.: Intelligence: computational versus artificial. IEEE Transactions on Neural Networks 4, 737–739 (1993)

McCulloch, W.S., Pitts, W.: A logical calculus of the ideas immanent in nervous activity. Bulletin of Mathematical Biophysics 5, 115–133 (1943)

Mendel, J.M.: Type-2 fuzzy sets and systems: an overview. IEEE Computational Intelligence Magazine 2, 20–29 (2007)

Mitra, S., De, R.K., Pal, S.K.: Knowledge-based fuzzy mlp for classification and rule generation. IEEE Transactions on Neural Networks 8, 1338–1350 (1997)

Moody, J., Darken, C.J.: Fast learning in networks of locally-tuned processing units. Neural Computation 1, 281–294 (1989)

Mou, S., Gao, H., Lam, J., Qiang, W.: A new criterion of delay-dependent asymptotic stability for hopfield neural networks with time delay. IEEE Transactions on Neural Networks 19, 532–535 (2008)

Nauck, D., Kruse, R.: A neuron-fuzzy method to learn fuzzy classification rules from data. Fuzzy Sets and Systems 89, 277–288 (1997)

Nauck, D., Kruse, R.: Neuro-fuzzy systems for function approximation. Fuzzy Sets and Systems 101, 261–271 (1999)

Pal, T., Pal, N.R.: SOGARG: A self-organized genetic algorithm-based rule generation scheme for fuzzy controllers. IEEE Transactions on Evolutionary Computations 7, 397–415 (2003)

Palmes, P.P., Hayasaka, T., Usui, S.: Mutation-based genetic neural network. IEEE Transactions on Neural Networks 16, 587–600 (2005)

Paul, S., Kumar, S.: Subsethood-product fuzzy neural inference system (SuPFuNIS). IEEE Transactions on Neural Networks 13, 578–599 (2002)

Pedrycz, W.: Computational intelligence: an introduction. In: Szczepaniak, P.S. (ed.) Computational Intelligence and Applications, pp. 3–17. Physical-Verlag, Heidelberg (1999)

Peña-Reyes, C.A., Sipper, M.: Fuzzy CoCo: A Cooperative-Coevolutionary Approach to Fuzzy Modeling. IEEE Transactions on Fuzzy Systems 9, 727–737 (2001)

Pitts, W., McCulloch, W.S.: How we know universals: the perception of auditory and visual forms. Bulletin of Mathematical Biophysics 9, 127–147 (1947)

Rechenberg, I.: Evolutionary strategy. In: Zurada, J.M., Marks II, R., Robinson, C. (eds.) Computational Intelligence: Imitating Life, pp. 147–159. IEEE Press, Los Alamitos (1994)

Rosenblatt, F.: The perceptron, a probabilistic model for information storage and organisation in the brain. Psychological Review 65, 386–408 (1958)

Rumelhart, D.E., Hinton, G.E., Williams, R.J.: Learning internal representation by error propagation. In: Rumelhart, D.E., McLelland, J.L. (eds.) Parallel Distributed Processing, vol. I, pp. 318–362. MIT Press, Cambridge (1986)

Schwefel, H.-P.: Numerical Optimization of Computer Models. John Wiley, Chichester (1981)

Song, Q., Kasabov, N.K.: NFI: A neuro-fuzzy inference method for transductive reasoning. IEEE Transactions on Fuzzy Systems 13, 799–808 (2005)

Song, T., Jamshidi, M.M., Lee, R.R., Huang, M.: A modified probabilistic neural network for partial volume segmentation in brain mr image. IEEE Transactions on Neural Networks 18, 1424–1432 (2007)

Su, S.-F., Yang, F.-Y.P.: On the dynamical modeling with neural fuzzy networks. IEEE Transactions on Neural Networks 13, 1548–1553 (2002)

Velayutham, S., Kumar, S.: Asymmetric subsethood-product fuzzy neural inference system (ASuPFuNIS). IEEE Transactions on Neural Networks 16, 160–174 (2005)

Wang, C.-H., Cheng, C.-S., Lee, T.-T.: Dynamical Optimal Training for Interval Type-2 Fuzzy Neural Network (T2FNN). IEEE Transactions on Systems, Man, and Cybernectics–B 34, 1462–1477 (2004)

Widrow, B., Hoff, M.E.: Adaptive switching circuits. In: IRE WESCON Convention Record, pp. 96–104 (1960)

Yamakawa, T., Uchino, E., Miki, T., Kusanagi, H.: A neo fuzzy neuron and its application to system identification and prediction of the system behaviour. In: Proc. of 2-nd International Conference on Fuzzy Logic and Neural Networks, pp. 477–483 (1992)

Yao, X.: Evolving artificial neural networks. Proceedings of IEEE 87, 1423–1447 (1999)

Yao, X., Liu, Y.: A new evolutionary system for evolving artificial neural networks. IEEE Transactions on Neural Networks 8, 694–713 (1997)

Zadeh, L.A.: Fuzzy Sets. Information and Control 8, 338–353 (1965)

Zadeh, L.A.: Outline of a new approach to the analysis of complex systems and decision processes. IEEE Transactions on System, Man and Cybernectics SMC-3, 28–44 (1973)

Zadeh, L.A.: The concept of a linguistic variable and its application to approximate reasoning–1. Information Sciences 8, 199–249 (1975)

Zhang, Y.-Q., Fraser, M.D., Gagliano, R.A., Kandel, A.: Granular neural networks for numerical-linguistic data fusion and knowledge discovery. IEEE Transactions on Neural Networks 11, 658–667 (2000)

A Quest for Adaptable and Interpretable Architectures of Computational Intelligence

Witold Pedrycz

Department pf Electrical and Computer Engineering, University of Alberta, Edmonton, T6R 2G7 Canada and Systems Research Institute, Polish Academy of Sciences, Warsaw, Poland
pedrycz@ee.ualberta.ca

Abstract. The agenda of fuzzy neurocomputing focuses on the development of artifacts that are both adaptable (so any learning pursuits could be carried out in an efficient manner) and interpretable (so that the results are easily understood by the user or designer). The logic is the language of interpretable constructs. Neural architectures offer a flexible and convenient setting for learning. The study conveys a message that a suitable combination of logic incorporated into the structure of a specialized neuron leads to interpretable and elastic processing units one can refer to as fuzzy neurons. We investigate the main categories of such neurons and elaborate on the ensuing topologies of the networks emphasizing a remarkably rich landscape of logic architectures associated with the use of the logic neurons.

1 Introductory Comments

Logic and logic constructs have been a conceptual and operational framework for computing, knowledge representation and numerous information systems. In the similar vein multivalued logic and fuzzy sets contribute to a variety of tasks of analysis and synthesis of intelligent systems. One of the fundamental quests of fuzzy systems is to construct architectures that are both flexible (and in this way can easily adjust to the dynamic environment) as well as easily interpretable and hence understood by users and designers. Interestingly enough, the current quite visible trend is concerned with the development of fuzzy systems (models) of high approximation capabilities while the interpretability issues seem to be addressed to a far lesser extent.

Our objective is to stress the need for forming structures that are both flexible and interpretable. This quest almost explicitly implies a direct need to exploit logic constructs of fuzzy logic. The concept of logic neurons becomes instrumental in the realization of transparent yet flexible structures of fuzzy logic.

The paper is arranged into 7 sections. We start by discussing generic types of logic connectives (logic operators) and show their use in the construction of basic processing units – fuzzy neurons. Here we distinguish between aggregative neurons (AND and OR neurons), referential neurons and unineurons. Next we show some main classes of logic networks emphasizing their diversity and processing functionality. Some discussion on learning strategies is also covered; in particular in the context of networks composed of unineurons. The role of such networks in fuzzy modeling and granular modeling [1][16][17], in general, is also discussed.

L.C. Jain et al. (Eds.): Comp. Intel. Para.: Innov. Applications, SCI 137, pp. 25–49, 2008.
springerlink.com

2 Generic Logic Processing Realized with the Aid of Logic Neurons

Logic operators (connectives) have been around since the inception of fuzzy sets as the integral part of the overall modeling environment built with the aid of information granules. The growing diversity of the operators both in terms of the main categories of logic connectives, their flexibility which manifests in an enormous parametric diversity of the detailed models have offered a wealth of opportunities to utilize them in constructs of logic neurons – computing elements that seamlessly combine the logic nature of operators with the parametric flexibility associated with their weights (connections). In what follows, we highlight the architectures of such neurons showing that the logic underpinnings offer a wealth of functionality. The key categories of neurons involve two classes. The first class consists of aggregative neurons completing AND and OR aggregation. A generalization of those comes under the rubric of uninorms which give rise to so-called unineurons. The second category concerns so-called referential neurons in which one realizes a suite of logic predicates such as dominance, inclusion, equality and difference. For the sake of completeness, we start with a brief recollection of the generic facts about triangular norms and uninorms.

2.1 Triangular Norms: t- Norms

When contemplating any realization of operations of intersection and union of fuzzy sets, we should require a satisfaction of the following set of intuitively appealing properties:

(a) commutativity,
(b) associativity,
(c) monotonicity, and
(d) identity.

The last requirement of identity takes on a different form depending on the operation. More specifically, in the case of intersection, we anticipate that an intersection of any fuzzy set with the universe of discourse **X** should return this fuzzy set. For the union operations, the identity implies that the union of any fuzzy set and an empty fuzy set returns the fuzzy set.

Thus any binary operator $[0,1] \times [0,1] \rightarrow [0,1]$ which satisfies the collection of the requirements outlined above can be regarded as a potential candidate to realize the intersection or union of fuzzy sets. Note also that identity acts as boundary conditions meaning that when confining to sets, the above stated operations return the same results as encountered in set theory. In general, idempotency is not required, however the realizations of union and intersection could be idempotent as this happens for the operations of minimum and maximum where (min(a, a) = a and max(a, a) = a).

In the theory of fuzzy sets, triangular norms [6][7][13][14][20] offer a general class of operators of intersection and union. For instance, t-norms generalize intersection of fuzzy sets. Given a t-norm, a dual operator called a t-conorm (or s-norm) can be derived using the relationship $x \text{ s } y = 1 - (1 - x) \text{ t } (1 - y)$, $\forall$ x,y $\in$ [0,1], the De Morgan law [(20], but t-conorm can also be described by an independent axiomatic definition [14] Triangular conorms provide generic models for the union of fuzzy sets.

Let us recall that a triangular norm, t-norm for brief, is a binary operation t: $[0,1] \times [0,1] \rightarrow [0,1]$ that satisfies the following properties

1. Commutativity: $a\ t\ b = b\ t\ a$
2. Associativity: $a\ t\ (b\ t\ c) = (a\ t\ b)\ t\ c$
3. Monotonicity: if $b \leq c$ then $a\ t\ b \leq a\ t\ c$
4. Boundary conditions: $a\ t\ 1 = a$
   $a\ t\ 0 = 0$

where $a,b,c \in [0,1]$.

Let us elaborate on the meaning of these requirements vis-à-vis the use of t-norms as models of operators of union and intersection of fuzzy sets. There is a one-to-one correspondence between the general requirements outlined above and the properties of t-norms. The first three reflect the general character of set operations. Boundary conditions stress the fact all t-norms attain the same values at boundaries of the unit square $[0,1] \times [0,1]$. Thus, for sets, any t-norm produces the same result that coincides with the one we could have expected in set theory when dealing with intersection of sets, that is $A \cap \mathbf{X} = A$, $A \cap \varnothing = \varnothing$. Some commonly encountered examples of t-norms include the following operations:

1. Minimum: $a\ t_m\ b = \min(a, b) = a \wedge b$
2. Product: $a\ t_p\ b = ab$
3. Lukasiewicz: $a\ t_l\ b = \max(a + b - 1, 0)$
4. Drastic product: $a\ t_d\ b = \begin{cases} a & \text{if } b=1 \\ b & \text{if } a=1 \\ 0 & \text{otherwise} \end{cases}$

The second class of logic operators concerns realizations of unions of fuzzy sets which lead to the concept of t-conorms. Triangular conorms are functions s: $[0,1] \times [0,1] \rightarrow [0,1]$. that serve as generic realizations of the union operator on fuzzy sets. Similarly as triangular norms, conorms provide the highly desirable modeling flexibility needed to construct fuzzy models.

2.2 Triangular Norms: t-Conorms

Triangular conorms can be viewed as dual operators to the t-norms and as such, explicitly defined with the use of De Morgan laws. We may characterize them in a fully independent manner by providing the following definition.

A triangular conorm (s-norm) is a binary operation s: $[0,1] \times [0,1] \rightarrow [0,1]$ that satisfies the following requirements

1. Commutativity: $a\ s\ b = b\ s\ a$
2. Associativity: $a\ s\ (b\ s\ c) = (a\ s\ b)\ s\ c$
3. Monotonicity: if $b \leq c$ then $a\ s\ b \leq a\ s\ c$
4. Boundary conditions: $a\ s\ 0 = a$
   $a\ s 1 = 1$

$a,b,c \in [0,1]$.

One can show that s: $[0,1] \times [0,1] \rightarrow [0,1]$ is a t-conorm if and only if there exists a t-norm (dual t-norm) such that for $\forall$ a,b $\in$ [0,1] we have

$$a\ s\ b = 1 - (1 - a)\ t\ (1 - b) \tag{1}$$

For the corresponding dual t-norm we have

$$a\ t\ b = 1 - (1 - a)\ s\ (1 - b) \tag{2}$$

The duality expressed by (1) and (2) can be viewed as an alternative definition of t-conorms. This duality allows us to deduce the properties of t-conorms on the basis of the analogous properties of t-norms. Notice that after rewriting (1) and (2), we obtain

$$(1 - a)\ t\ (1 - b) = 1 - a\ s\ b$$

$$(1 - a)\ s\ (1 - b) = 1 - a\ t\ b$$

These two relationships can be expressed symbolically as

$$\overline{A} \cap \overline{B} = \overline{A \cup B}$$

$$\overline{A} \cup \overline{B} = \overline{A \cap B}$$

that are nothing but the De Morgan laws.

The boundary conditions mean that all t-conorms behave similarly at the boundary of the unit square $[0,1] \times [0,1]$. Thus, for sets, any t-conorm returns the same result as encountered in set theory

A list of commonly used t-conorms includes the following

Maximum:	$a\ s_m\ b = \max(a, b) = a \vee b$
Probabilistic sum:	$a\ s_p\ b = a + b - ab$
Lukasiewicz:	$a\ s_l\ b = \min(a + b, 1)$
Drastic sum:	$a\ s_d\ b = \begin{cases} a & \text{if } b=0 \\ b & \text{if } a=0 \\ 1 & \text{otherwise} \end{cases}$

2.3 Uninorms – A Hybrid Structure of Logic Operators

Uninorms form an interesting generalization of t- norms and t-conorms as they bind these two constructs. The switching between the "and" and "or" type of aggregation occurs on a basis of the numeric values of its arguments.

Formally speaking, a uninorm is a two argument mapping [2][3][9][10][11][22] [23] u: $[0,1]^2 \rightarrow [0,1]$ satisfying the following properties

Commutativity	$u(x, y) = u(y, x)$
Monotonicity	$u(x,y) \geq u(z, v)$ for $x > z$ and $y > v$
Associativity	$u(x, u(y, z)) = u(u(x,y), z)$

Existence of identity (neutral) element (g):

there exists some $g \in [0,1]$, called identity, such that $u(x,g) = x$

Uninorms differ from t-norms or t-conorms (s-norms) in the last property shown above. In the definition, we allow the values of the identity element "g" to vary in-between 0 and 1. As a result of this, we can implement switching between pure *and* and *or* properties of the logic operators occurring in this construct. Evidently when g = 0 we end up with the "or" type of aggregation, u(x, 0) = x. The identity element such that g =1 returns the "and" type of aggregation, namely u(x, 1) = x.

In the existing literature we can encounter different realizations of uninorms, see [11][21]. In this study, we confine ourselves to the following family of constructs that seem to be highly readable and in this sense intuitively appealing

$$u(x,y)=\begin{cases} g\,t(\frac{x}{g},\frac{y}{g}) & \text{if } x,y\in[0,g] \\ g+(1-g)s(\frac{x-g}{1-g},\frac{y-g}{1-g}) & \text{if } x,y\in[g,1] \\ \min(x,y) \text{ or } \max(x,y), & \text{otherwise} \end{cases} \tag{3}$$

In the above expression, "t" denotes a certain t-norm while "s" stands for some t-conorm (s-norm). Graphically, we can portray this uninorm, including its two fundamental variants (based on the max and min operations), in the way visualized in Figure 1. In general, we do not require any relationship (e.g., duality) to be satisfied between the specific t- norm and t-conorm being used in this construct.

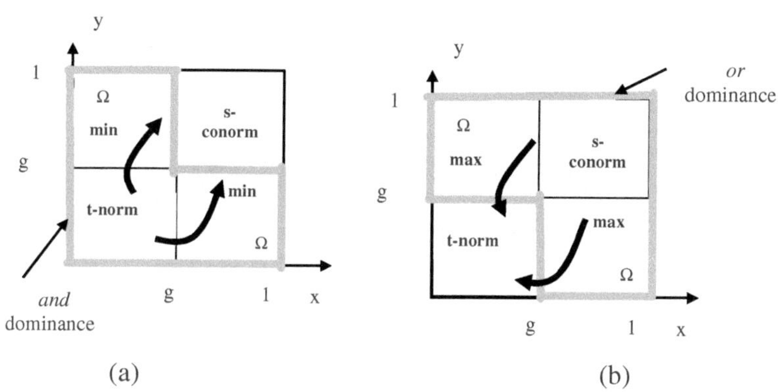

Fig. 1. Uninorm development – two alternative realizations: *and*-dominated (a), and *or*-dominated (b). The pertinent dominance regions (*and* or *or*, respectively) are highlighted.

In virtue of definition (3), we observe that the two intermediate regions (denoted here by Ω) deliver some flexibility to the specific realization of the uninorm. In the first case, we encounter the minimum operation which in conjunction with the region positioned below the identity element can be treated as the extension of the *and*-like aggregation (note that the minimum is the largest t-norm and in this sense could be sought as the extreme realization of the *and* connective). We may refer to this construct as the *and*-dominated uninorm. In the second case, we prefer extending the *or*-type of aggregation by accepting the maximum operation in the intermediate regions. This leads to the emergence of the *or* type of dominance. Note that the maximum operator is the smallest among all t-conorms. This brings up a concept of the *or*-dominated uninorm. Conceptually, these two realizations are distinct as they favor different slant of the logic aggregation with the dominant nature of the *and* or *or* character. As indicated, a certain value of the identity element implies the corresponding dominance mix of the logic operations. Considering the area occupied by some t-norm (and minimum) and t-conorm (and maximum, respectively), we can quantify the *or*-ness and *and*-ness character of the uninorm by computing the following integrals (those are the corresponding areas of the unit square),

or-ness degree: $(1\text{-}g)^2$ for *and* dominance and $1\text{-}g^2$ for *or* dominance
and-ness degree : $1\text{-} (1\text{-}g)^2$ for *and* dominance and $1\text{-}g^2$ for *or* dominance

This helps us understand the character of the logic aggregation completed by some uninorm and being expressed in terms of the generic logic connectives. The case of g = 0.5 places the uninorm in a somewhat neutral position with an equal balance of the *or* and *and* character of processing (when being expressed by the degrees formulated above).

In addition to the previous schematic view of the uninorms, the detailed plots of the characteristics for some specific t-norms and t-conorms are displayed in Figure 2.

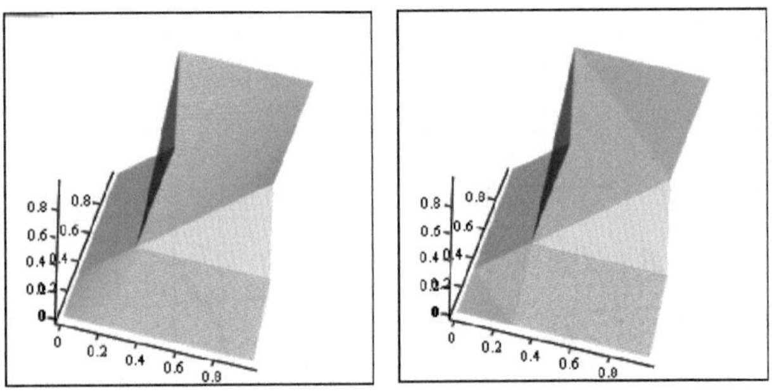

Fig. 2. Two-dimensional plots of the uninorms (*and* dominated) with two realizations: product and probabilistic sum, (b) Lukasiewicz connectives. The value of "g" is equal to 0.3.

3 Logic Neurons

Formally, these neurons realize a logic mapping from $[0,1]^n$ to $[0,1]$ cf. [8][12][15][16][18]. Two main classes of the processing units exist in this category, namely aggregative and referential neurons.

3.1 Aggregative Neurons: OR and AND Neurons

OR neuron: realizes an *and* logic aggregation of inputs $\mathbf{x} = [x_1\ x_2 \ldots x_n]$ with the corresponding connections (weights) $\mathbf{w} = [w_1\ w_2 \ldots\ w_n]$ and then summarizes the partial results in an *or*-wise manner (hence the name of the neuron). The concise notation underlines this flow of computing, $y = \mathrm{OR}(\mathbf{x}; \mathbf{w})$ while the realization of the logic operations gives rise to the expression (we commonly refer to it as an s-t combination or s-t aggregation)

$$y = \mathop{S}_{i=1}^{n} (x_i t w_i) \tag{4}$$

Bearing in mind the interpretation of the logic connectives (t-norms and t-conorms), the OR neuron realizes the following logic expression being viewed as an underlying logic description of the processing of the input signals

$$(x_1 \textit{ and } w_1) \textit{ or } (x_2 \textit{ and } w_2) \textit{ or } \ldots \textit{ or } (x_n \textit{ and } w_n) \tag{5}$$

Apparently the inputs are logically "weighted" by the values of the connections before producing the final result. In other words we can treat "y" as a truth value of the above statement where the truth values of the inputs are affected by the corresponding weights. Noticeably, lower values of w_i discount the impact of the corresponding inputs; higher values of the connections (especially those being positioned close to 1) do not affect the original truth values of the inputs resulting in the logic formula. In limit, if all connections w_i , i =1, 2,…,n are set to 1 then the neuron produces a plain *or*-combination of the inputs, $y = x_1$ *or* x_2 *or* … *or* x_n. The values of the connections set to zero eliminate the corresponding inputs. Computationally, the OR neuron exhibits nonlinear characteristics (that are inherently implied by the use of the t- and t-conorms (that are evidently nonlinear mappings). The plots of the characteristics of the OR neuron shown in Figure 3 shows this effect (note that the characteristics are affected by the use of some triangular norms). The connections of the neuron contribute to its adaptive character; the changes in their values form the crux of the parametric learning.

AND neuron:the neurons in the category, denoted by y = AND(**x**; **w**) with **x** and **w** being defined as in case of the OR neuron, are governed by the expression

$$y = \mathop{T}_{i=1}^{n} (x_i s w_i) \tag{6}$$

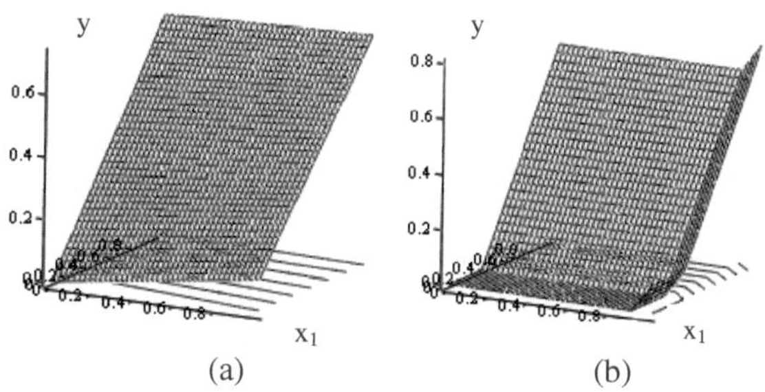

Fig. 3. Characteristics of the OR neuron for selected pairs of t- and t-conorms. In all cases the corresponding connections are set to 0.1 and 0.7 with intent to visualize their effect on the input-output characteristics of the neuron: (a) product and probabilistic sum, (b) Lukasiewicz *and* and *or* connectives (b).

Here the *or* and *and* connectives are used in a reversed order: first the inputs are combined with the use of the t-conorm and the partial results produced in this way are aggregated *and*-wise. Higher values of the connections reduce impact of the corresponding inputs. In limit $w_i = 1$ eliminates the relevance of x_i. With all values of w_i being set to 0, the output of the AND neuron is just an *and* aggregation of the inputs

$$y = x_1 \text{ and } x_2 \text{ and } \ldots \text{ and } x_n \quad (7)$$

The characteristics of the AND neuron are shown in Figure 4; note the influence of the connections and the specific realization of the triangular norms on the mapping completed by the neuron.

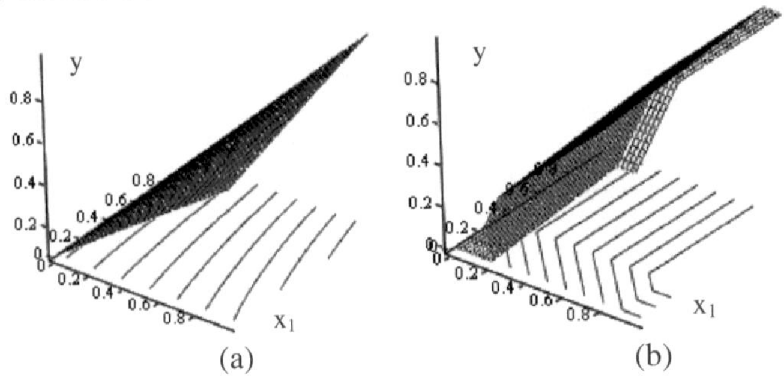

Fig. 4. Characteristics of AND neurons for selected pairs of t- and t-conorms. In all cases the connections are set to 0.1 and 0.7 with intent to visualize their effect on the characteristics of the neuron: (a) product and probabilistic sum, (b) Lukasiewicz logic connectives.

Let us conclude that the neurons are highly nonlinear processing units whose nonlinear mapping depends upon the specific realizations of the logic connectives. They also come with potential plasticity whose usage becomes critical when learning the networks including such neurons.

3.2 Referential (Reference) Neurons

The essence of referential computing deals with processing logic predicates. The two-argument (or generally multivariable) predicates such as *similar*, *included in*, *dominates* [18] are essential components of any logic description of a system. In general, the truth value of the predicate is a degree of satisfaction of the expression P(x, a) where "a" is a certain reference value (reference point). Depending upon the meaning of the predicate (P), the expression P(x, a) reads as " x is similar to a", " x is included in a", "x dominates a", etc. In case of many variables, the compound predicate comes in the form $P(x_1, x_2, .., x_n, a_1, a_2, .., a_n)$ or more concisely $P(\mathbf{x}; \mathbf{a})$ where **x** and **a** are vectors in the n-dimensional unit hypercube. We envision the following realization of $P(\mathbf{x}; \mathbf{a})$

$$P(\mathbf{x}; \mathbf{a}) = P(x_1, a_1) \text{ and } P(x_2, a_2) \text{ and } \ldots \text{ and } P(x_n, a_n) \quad (8)$$

meaning that the satisfaction of the multivariable predicate relies on the satisfaction realized for each variable separately. As the variables could come with different levels of relevance as to the overall satisfaction of the predicates, we represent this effect by some weights (connections) $w_1, w_2, \ldots, w_n$ so that (8) can be expressed in the following form

$$P(\mathbf{x}; \mathbf{a}, \mathbf{w}) = [P(x_1,a_1) \textit{ or } w_1] \textit{ and } [P(x_2,a_2) \textit{ or } w_2] \textit{ and } \ldots \textit{ and } [P(x_n,a_n) \textit{ or } w_n] \tag{9}$$

Taking another look at the above expression and using a notation $z_i = P(x_i,a_i)$, it corresponds to a certain AND neuron y =AND(**z**; **w**) with the vector of inputs **z** being the result of the referential computations done for the logic predicate. Then the general notation to be used reads as REF(**x**; **w**, **a**). In the notation below, we explicitly articulate the role of the connections

$$y = \mathop{\mathrm{T}}_{i=1}^{n} (\mathrm{REF}(x_i, a_i) s w_i) \tag{10}$$

In essence, as visualized in Figure 5, we may conclude that the reference neuron is realized as a two-stage construct where first we determine the truth values of the predicate (with **a** being treated as a reference point) and then treat these results as the inputs to the AND neuron.

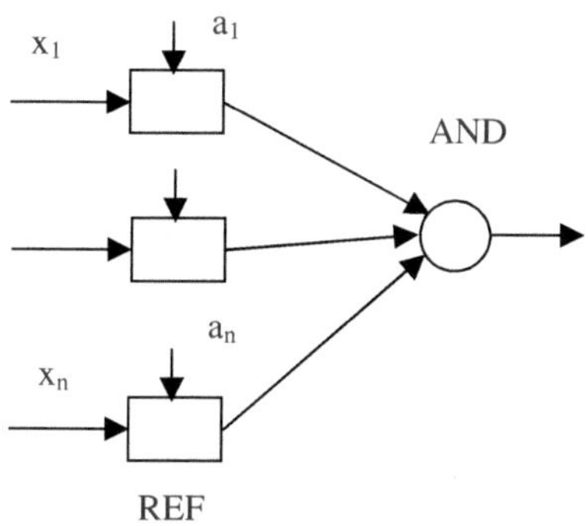

Fig. 5. A schematic view of computing realized by a reference neuron an involving two processing phases (referential computing and aggregation)

So far we have used the general term of predicate-based computing not confining ourselves to any specific nature of the predicate itself. Among a number of available possibilities of such predicates, we discuss the three of them, which tend to occupy an important place in logic processing. Those are inclusion, dominance and match (similarity) predicates. As the names stipulate, the predicates return truth values of satisfaction of the relationship of inclusion, dominance and similarity of a certain argument "x" with respect to the given reference "a". The essence of all these calculations is in the determination of the given truth values and this is done in the carefully developed logic framework so that the operations retain their semantics and interpretability. What makes our discussion coherent is the fact that the proposed operations originate from triangular norms. The inclusion operation, as discussed earlier, denoted by $\subset$, is modeled by an implication $\Rightarrow$ that is induced by a certain left continuous t-norm [16]

$$a \Rightarrow b = \sup\{c \in [0,1] \mid atc \le b\}, \; a, b \in [0,1] \tag{12}$$

For instance, for the product the inclusion takes on the form $a \Rightarrow b = \min(1, b/a)$. The intuitive form of this predicate is self-evident: the statement "x is included in a" and modeled as INCL(x, a) = x ⇒ a comes with the truth value equal to 1 if x is less or equal to a (which in other words means that x is included in a) and produces lower truth values once x starts exceeding the truth values of "a". Higher values of "x" (those above the values of the reference point "a") start generating lower truth values of the predicate. The dominance predicate acts in a dual manner when compared with the predicate of inclusion. It returns 1 once "x" dominates "a" (so that its values exceeds "a") and values below 1 for x lower than the given threshold. The formal model can be realized as DOM(x, a) = a ⇒x. With regard to the reference neuron, the notation is equivalent to the one being used in the previous case, that is DOM(**x**; **w**, **a**) with the same meaning of **a** and **w**.

The similarity (match) operation is an aggregate of these two, SIM(x,a) = INCL(x,a) t DOM(x,a) which is appealing from the intuitive standpoint: we say that x is similar to a if x is included in a *and* x dominates a. Noticeably, if x =a the predicate returns 1; if x moves apart from "a" the truth value of the predicate becomes reduced. The resulting similarity neuron is denoted by SIM(**x**; **w**, **a**) and reads as

$$y = \mathop{T}_{i=1}^{n} (\mathrm{SIM}(x_i, a_i) s w_i) \tag{13}$$

The reference operations form an interesting generalization of the threshold operations. Consider that we are viewing "x" as a temporal signal (that changes over time) whose behavior needs to be monitored with respect to some bounds (α and β). If the signal does not exceed some predefined threshold α then the acceptance signal should go off. Likewise we require another acceptance mechanism that indicates a situation where the signal does not go below another threshold value of β. In the case of fuzzy predicates, the level of acceptance assumes values in the unit interval rather than being a Boolean variable. Furthermore the strength of acceptance reflects how much the signal adheres to the assumed thresholds.

The plots of the referential neurons with two input variables are shown in Figure 6 and 7; here we have included two realizations of the t-norms to illustrate their effect on the nonlinear characteristics of the processing units.

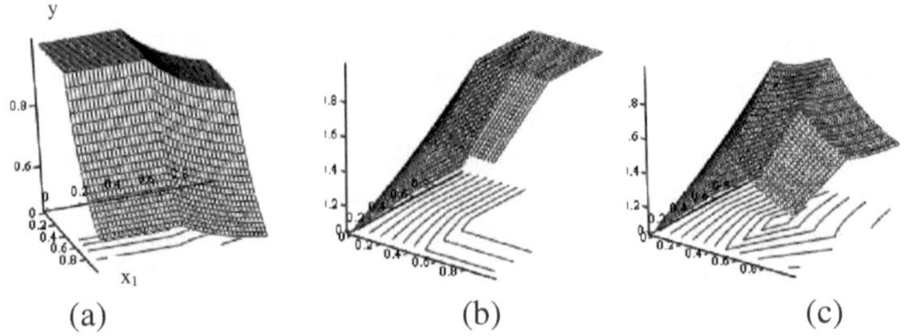

Fig. 6. Characteristics of the reference neurons for the product (t-norm) and probabilistic sum (t-conorm). In all cases the connections are set to 0.1 and 0.7 with intent to visualize the effect of the weights on the relationships produced by the neuron. The point of reference is set to (0.5, 0.5): inclusion neuron (a), dominance neuron (b), similarity neuron (c).

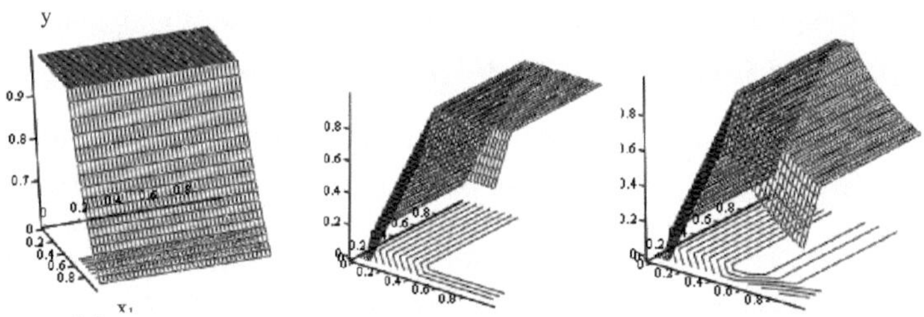

(a) (b) (c)

Fig. 7. Characteristics of the reference neurons for the Lukasiewicz t-norm and t-conorm (that is a t b = max(0, a+b-1) and a s b = min (1, a+b)). In all cases the connections are set to 0.1 and 0.7 with intent to visualize the effect of the weights. The point of reference is set to (0.5, 0.5): inclusion neuron (a), dominance neuron (b), similarity neuron (c).

It is worth noting that by moving the reference point to the origin or the **1**-vertex of the unit hypercube (with all its coordinates being set up to 1), the referential neuron starts resembling the aggregative neuron. In particular, we have

- for **a** = **1** = [1 1 1… 1] the inclusion neuron reduces to the AND neuron
- for **a** = **0** = [0 0 0… 0] the dominance neuron reduces to the AND neuron

One can draw a loose analogy between some types of the referential neurons and the two categories of processing units encountered in neurocomputing. The analogy is based upon the *local* versus *global* character of processing realized therein. Perceptrons come with the global character of processing. Radial basis functions realize a local character of processing as focused on receptive fields. In the same vein, the inclusion and dominance neurons are after the global nature of processing while the similarity neuron carries more confined and local processing.

3.3 Uninorm-Based Logic Neurons

In the previous studies carried out in the realm of logic-based neurocomputing, we have distinguished between two general categories of neurons that is OR and AND neuron. Let **x** be a vector in the unit hypercube, $\mathbf{x} \in [0,1]^n$ and y denote an element in [0,1]. Formally speaking, their underlying logic processing is governed by a composition of some logic operation occurring at the level of processing of the individual inputs and their corresponding connections (weights) $\mathbf{w} \in [0,1]^n$ that is L_1: $(x_i, w_i) \rightarrow [0,1]$ followed by some overall logic aggregation L_2 giving rise to the output y, that is

$$y = L_2\,[\,L_1(x_1,w_1), L_1(x_2,w_2),\ldots, L_1(x_n, w_n)] \tag{14}$$

In the OR neuron, the semantics of L_1 is concerned with any realization of the *and* operator (t-norm), L_1 = *and*. L_2 concerns the realization of the *or* operator (t-conorm), L_2 = *or*. In contrast to this processing, in the AND neuron we have these two operations reversed, that is L_1 = *or* and L_2 =*and*. To emphasize the functionality of the neurons as well as to highlight the parametric flexibility residing with the connections **w** that are necessary to support all learning pursuits, we use the following concise notation y = OR(**x**; **w**) and y = AND(**x**; **w**). The semantics of the operations governing the logic mapping realized by the neurons fully justifies the names of these logic neurons.

Uninorms offer several quite general avenues of architectural augmentations of the logic neurons. Given the allocation of the uninorms across the composition operation realizing the logic neuron , the two major possibilities seem to be of particular interest. Here, we consider L_1 or L_2 to be implemented by means of some uninorms. We envision here two general alternatives such as: (a) a uninorm–based aggregation realized at the *global* level, and (b) a uninorm-based aggregation completed at the *local* level involving individual inputs of the neuron.

Following the taxonomy introduced above, we place the uninorm-based processing at the global level of aggregation of the logic transformations of the individual inputs. In line of the notation already used in (13), we arrive at the following descriptions

$$y = u\,[\, L_1(x_1,w_1), L_1(x_2,w_2),\ldots, L_1(x_n, w_n), g] \tag{15}$$

and

$$y = u\,[\, L_2(x_1,w_1), L_2(x_2,w_2),\ldots, L_2(x_n, w_n), g] \tag{16}$$

Depending upon the mechanism of the aggregation occurring at the level of the individual inputs and their corresponding connections, we encounter two categories of processing. Subsequently, we may talk about *and*-type and *or*-type unineuron, UNI-AND and UNI-OR, respectively [18][16]. The concise notation reads as follows

for (14)

$$y = \text{UNI-AND}(\mathbf{x};\, \mathbf{w}, g) \tag{17}$$

for (15)

$$y = \text{UNI-OR}(\mathbf{x};\, \mathbf{w}, g) \tag{18}$$

In both these situations we fully acknowledge the role of the identity element (g) whose value impacts the nature of the overall aggregation completed by the neurons. For the UNI-AND neuron, we consider the use of the min operation in the Ω region of the unit square. Likewise for the UNI-OR the uninorm in the same region (Ω) uses the maximum operator. The generalized character of the unineuron is obvious: for g = 1

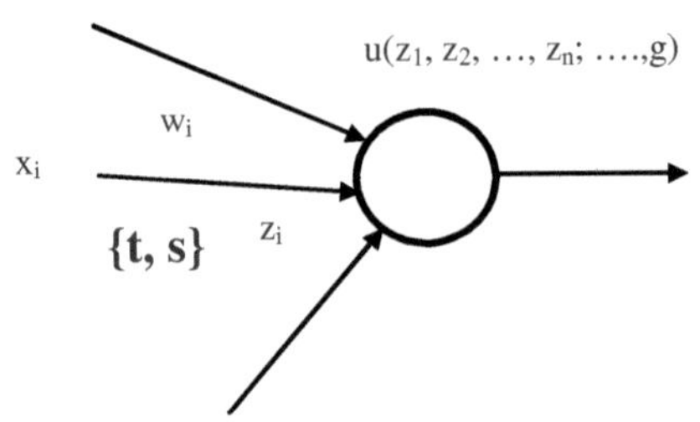

Fig. 8. Schematics of a unineuron where a uninorm is used at the global level of aggregation; at the level of the individual inputs we consider the use of t- norms or t-conorms (s-norms)

and $L = L_1$ we end up with the AND neuron, that is $y = t(s(x_1,w_1), s(x_2, w_2),\ldots, s(x_n, w_n))$. The combination $g = 0$ and $L = L_2$ returns the OR neuron, $y = s(t(x_1,w_1), t(x_2, w_2),\ldots, t(x_n, w_n))$.

We illustrate the characteristics of the unineurons developed so far considering some two-input constructs, Figure 9 and 10. Here we use some selected combinations of the values of the parameters of the neuron. The graphs deliver an interesting insight into the nature of processing. First, the neurons produce nonlinear characteristics. Second, the form of these input-output dependencies could be controlled by the choice of specific numeric values of the parameters (connections and the identity element). Similarly, by choosing various t- norms and conorms, one could influence some nonlinear dependencies captured by the neuron.

In this case, the uninorm is positioned at the level of the individual inputs of the neurons. This gives rise to the following expressions describing a way in which the logic processing takes place. For the i-th input we come up with the expression $u(x_i, w_i; g_i)$ and this underlines that each input (x_i) comes with the two modifiable parameters (viz. the connection and the identity element). Their values are specific for the given input and could be adjusted during the learning of the neuron. The overall aggregation is completed through the use of L_1 or L_2. We then have

$$y = L_1 [\, u(x_1, w_1; g_1), u(x_2, w_2; g_2),\ldots, u(x_n, w_n; g_n)] \tag{19}$$

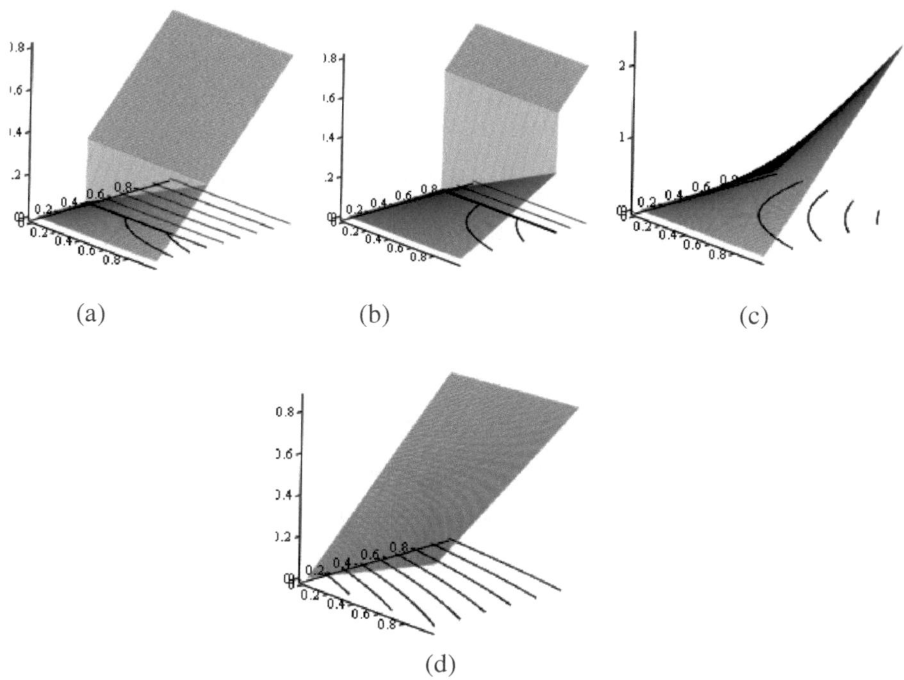

Fig. 9. Input-output characteristics of the UNI-AND neuron; **w** =[0.3 0.8]; g = 0.3 (a) and g= 0.6 (b) Shown are also boundary cases with g =1 (c) and g = 0 (d). t-norm and t-conorm: product and probabilistic sum

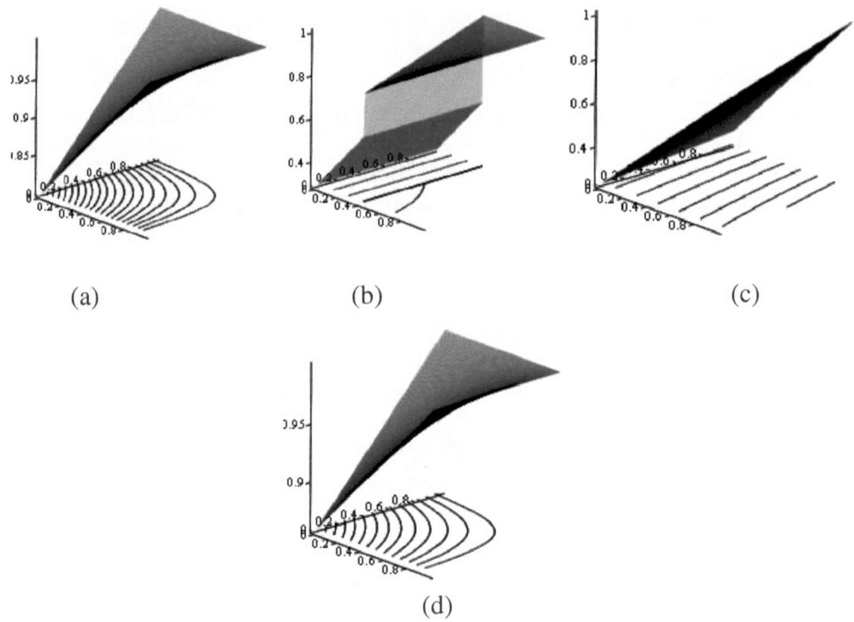

Fig. 10. Input-output characteristics of the UNI-OR neuron; **w** =[0.3 0.8] g = 0.3 (a) and g= 0.6 (b) Included are boundary cases with g =1 (c) and g = 0 (d). t-norm: product; t-conorm: probabilistic sum

and

$$y = L_2\,[\,u(x_1, w_1; g_1), u(x_2, w_2; g_2),..., u(x_n, w_n; g_n)] \tag{20}$$

Given the form of the logic operator, we end up with the *and*-type or *or*-type of aggregation; y = AND $[u(x_1, w_1; g_1), u(x_2, w_2; g_2),..., u(x_n, w_n; g_n)]$ and y = OR $[u(x_1, w_1; g_1), u(x_2, w_2; g_2),..., u(x_n, w_n; g_n)]$. We may refer to these two categories of the neurons as AND_U or OR_U logic neurons. Schematically, such unineurons are portrayed in Figure 11.

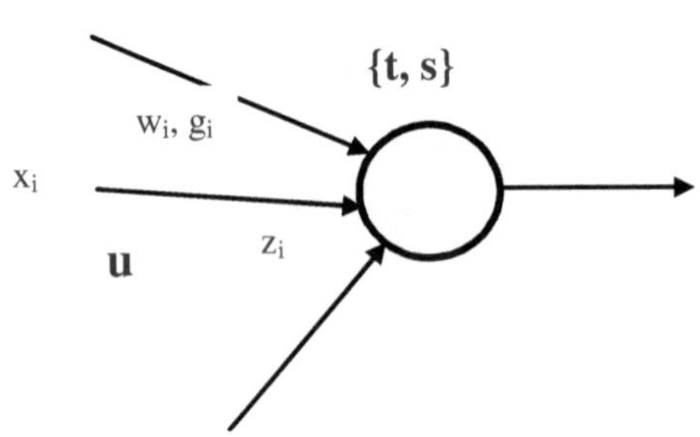

Fig. 11. A schematic view of a unineuron in which uninorms are used at the local level of processing (viz. associated with each input) and the partial results produced in this way are aggregated using the *and* or *or* type of logic aggregation

From the standpoint of the available parametric flexibility of this neuron, we note that it becomes concentrated at the level of the inputs; hence most learning will need to be focused there. To illustrate the form of dependencies produced by the neuron, we look at the expression u(x, w; g) being treated as a function of the connection (**w**) and the identity element (g). Selecting some values of "x", the corresponding plots are shown in Figure 12. A remarkable is a diversity of the nonlinear characteristics whose shape could be efficiently affected by adjusting the numeric values of these parameters.

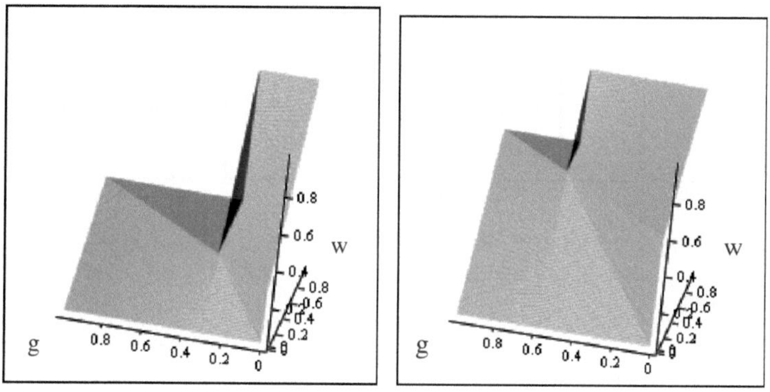

Fig. 12. Plots of u(x, w; g) for selected values of the input (x); t-norm: product; t-conorm: probabilistic sum: (a) x =0.3, (b) x =0.6

Given this architecture, the logic aggregation for each input could be done independently and because of the adjustable value of the identity element we encounter a different way of logic processing.

We may also envision another topology of the neuron in which uninorms are used at both levels (viz. locally and globally). This leads to the following expression

$$y = u\,[\,u(x_1, w_1; g_1), u(x_2, w_2; g_2),\ldots, u(x_n, w_n; g_n); g] \tag{21}$$

This architecture of the neuron exhibits a high level of flexibility yet its interpretation could be more complicated given the semantics of the uninorms. Likewise its development (learning) might cause more difficulties.

In our discussion so far we have focused on a single n-input unineuron, which in essence forms a multi-input single-output computing structure. The multi-output network can be constructed by putting together several unineurons with each of them contributing to a specific output. The multilevel structure usually encountered in neurocomputing is not necessary here because of the fact that each unineuron realizes a fairly complete logic function. There are four modes of logic computing that become predominantly visible depending upon the numeric values of the connections and the value of "g".Let us concentrate on the UNI-AND and UNI-OR neurons; refer to Table 1. The inherent character of logic processing is beneficial when pruning connections established during the learning. In essence, cf. [15][16] if the connection is processed *and*–wise (using any t-norm), its low value (close to 0) indicates that it is a candidate for pruning. An opposite situation happens if we consider a combination realized via any t-conorm: here high values (around 1) make the corresponding input a candidate for potential pruning.

Table 1. Four fundamental modes of logic processing realized by unineurons

	Individual inputs of unineuron (*and* and *or*)	
Global aggregation $g \rightarrow 0$	*or*-like summarization of calibrated conditions; *and* calibration of conditions	*or*-like summarization of calibrated conditions; *or* calibration of conditions
$g \rightarrow 1$	*and*-like summarization of calibrated conditions; *and* calibration of conditions	*and*-like summarization of calibrated conditions; *or* calibration of conditions

In a nutshell, any collection of uninorms returns a certain specialized family of rules and in this manner gives rise to a transparent yet highly adjustable description of data.

4 Architectures of Logic Networks

The logic neurons (aggregative and referential) can serve as building blocks of more comprehensive and functionally appealing architectures. The diversity of the topologies one can construct with the aid of the proposed neurons is surprisingly high. This architectural multiplicity is important from the application point of view as we can fully reflect the nature of the problem in a flexible manner. It is essential to capture the problem in a logic format and then set up the logic skeleton – conceptual blueprint (by forming the and finally refine it parametrically through a thorough optimization of the connections). Throughout the entire development process we are positioned quite comfortably by monitoring the optimization of the network as well as interpreting its meaning.

4.1 Logic Processor in the Processing of Fuzzy Logic Functions: A Canonical Realization

The typical logic network that is at the center of logic processing originates from the two-valued logic and comes in the form of the famous Shannon theorem of decomposition of Boolean functions. Let us recall that any Boolean function $\{0,1\}^n \rightarrow \{0,1\}$ can be represented as a logic sum of its corresponding miniterms or a logic product of maxterms. By a minterm of "n" logic variables $x_1, x_2, \ldots, x_n$ we mean a logic product involving all these variables either in direct or complemented form. Having "n" variables we end up with 2^n minterms starting from the one involving all complemented variables and ending up at the logic product with all direct variables. Likewise by a maxterm we mean a logic sum of all variables or their complements. Now in virtue of the decomposition theorem, we note that the first representation scheme involves a two-layer network where the first layer consists of AND gates whose outputs are combined in a single OR gate. The converse topology occurs for the second decomposition mode: there is a single layer of OR gates followed by a single AND gate aggregating *or*-wise all partial results.

The proposed network (referred here as a logic processor) generalizes this concept as shown in Figure 13. The OR-AND mode of the logic processor comes with the two types of aggregative neurons being swapped between the layers. Here the first (hidden) layer is composed of the OR neuron and is followed by the output realized by means of the AND neuron.

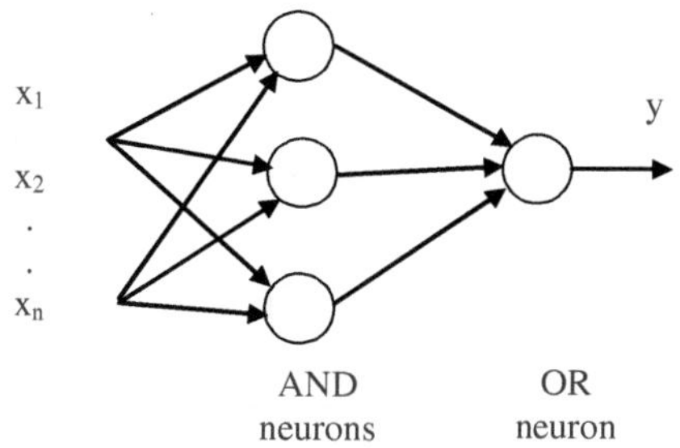

Fig. 13. A topology of the logic processor in its AND-OR mode

The logic neurons generalize digital gates. The design of the network (viz. any fuzzy function) is realized through learning. If we confine ourselves to {0,1} values, the network's learning becomes an alternative to a standard digital design, especially a minimization of logic functions. The logic processor translates into a compound logic statement (we skip the connections of the neurons to underline the underlying logic content of the statement)

- if (input_1 *and... and* input_j) *or* (input_d *and ...and* input_f) then *output*

The logic processor's topology (and underlying interpretation) is standard. Two LPs can vary in terms of the number of AND neurons, their connections but the format of the resulting logic expression is quite uniform (as a sum of generalized minterms).

As an example, let us consider a simple fuzzy neural network network in which the hidden layer includes two AND neurons whose outputs are combined through a single OR neuron located in the output layer. The connections of the first AND neuron are equal to 0.3 and 0.7. For the second AND neuron we have the values of the connections equal to 0.8 and 0.2. The connections of the OR neuron are equal to 0.5 and 0.7, respectively. The input-output characteristics of the network are illustrated in Figure 14; to demonstrate the flexibility of the architecture, we included several combinations of the connections as well as used alternative realizations of the triangular norms and conorms.

The resulting networks exhibit a significant diversity in terms of the resulting nonlinear dependencies. More importantly, we note that by choosing certain logic connectives (triangular norms) and adjusting the values of the connections, we could substantially affect the behavior (input-output characteristics) of the corresponding network. This plasticity becomes an important feature that plays a paramount role in the overall learning process.

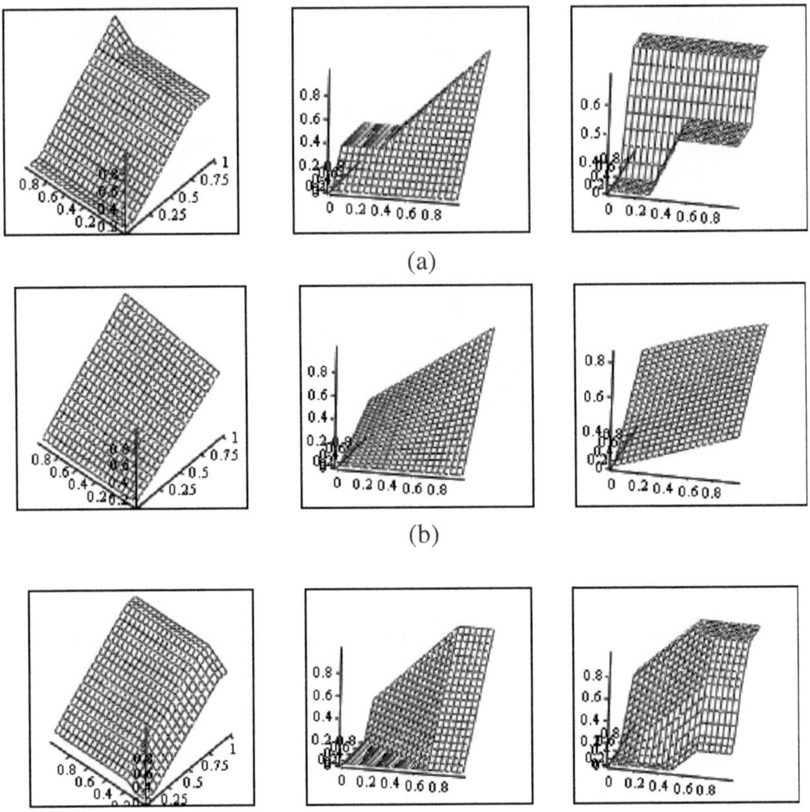

(a)

(b)

(c)

Fig. 14. Plots of the characteristics of the fuzzy neural network: output of the two AND neurons and the output of the network (from left to right) for different realization of the logic operators: (a) min and max, (b) product and probabilistic sum, and (c) Lukasiewicz logic operators

4.2 Heterogeneous Logic Networks

Logic processing with fuzzy predicates is an important endeavor when dealing with decision-making processes, sensor aggregation where we encounter a number of constraints which are articulated as a part of domain knowledge about the problem. The domain knowledge conveyed in the form of constraints gives rise to the structure of the logic network and whose connections are afterwards adjusted when dealing with the calibration of the network realized in presence of some numeric data. For instance, the following statement

> " decision "d" is contemplated when inputs x_1 and x_2 are similar (close to) $\mathbf{r}_1$ (= $[r_{11}\ r_{12}]$) and input x_2 and x_3 are not greater than threshold $\mathbf{r}_2$ (=$[r_{21}\ r_{22}]$) or inputs x_1, x_2, and x_3 dominate threshold $\mathbf{r}_3$"

is directly transformed (mapped) into the structure of the logic network shown in Figure 15. Here the output "d" pertains to the level of confidence (level of support) in

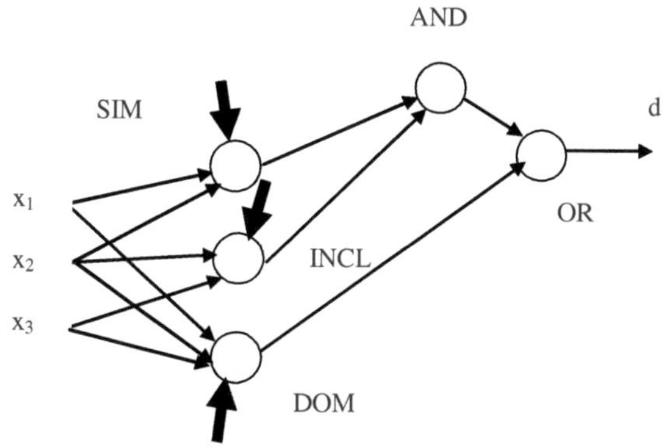

Fig. 15. An example of a logic network whose topology is induced by some prior domain knowledge conveyed by a logic description of the problem involving some reference computing

favor of the specific decision. The network has 3 inputs and they are fed into the corresponding reference neurons whose reference points are formed by the threshold vectors (r_1, r_2, and r_3, respectively).

The learning of the connections of the neurons becomes a part of the parametric learning which is accomplished by taking into account some training data.

4.3 Unineuron-Based Topologies of Logic Networks

Unineurons are by far more flexible computing elements (neurons) in comparison with the generic AND and OR neurons. The generic topology of the network could then mimic the two-layered structure as developed for the logic processor. Here the first layer is composed of UNI_AND neurons which is followed by the output layer built by means of the UNI_OR neurons, see Figure 16.

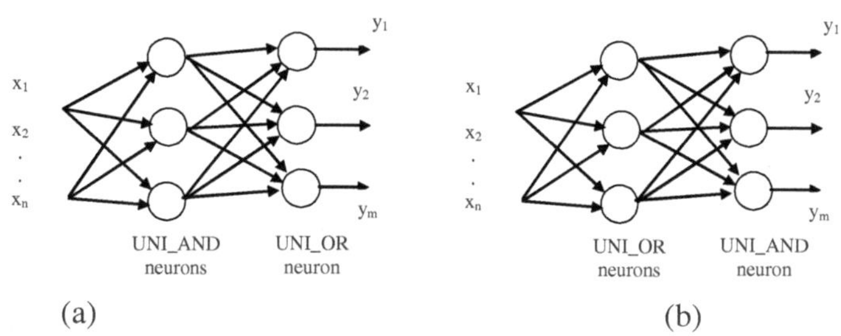

Fig. 16. A two-layered structure of logic processing constructed by unineurons: (a) formed by the UNI_AND and UNI_OR neurons, and (b) constructed by means of UNI_OR and UNI_AND neurons

The logic dominance (either *or* or *and* character) of the unineurons is determined by the selected numeric values of the parameters. By choosing the topology displayed in Figure 16 (a), we may anticipate that for some specific numeric values of the connections and identity points, we arrive at the logic processor with the layer of

AND and OR neurons. The dual topology of the network is obtained by positioning the unineurons in a reverse order (viz. the layer of UNI_OR followed by the UNI_AND).

5 Learning of Unineurons and Networks of Unineurons

While for the networks composed of AND and OR neurons, the learning could rely to a significant extent on gradient-based optimization, in case of unineurons more attention is required given the nature of the processing and a role being played by the individual parameters.

Any unineuron is equipped with a significant level of parametric flexibility that resides within its connections (**w**) as well as the identity element (g). The role of the latter, as illustrated in the previous section, is very evident. Any learning has to capitalize on this flexibility inbuilt into the structure. By modifying the values of the parameters (parametric learning) we allow the neuron to adhere to the required (target) values. If we are concerned with the supervised learning in which we are provided with a series of input-output data pairs (**x**(k), target(k)), k=1,2, …,N then the typical learning invokes a series of iterations in which the parameters of the unineuron are adjusted following the gradient of a certain minimized objective function (performance index) Q. Thus we have

$$\mathbf{w}(\text{iter}+1) = \mathbf{w}(\text{iter}) - \alpha \nabla_{\mathbf{w}} Q \tag{22}$$

with some positive learning rate α (we assume that the values of the connections are always retained in the unit interval and clipped, if necessary). The details of the computations of the gradient depend upon the specific realizations of t-norms and t-conorms being used in the construct. Once all structural elements have been sorted out, the computing becomes quite standard.

The optimization of a certain connection, say w_j requires some regrouping of the results of aggregation at the level of the individual inputs. Let us consider the UNI-AND neuron and use the shorthand notation that helps us capture all inputs,

$$y = \text{UNI\_AND}(\mathbf{x}; \mathbf{w}, g) = \mathop{U}_{i=1}^{n} (x_i t w_i; g) \tag{23}$$

The objective is to make the output of the unineuron, y, equal to the target value given in the data set. The learning is realized in the on-line mode so we discuss each pair of data (**x**, target) at a time (the index of the pair, k=1, 2, …, N could then be dropped). The objective is defined as $Q = (\text{target} - y)^2$. Making use of the uninorm type of aggregation used in the neuron, we introduce the notation $A_i = \mathop{U}_{\substack{j=1 \\ j\neq i}}^{n} (x_j t w_j; g)$ to isolate the connection of interest, that is UNI_AND(**x**; **w**, g) = $u_g(x_i t w_i, A_i)$. Subsequently we have

$$w_i = w_i - \alpha \frac{\partial Q}{\partial w_i} = w_i + 2\alpha(\text{target} - y)\frac{\partial y}{\partial w_i} \tag{24}$$

i=1, 2,…,n. As A_i does not depend on w_i, the ensuing computations invoke explicitly the definition of the uninorm. For instance, assuming that the uninorm uses the maximum operation defined over the off-diagonal regions of the unit square of its arguments, we obtain the derivative of the output of the unineuron with respect to its connection,

$$\frac{\partial y}{\partial w_i} = \frac{\partial}{\partial w_i}\begin{cases} g\,t(\frac{x_i t w_i}{g}, \frac{A_i}{g}) & \text{if } x_i t w_i, A_i \in [0, g] \\ g + (1-g)s(\frac{x_i t w_i - g}{1-g}, \frac{A_i - g}{1-g}) & \text{if } x_i t w_i, A_i \in [g, 1] \\ \max(x_i t w_i, A_i), \text{otherwise} \end{cases} \tag{25}$$

(formally speaking, the derivative does not exist at the lines where the regions meet however the probability of reaching such an event is zero; in the calculations above we have made appropriate provisions by defining the values of such derivative). Further calculations of the derivative (24) can be completed once we have decided upon the specific form of the t- and t-conorm.

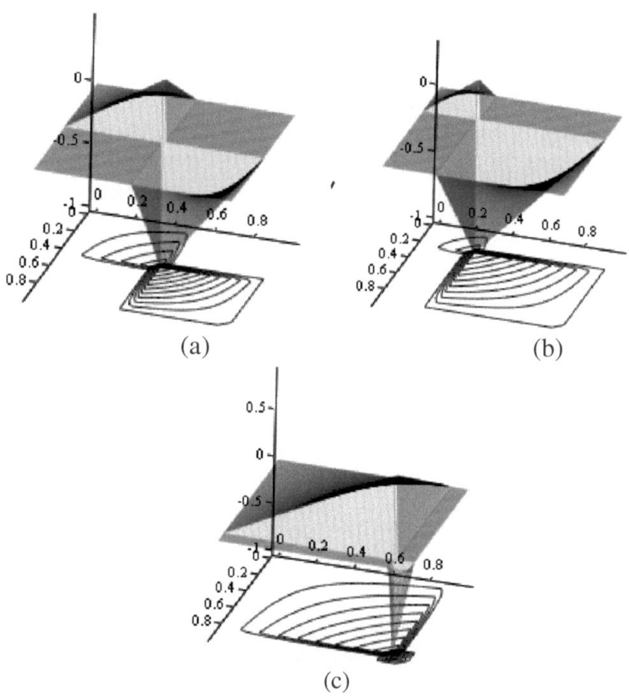

Fig. 17. Derivative of Q with respect to g treated as a function of x and y: (a) g = 0.3, (b) g = 0.5, and (c) g = 0.9

As far as the optimization of the identity value of the uninorm (g) is concerned, we should exercise caution when proceeding with the gradient-based optimization. The derivative is quite convoluted; see Figure 17 with significant regions of the unit hypercube over which it assumes zero (which means that the learning could be easily trapped in the region where there is no minimum).

Given all of these, it would be prudent to look at some simpler optimization scenario. Even a straightforward sweep through the values of "g" during which we record the lowest value of the performance index (with the connections of the unineuron being optimized gradient-wise) sounds like a viable optimization alternative option. While being simple, this approach leads to a comprehensive visualization of the relationship Q=Q(g) and in this sense helps gain a better view at the *and-or* tradeoff of the aggregation. Likewise we can show how much the performance index changes with respect to "g" and in this sense learn about the sensitivity of the logic operator.

6 Linguistic Modeling of Relationships between Information Granules: The Paradigm of Granular (Fuzzy) Models

Granular Computing [1] [[15][17] is concerned with processing information granules. In particular, information granules [20] are realized in the form of fuzzy sets and fuzzy relations. The key advantage of computing at the level of information granules is that these constructs are the key building blocks [16] that we use to form an abstract, user-driven view of the world in the form of some model. This also contributes to the interpretability of the resulting models. The essence of the ensuing modeling is to form some logic expressions (formulas) over the granules where such formulas are instrumental in capturing the essence of the data (being cast in the logic setting) and reveal no relationships between the granules. The transparency of such constructs is evident. Before moving to the architectures of the modeling pursuits, one should elaborate a bit on the origin of information granules. Typically those are formed through abstraction of data through building a collection of clusters. When considering the Fuzzy C-Means algorithm (FCM) [4] the membership grades of the individual information granules (clusters) are represented in the successive rows of the partition matrix U. Thus the rows of U collect membership functions of the corresponding fuzzy sets, say A_1, A_2, …, A_c (assuming that "c" denotes the number of clusters being investigated in the dataset).

When dealing with information granules, the computing facet usually requires more sophisticated models of logic operations supporting the development of the mapping between fuzzy sets or fuzzy relations. This is a direct motivation behind the use of unineurons.

As there are a vast number of possible modeling tasks, it is instructive to concentrate on the two quite general categories of problems as visualized in Figure 18. In the first one, Fig. 18(a), we are interested in revealing associations between information granules formed for the same dataset **D**. In the second one, Fig. 18 (b), we form a logic mapping between information granules constructed in the input and output spaces; this leads us to the idea of the logic input-output model.

Let us briefly summarize the essence of these two major streams of processing. The first one, which could be referred to as building associations between two granular descriptions of the same data, can be succinctly described in the following way. A given data set **D** is captured through a collection of information granules **A**, card(**A**) = n. The same dataset while processed in some other way (e.g., using different mechanisms of information granulation that is specifying a different number of clusters) leads to another finite collection of information granules, **B**, card (**B**) = m. The granularity (specificity) of these new granules could be lower or higher depending upon the relationship between "n" and "m". We are interested in expressing B_j's in term of A_is. To do that we require an n-input m-output network composed of 'm" unineurons. The ultimate role of the model here is to associate two descriptions of data completed at different levels of granularity. The more advanced alternative that is a certain more abstract version of the problem discussed so far involves a collection of logic expressions built upon the original information granules. More specifically, consider that A_is give rise to some logic expressions ϕ_i with arguments involving membership grades of A_i, say $\phi_i(A_1, A_2, \ldots, A_n)$, i=1, 2, …, n' Likewise there are some formulas built upon B_js, say $\psi_j(B_1, B_2, \ldots, B_m)$, j=1, 2,…, m'. Then the logic network realizes some linguistic modeling between such formulas, refer to Figure 18 (a).

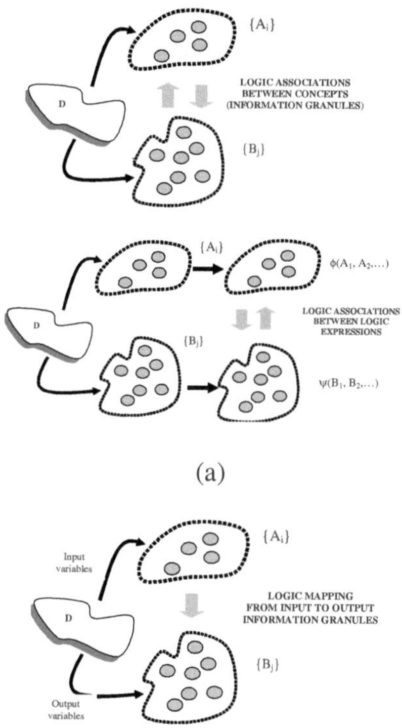

Fig. 18. Two main modes of modeling at the level of information granules: (a) revealing associations between information granules formed for the same dataset, and (b) building a logic model

The second stream of logic is concerned with the logic mode of modeling. In this case, refer to Figure 18 (b), the clusters are formed independently in the input and output spaces (note that in the first category of tasks we did not make any distinction between the input and outputs—the problem was simply made direction free as we were involved in the formation of the associations with the data and any distinction between input and output variables was not necessary). The discrimination between input and output variables is commonly encountered in fuzzy modeling. Now given a family of information granules $\{A_i\}$ defined in the input space and another family of granules (clusters) $\{B_j\}$ formed in the output space, the role of the model is to form a logic mapping from A_i s to B_j (viz. to express B_j in terms of A_is).

7 Conclusions

The logic-driven fuzzy constructs contribute to the paradigm of fuzzy or granular modeling in which we strive for transparency and abstraction (generality) of the resulting constructs. In this study, we have emphasized the genuine need to embark on logic processing which becomes an ultimate prerequisite when developing interpretable structures. The richness of the landscape of logic processing units – fuzzy neurons is remarkable and this diversity becomes highly beneficial when capturing logic operations and their calibration as well as computing with logic predicates. The study offered an overview of the main categories of logic neurons and presented taxonomy of the ensuing architectures including logic processors as well as more advanced and heterogeneous topologies. The issues of transparency of logic networks are crucial to an effective accommodation of prior domain knowledge. Owing to the very transparency any knowledge about the problem we have at a very initial phase of the development of the granular model could be easily "downloaded" onto the corresponding structure of the network. The ensuing learning becomes then more focused and hence far more effective than the one when we proceed with any initial random configuration of the connections.

References

1. Bargiela, A., Pedrycz, W.: Granular Computing: An Introduction. Kluwer Academic Publishers, Dordercht (2003)
2. De Baets, B.: Idempotent Uninorms. European Journal of Operational Research 118, 631–642 (1999)
3. De Baets, B., Fodor, F.: van Melle's Combining Function in MYCIN is a Representable Uninorm: An Alternative Proof. Fuzzy Sets and Systems 104, 133–136 (1999)
4. Bezdek, J.C.: Pattern Recognition with Fuzzy Objective Function Algorithms. Plenum Press, N. York (1981)
5. Calvo, T., De Baets, B., Fodor, J.: The Functional Equations of Frank and Alsina for Uninorms and Nullnorms. Fuzzy Sets and Systems 120, 385–394 (2001)
6. Calvo, T., Kolesárová, A., Komorníková, M., Mesiar, R.: Aggregation Operators: Properties, Classes and Construction Methods. In: Aggregation Operators: New Trends and Applications, pp. 1–104. Physica-Verlag, Heildelberg (2002)
7. Calvo, T., Mesiar, R.: Continuous Generated Associative Aggregation Operators. Fuzzy Sets & Systems 126, 191–197 (2002)

8. Ciaramella, A., Tagliaferri, R., Pedrycz, W.: The Genetic Development of Ordinal Sums. Fuzzy Sets & Systems 151, 303–325 (2005)
9. Feng, Q.: Uninorm Solutions and Nullnorm Solutions to the Modularity Condition Equations. Fuzzy Sets & Systems 148, 231–242 (2004)
10. Fodor, J., Yager, R.R., Rybalov, A.: Structure of Uninorms. Int. J. of Uncertainty, Fuzziness and Knowledge-Based Systems 5, 113–126 (1997)
11. Fodor, J., Yager, R.R., Rybalov, A.: Structure of Uninorms. Int. J. of Uncertainty, Fuzziness and Knowledge-Based Systems 5, 411–427 (1997)
12. Hirota, K., Pedrycz, W.: OR/AND Neuron in Modeling Fuzzy Set Connectives. IEEE Trans. on Fuzzy Systems 2, 151–161 (1994)
13. Klement, E., Navara, M.: A Survey on Different Triangular Norm-Based fuzzy Logics. Fuzzy Sets and Systems 101, 241–251 (1999)
14. Klement, E., Mesiar, R., Pap, E.: Triangular Norms. Kluwer Academic Publishers, Dordrecht (2000)
15. Pedrycz, W.: Fuzzy Neural Networks and Neurocomputations. Fuzzy Sets & Systems 56, 1–28 (1993)
16. Pedrycz, W., Gomide, F.: Fuzzy Systems Engineering. J. Wiley, Hoboken (2007)
17. Pedrycz, W. (ed.): Granular Computing: An Emerging Paradigm. Physica Verlag, Heidelberg (2001)
18. Pedrycz, W.: Heterogeneous Fuzzy Logic Networks: Fundamentals and Development Studies. IEEE Trans. on Neural Networks 15, 1466–1481 (2004)
19. Pedrycz, W.: Knowledge-Based Clustering. J. Wiley & Sons, N. York (2005)
20. Schweizer, B., Sklar, A.: Probabilistic Metric Spaces. North-Holland, New York (1983)
21. Tasdiras, A., Mazrgaritas, K.: The MYCIN Certainty Factor Handling Function as Uninorm Operator and Its use as Threshold Function in Artificial Neurons. Fuzzy Sets and Systems 93, 263–274 (1998)
22. Yager, R.R.: On Weighted Median Aggregation. Int. J. of Uncertainty, Fuzziness and Knowledge-Based Systems 2, 101–114 (1994)
23. Yager, R.R., Rybalov, A.: Uninorm Aggregation Operators. Fuzzy Sets & Systems 80, 111–120 (1996)
24. Yager, R.R.: Uninorms in Fuzzy Systems Modeling. Fuzzy Sets & Systems 122, 167–175 (2001)
25. Yager, R.R.: Defending Against Strategic Manipulation in Uninorm-based Multi-agent Decision Making. European Journal of Operational Research 141, 217–232 (2002)
26. Zadeh, L.A.: Toward Theory of Fuzzy Information Granulation and Its Centrality in Human Reasoning and Fuzzy Logic. Fuzzy Sets & Systems 90, 111–127 (1997)

MembershipMap: A Data Transformation for Knowledge Discovery Based on Granulation and Fuzzy Membership Aggregation

Hichem Frigui

Multimedia Research Lab
Computer Engineering & Computer Science Dept.
University of Louisville
h.frigui@louisville.edu

Abstract. In this chapter, we describe a new data-driven transformation that facilitates many data mining, interpretation, and analysis tasks. This approach, called MembershipMap, strives to granulate and extract the underlying sub-concepts of each raw attribute. The orthogonal union of these sub-concepts are then used to define a new membership space. The sub-concept soft labels of each point in the original space determine the position of that point in the new space. Since sub-concept labels are prone to uncertainty inherent in the original data and in the initial extraction process, a combination of labeling schemes that are based on different measures of uncertainty will be presented. In particular, we introduce the CrispMap, the FuzzyMap, and the PossibilisticMap. We outline the advantages and disadvantages of each mapping scheme, and we show that the three transformed spaces are complementary. We also show that in addition to improving the performance of clustering by taking advantage of the richer information content, the MembershipMap can be used as a flexible pre-processing tool to support such tasks as: sampling, data cleaning, and outlier detection.

Keywords: Preprocessing, Transformations, Clustering, Labeling, Fuzzy sets.

1 Introduction

1.1 Background

Knowledge Discovery in Databases (KDD) aims at discovering interesting and useful knowledge from large amounts of data, such as patterns, trends, and associations. The KDD process consists of the following main phases[1]:

- **Data preparation and preprocessing:** This phase generally consists of cleaning of imperfect or noisy data, integration of data from multiple sources, selection of relevant data, and transformation of data.
- **Data Mining:** This phase consists of the application of intelligent methods to extract knowledge from data. Data Mining relies on several essential tools, such as clustering, classification, and association rules.
- **Pattern evaluation and presentation:** This phase involves the identification and evaluation of interesting and useful knowledge from the mined information and its presentation to the user in a suitable format.

L.C. Jain et al. (Eds.): Comp. Intel. Para.: Innov. Applications, SCI 137, pp. 51–76, 2008.
springerlink.com

The preprocessing phase, is an essential but compared to the other phases, largely ignored subject. Traditionally, it has taken a back seat to the data mining algorithms. However, without adequate preparation of the data, the outcome of any data mining algorithm can be disappointing, hence the saying "garbage in garbage out".

In this chapter, we propose a new data pre-processing approach that facilitates many data mining, interpretation, and analysis tasks for the case of unlabeled data with limited or no prior knowledge. Our approach, called MembershipMap, is different from other techniques because it takes into account the intricate nature of each attribute's distribution *independently* of the others and then aggregates the cluster labels to create a new membership space. This separate treatment of the attributes allows the complete freedom to adapt different ways to compute the concept membership functions for different attributes depending on prior knowledge and attribute type such as categorical versus numerical. The MembershipMap has several advantages including:

1. The mapping can be exploited in any of the three complementary spaces: crisp, fuzzy, or possibilistic.
2. According to their location in the MembershipMap, data can be labeled as noise, boundary, or seeds in an unsupervised way.
3. Suitable for data sets with features that vary widely in scale and type.
4. Can be updated incrementally.

Even though the attributes are treated independently, they do get combined at a later stage. Hence, information such as correlations is not lost in the transformed space. Treating the attributes independently in the initial phase is analogous to the standard mean-variance normalization. In fact, if the data has a simple unimodal distribution along each dimension, the two approaches are equivalent. However, since in general, multiple clusters are sought in each dimension, the MembershipMap is essentially a generalization of the simple normalization that performs local scaling that is optimal for each sub-group or cluster.

The organization of the rest of this chapter is as follows. In section 2, we briefly review different pre-processing and data partitioning and labeling schemes. In section 3, we describe our transformation approach, and in section 4, we analyze the properties of the different types of MembershipMaps. In section 5, we describe methods to explore the MembershipMaps and identify regions of interest. In section 6, we illustrate the application of the proposed mapping to extract and label candidate samples for semi-supervised clustering. In section 7, we apply our approach to data clustering and image segmentation. Finally, section 8 contains the conclusions.

2 Related Work

2.1 Data Preprocessing

The data preprocessing phase can take many forms including normalization/scaling, transformation/projection, cleaning, and data reduction [2, 3].

2.1.1 Data Normalization

Data normalization involves scaling the attribute values to make them lie numerically in the same interval/scale, and thus have the same importance. This step is particularly useful for distance-based methods such as clustering and nearest neighbor classification. Some data normalization techniques can be found in [4]. In general, a single successful linear transformation of the data to a space where all dimensions are normalized is reliable only when the data has a unimodal distribution. If this is not the case, then theoretically, several distinct transformations are optimal (each one tailored to one of the modes). Performing such an ideal normalization is impossible when the data is *not labeled*. In this case, "blind" normalization which may not preserve the distribution of the data is the only option. For instance, z-score or max-min scaling can destroy the structure (shape, size, density) of the clusters, and can even create "ghost" clusters (when scattered data is normalized within a small interval). Other conditions that can make normalization even more difficult and less satisfactory are the presence of noise, and the presence of categorical attributes.

2.1.2 Data Transformation

Data transformation is a common preprocessing technique that is used to map data to an equal or lower dimensional space. In general, the goals are to facilitate visualization, select a set of relevant attributes, or map object data to an Euclidean space. Energy preserving techniques such as the Karhunen-Loeve (KL) transform map the n-dimensional points to a lower k-dimensional space using linear projections via Principal Component Analysis (PCA)[5].

Another class of transformations attempts to preserve the topology of the data by maintaining the pair-wise dissimilarities (or their order) between objects. Multi-Dimensional Scaling (MDS)[6], Self-Organizing maps [7], and Sammon's mapping [8] fall into this category. The sharpest criticism against MDS consists of their high computational and storage requirements ($O(N^2)$). Jagadish[9] suggested relying on domain experts to derive k feature extraction functions to be used to map each object into a point in k-dimensional space. Faloutsos and Lin [10] generalized this approach by requiring experts to define only the objects' dissimilarity values.

2.1.3 Data Cleaning

Data cleaning includes smoothing out noise, identification of outliers, irrelevant features, and identification and mending of missing data. Outlier detection is important both as a prerequisite to later discovery tasks that cannot handle outliers gracefully, or even on its own, if outliers represent the knowledge nuggets to be discovered. Smoothing techniques have predominantly relied on histogram binning [4].

In statistics, missing data has been handled mainly by data deletion or substitution. The former method ignores records containing missing attributes. This approach may forfeit a proportion of data that is unacceptably high, and can yield a biased data set. The substitution method fills in the missing attributes using mean (or median) substitution, regression methods, hot deck imputation, or the EM algorithm [11].

In data mining, the problem of missing data has been addressed using **(i)** stand alone techniques that fill the missing data with the mean or median of related values, **(ii)** regression methods to estimate the missing value [12], and **(iii)** rule induction techniques that predict the missing value based on the value of other attributes in that record. Most of these methods were designed for categorical attribute with low attribute numerosity.

2.1.4 Visualization

Preprocessing can also involve certain transformations and aggregations that allow better visualization, interpretation, and analysis of the data. For instance, the data cube [13] is a widely used model for OLAP [14]. Data transformation techniques (described earlier) can also be used for visualization purposes.

2.1.5 Data Reduction

Data reduction includes reducing the number of attributes (feature reduction), the number of records (data sampling), or the number of possible values taken by an attribute (numerosity reduction). Feature Reduction has been addressed through feature selection techniques [15]. Unsupervised attribute selection was introduced in [16], and an unsupervised attribute weighting was introduced in [17].

Data sampling may be necessary to reduce the number of items to be mined when the data size is very large. Sampling techniques are discussed in the context of KDD in [18]. Sampling cannot be done blindly if the data is not uniformly distributed. A typical side effect of sampling when the data is distributed into clusters, is losing small clusters. A similar side effect occurs when mining frequent item sets and association rules.

Attribute numerosity reduction is a technique that reduces the number of values taken by an attribute. It can be achieved by several techniques such as numerical attribute discretization by histogram binning [19, 20, 21, 22].

2.2 Data Partitioning and Labeling

Clustering aims at classifying a set of N unlabeled data points $\mathcal{X}=\{\mathbf{x}_j | j = 1, \ldots, N\}$ into C clusters $G_1, \cdots, G_C$, and assigning a label to each point to represent information about its belongingness. In general, there are four types of cluster labels depending on the uncertainty framework: *crisp*, *fuzzy*, *probabilistic*, and *possibilistic*. Each labeling paradigm uses a different uncertainty model, and thus, has a different semantic interpretation.

In crisp labeling, each data point $\mathbf{x}_j$ is assigned a binary membership value,$(u_c)_{ij}$, in each cluster G_i according to the following characteristic function:

$$(u_c)_{ij} = \begin{cases} 1 \text{ if } \mathbf{x}_j \in G_i \\ 0 \text{ otherwise} \end{cases} \tag{1}$$

The K-Means [23], K-Medoids [24], and other crisp partitional clustering algorithms assign this type of labels to each data sample in each iteration of the algorithm.

Fuzzy and probabilistic labeling allow for partial or gradual membership values. This type of labeling offers a richer representation of belongingness and can handle uncertain cases[25]. Fuzzy and probabilistic memberships can have a different interpretation at high level reasoning. However, for the purpose of assigning labels, they can be treated

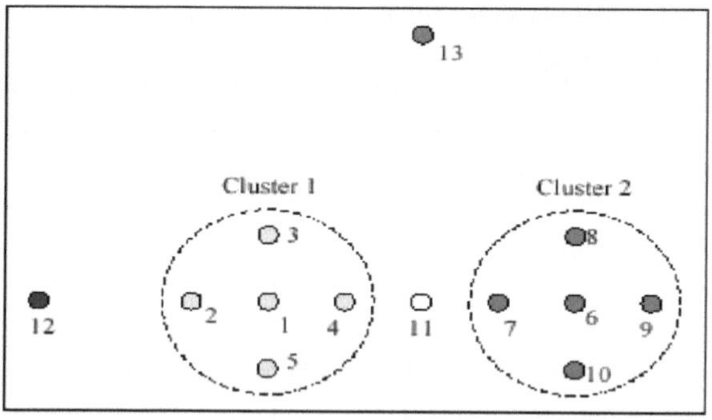

Fig. 1. Two simple clusters with noise points

Table 1. Crisp, Fuzzy, and Possibilistic membership degrees of the data in Fig. 1

Point	Crisp		Fuzzy		Possibilistic	
	$(u_c)_1$	$(u_c)_2$	$(u_f)_1$	$(u_f)_2$	$(u_p)_1$	$(u_p)_2$
1	1	0	1.00	0.00	1.00	0.06
2	1	0	0.96	0.04	0.50	0.04
3	1	0	0.94	0.06	0.50	0.06
4	1	0	0.90	0.10	0.50	0.10
5	1	0	0.94	0.06	0.50	0.06
6	0	1	0.00	1.00	0.06	1.00
7	0	1	0.10	0.90	0.10	0.50
8	0	1	0.06	0.94	0.06	0.50
9	0	1	0.04	0.96	0.04	0.50
10	0	1	0.06	0.94	0.06	0.50
11	1*	0	0.50	0.50	0.20	0.20
12	1	0	0.84	0.16	0.10	0.02
13	1*	0	0.50	0.50	0.05	0.05

the same way. We will refer to both of them as *(constrained) fuzzy* labels. Fuzzy labels, (u_f), satisfy the constraints:

$$0 \leq (u_f)_{ij} \leq 1 \qquad and \quad \sum_{i=1}^{C} (u_f)_{ij} = 1 \quad \forall j = 1 \cdots N. \tag{2}$$

Many fuzzy partitional clustering algorithms assign this type of labels in each iteration. For instance, the Fuzzy C-Means (FCM) [26] assigns the labels:

$$(u_f)_{ij} = \frac{(1/d_{ij})^{\frac{2}{m-1}}}{\sum_{k=1}^{C} (1/d_{kj})^{\frac{2}{m-1}}}, \tag{3}$$

where d_{ij} is the distance between cluster i and data point $\mathbf{x}_j$, and $m \in (1, \infty)$ is a constant that controls the degree of fuzziness. The EM algorithm [27] uses a similar equation where $m=2$, and the $(1/d_{ij})$ are replaced by probabilities (these labels correspond to the posteriori probabilities). Due to the constraint in (2), the membership degrees computed using eq. (3) represent relative degrees of sharing of a given point among several clusters. In section 5.3, we will show that these memberships could be used to identify boundary points.

Possibilistic labeling relaxes the constraint that the memberships must sum to one. It assigns *"typicality"* values, (u_p), that do not consider the *relative* position of the point to all clusters. As a result, if $\mathbf{x}_j$ is a noise point, then $\sum_{i=1}^{C}(u_p)_{ij} << 1$, and if $\mathbf{x}_j$ is typical of more than one cluster, we can have $\sum_{i=1}^{C}(u_p)_{ij} > 1$. Many robust partitional clustering algorithms [28, 29, 30] use this type of labeling in each iteration. For instance, the Possibilistic C-Means [29] assigns labels using

$$(u_p)_{ij} = \frac{1}{1 + (d_{ij}/\eta_i)^{\frac{2}{m-1}}}, \tag{4}$$

where η_i is a cluster-dependent resolution/scale parameter [29]. Robust statistical estimators, such M-estimators and W-estimators [31] use this type of memberships to reduce the effect of noise and outliers.

Unlike the memberships degrees in eq. (3), $(u_p)_{ij}$ are not constrained to add to one and represent absolute degrees of typicality to a given cluster regardless of the proximity of the points to the other clusters. In sections 5.1 and 5.2, we show that these memberships could be used to identify seed and noise points.

The example in Fig.1 and Table 1 illustrates the three labeling paradigms. In Table 1, The '*' indicates that the point is equally distant from both clusters, and labels were assigned arbitrarily. Notice how fuzzy and possibilistic labeling treat noise (points 12 & 13) and points at the boundaries (point 11) differently. Finally, we shall note here that labeling can be either integrated into the clustering process, or it can be applied as a post-clustering step using equations similar to those in (1), (3), or (4).

3 MembershipMap Generation

In this paper, we propose a semantic preprocessing that facilitates a variety of data interpretation, analysis, and mining tasks. Here, we use the term semantic to indicate that the coordinates of each point in the transformed space could reveal additional information that cannot be easily extracted from the original feature space. Our approach maps the original feature space into a unit hypercube with richer information content. The proposed mapping starts by clustering each attribute independently. Each cluster, thus obtained, can be considered to form a subspace that reflects a "more specific" concept along that dimension. The orthogonal union of all the subspaces obtained for each attribute compose the transformed space. The class labels (crisp, fuzzy, or possibilistic) determine the position of data points in the new space. Hence, the proposed preprocessing can be considered as a special type of "semantic mapping" that maps the data

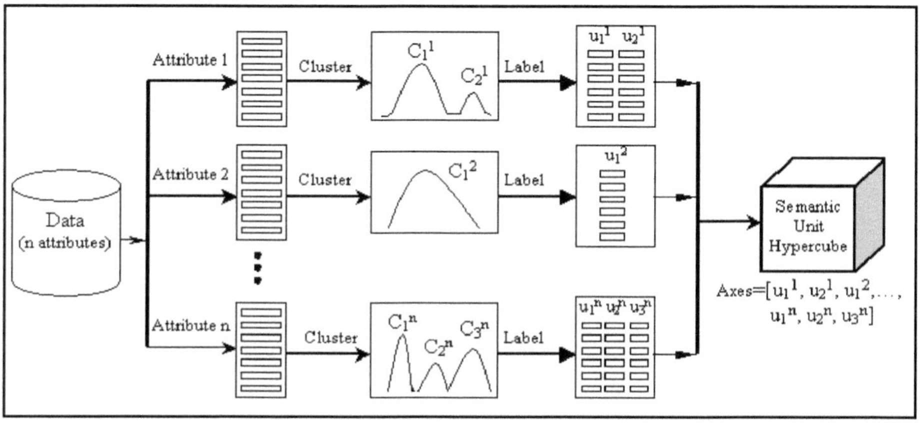

Fig. 2. Illustration of the MembershipMap

from the "raw" feature space onto a new space characterized by a semantically richer dichotomization of the original features. Our approach is outlined in Fig.2 and is summarized below.

MembershipMap Generation

FOR each dimension k, $1 \leq k \leq n$ **Do** {

Let $\mathcal{X}_k$= *projection of the data onto the* k^{th} *dimension;*

Partition $\mathcal{X}_k$ *into* C_k *clusters/components;*

For each $\mathbf{x}_j \in \mathcal{X}_k$, *compute any or all of the following memberships: crisp* $(u_c)_{ij}^k$, *fuzzy* $(u_f)_{ij}^k$, *or possibilistic* $(u_p)_{ij}^k$ *memberships; for* $1 \leq i \leq C_k$;

}

Dimension of the MembershipMap: $n_H = \sum_{k=1}^{n} C_k$;

Map each data sample $\mathbf{x}_j$ *to a new space,* $\mathcal{S}$, *using*

$\mathcal{M} : \mathcal{X} \rightarrow \mathcal{S}$;

$\mathcal{M}(\mathbf{x}_j) = [u_{1j}^1, \cdots, u_{c_1 j}^1,\ u_{1j}^2, \cdots, u_{c_2 j}^2,\ \cdots,\ u_{1j}^k, \cdots, u_{c_k j}^k]$;

The first step needed to generate the MembershipMap consists of performing feature wise clustering and assigning class memberships. This idea is not new as it has been used extensively in several pattern recognition and data mining techniques. For instance, the first step in computing with words (or granular computing) [32, 33] consists of granulating the data space into concept space. This is usually achieved by partitioning each feature into granules (i.e., clusters). Each granule would be assigned a meaningful name to represent its information content. Feature wise clustering is also a common technique used to derive fuzzy rules from data for system modeling and classification [34, 35, 36]. The main contribution of the proposed approach is the aggregation of the clusters' labels to form a new space that can be explored to facilitate many KDD tasks.

The MembershipMap approach is also different from other data pre-processing techniques such as z-score normalization, max-min scaling, and PCA [5]. This is because it takes into account the intricate nature of each attribute's distribution *independently* of the others and uses *mixed* cluster labels to create a new membership space. Thus, it is suitable for data sets with features that vary widely in scale and type. Moreover, our approach differs from techniques that impose a grid or a hyperbox structure on the data space. This is because the conceptual units in the semantic space are not constrained to take any specific shape, size, or number. This information is automatically derived from the data distribution. In fact, the MembershipMap can support grid-based techniques by providing a more accurate dichotomization of the data space.

The 1-dimensional clustering and membership assignment can be computed using many techniques. This task is generally trivial when performed on a univariate projection of the data (compared to the entire data set in the original space). For instance, the competitive agglomeration (CA) algorithm [37], or SOON[38] could be used to cluster each attribute and identify the optimal number of clusters. Also, these memberships can be estimated automatically based on statistical distributions or histogram density. For instance, the values of each attribute could be represented by a histogram. This histogram would then be thresholded using the discrimination criterion of Otsu [39]. The attribute clusters could even be user-defined depending on the conceptual meaning attached to the particular attribute. Most importantly, separate treatment of the attributes allows the complete freedom to adapt different ways to compute the membership functions for different attributes depending on prior knowledge and attribute type such as categorical versus numerical.

Illustrative Example

Figure 3 illustrates our approach using a simple 2-D (n=2) data consisting of three clusters. The 2 attributes ($x-y$ positions) are distributed between 0 and 200, and the upper left corner is the origin of the $x-y$ space. The histograms of the data's projection on each axis are also shown parallel to the corresponding axis. The two clusters of each dimension are displayed with different gray values and the '+' signs indicates the centers. The crisp, fuzzy, and possibilistic labels are generated using these cluster centers and equations (1), (3), and (4) respectively. Table 2 lists these labels for the centers and four other points in each group. These points were manually selected as points located on the boundaries to highlight the difference between crisp, fuzzy, and possibilistic labeling methods. In table 2, u_i^x (u_i^y) refer to the memberships of the projection of the data onto the x-axis (y-axis) in cluster i. Hence, for any of the selected data samples, these four membership values in the corresponding row of the table form the components of a new vector in the new 4-D MembershipMap.

4 Properties of the Crisp, Fuzzy, and Possibilistic Maps

4.1 The Crisp MembershipMap

Crisp labeling assigns membership values in $\{0,1\}$. Thus, the transformed space, called *CrispMap*, consists of vecores with binary attributes. This simple representation has several advantages:

1. **Comparison of two data samples:** The similarity between two data samples can be easily assessed using their mapped vectors. This is because the number of *common bits* between their two corresponding binary vectors represents the *number of shared concepts*, i.e., the number of attributes where the two points lie in the same cluster as identified by the preprocessing step. For instance, all points in the top left cluster in Fig. 3 are mapped to [1 0 0 1]. Thus, all intra-cluster distances will be 0.
2. **Querying the CrispMap:** The transformed data can be queried easily to retrieve samples that satisfy certain criteria. For instance, "data samples that are similar from the point of view of attribute $k1$ and dissimilar from the point of view of Attribute $k2$" or "data samples that are similar in at least N_k attributes". These queries can be implemented very efficiently using binary masking operators.
3. **Using Rules to cluster or to retrieve:** The hamming (or edit) distance between the points in the CrispMap can be used to cluster the original data. Clustering can be simpler in this new space because all the vectors are binary and each bit has a special meaning. Techniques such as hierarchical clustering [23] can be used to produce a dendrogram to show the structure of the data. Unlike a generic dendrogram, in this case, the levels of the dendrogram have a rich semantic meaning corresponding to the number of attributes that share the same concept/cluster.

The limitations of the CrispMap are a direct consequence of their failure to model areas of overlap and uncertainty, as well as their inability to model outliers or noise.

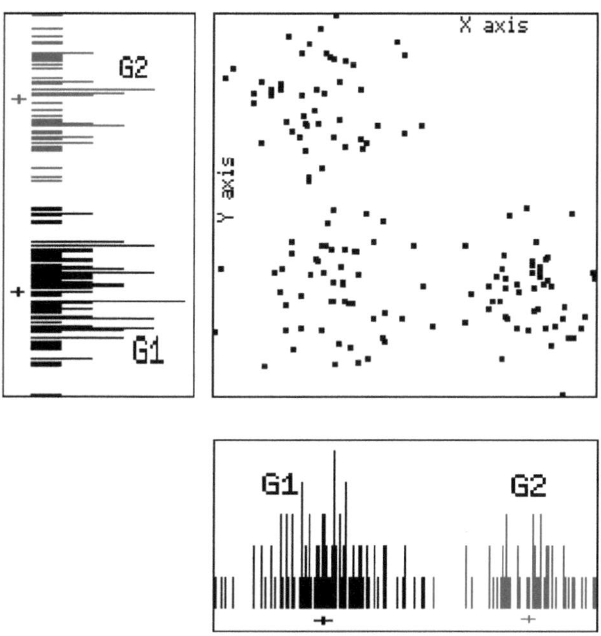

Fig. 3. A 2-D data and its corresponding histograms

Table 2. Transformation of the data in Fig. 3 into 4-Dim Crisp, Soft, and Possibilistic SemanticMaps

Cluster	x-y space		Crisp Map				Fuzzy Map				Possibilistic Map			
	x	y	u_1^x	u_2^x	u_1^y	u_2^y	u_1^x	u_2^x	u_1^y	u_2^y	u_1^x	u_2^x	u_1^y	u_2^y
1	50	52	1	0	0	1	1.00	0.00	0.00	1.00	0.97	0.03	0.04	0.95
1	7	55	1	0	0	1	1.00	0.00	0.07	0.93	1.00	0.03	0.02	0.25
1	87	49	1	0	0	1	1.00	0.00	0.34	0.66	0.91	0.03	0.10	0.21
1	58	108	1	0	0	1	0.53	0.47	0.02	0.98	0.13	0.11	0.05	0.72
1	33	5	1	0	0	1	0.91	0.09	0.01	0.99	0.14	0.02	0.03	0.78
⋮	⋮	⋮	⋮	⋮	⋮	⋮	⋮	⋮	⋮	⋮	⋮	⋮	⋮	⋮
2	140	62	1	0	1	0	1.00	0.00	1.00	0.00	0.91	0.04	0.93	0.05
2	109	59	1	0	1	0	1.00	0.00	0.74	0.26	0.97	0.03	0.21	0.10
2	142	109	1	0	1	0	0.51	0.49	1.00	0.00	0.13	0.12	0.96	0.05
2	182	75	1	0	1	0	0.96	0.04	0.94	0.06	0.54	0.05	0.23	0.02
2	133	3	1	0	1	0	0.90	0.10	0.98	0.02	0.13	0.02	0.71	0.06
⋮	⋮	⋮	⋮	⋮	⋮	⋮	⋮	⋮	⋮	⋮	⋮	⋮	⋮	⋮
3	141	158	0	1	1	0	0.00	1.00	1.00	0.00	0.04	0.92	0.95	0.05
3	101	163	0	1	1	0	0.00	1.00	0.62	0.38	0.04	1.00	0.16	0.13
3	146	134	0	1	1	0	0.13	0.87	1.00	0.00	0.07	0.30	1.00	0.04
3	199	182	0	1	1	0	0.02	0.98	0.90	0.10	0.03	0.55	0.12	0.02
3	138	199	0	1	1	0	0.06	0.94	0.99	0.01	0.02	0.25	0.85	0.05

1. **Points on areas of overlap between clusters:** These points will be mapped to the same image in the new space as points at the very core of either one of the overlapping clusters. Hence, it is not possible to distinguish between these points and those at the core.
2. **Outliers and border points:** No matter how distant an original data sample is from its cluster, it will be mapped to the same point as any point at the core of the cluster. This means that outliers cannot be detected, and that points on the outskirts of a cluster are indistinguishable from points at its center.

4.2 The Fuzzy MembershipMap

Fuzzy memberships are confined to $[0, 1]$. Hence, the transformed space, called *FuzzyMap*, consists of vectors with continuous values, and may embrace more space in the MembershipMap. The constraint that the memberships along each dimension must sum to one will constrain the placement of the transformed data. As a result, the FuzzyMap offers the following advantages:

1. **Identifying interesting regions via simple queries:** Without further processing, it is possible to use the transformed space to query and retrieve data that satisfy certain properties. For example, it is trivial to identify data samples that:

a) *Have low memberships in all dimensions:* These are boundary points. Once identified, they can be filtered out, resulting in a data set with well-separated clusters.
b) *Have strong memberships in all dimensions:* These are the most representative samples of the data. Once identified, they can be used for the purpose of instance selection or data summarization.
c) *Have no strong commitment to a specific attribute:* This query could be used to identify relevant (or irrelevant) attributes to different subsets of the data.

2. **Mining the Transformed Space:** Mining in the fuzzy membership space can be simpler than mining in the original space because all the feature vectors are normalized and each feature value has a special meaning: strength of matching between the data and one of the 1-D clusters/concepts. Moreover, fuzzy set union, intersection, and other aggregation operators [40] can be used to develop semantic similarity measures. This can be very helpful in case the original data has too many attributes, making the Euclidean distance unreliable.

The limitations of the FuzzyMap stem from the *constrained* fuzzy partition that characterizes them, and consist of the inability of fuzzy memberships to model outliers, as well as their being insufficient to convey the maximum amount of semantic information about the underlying structure. For instance, in Fig. 1 (and Table 1), unlike the possibilistic memberships, the fuzzy membership values do not differentiate between point 11 which is close to two clusters (memberships 0.5 and 0.5) and point 13 (noise) which is equally far from both clusters. We refer the reader to [29] for a more detailed comparison between fuzzy and possibilistic membership values in the context of clustering or labeling.

4.3 The Possibilistic MembershipMap

Possibilistic memberships are also confined to [0,1], but they are not constrained to sum to one like fuzzy memberships. Hence, the transformed space, called *PossibilisticMap*, consists of vectors that span the *entire* transformed space. Possibilistic labels have almost the same advantages as fuzzy labels. The only difference is that constrained fuzzy labels can capture the notion of *sharing* but not noise points, while *possibilistic* labels can identify noise but not the notion of sharing.

5 Exploring the Membership Maps

The MembershipMap can be mined just like the original data space, for clusters, classification models, association rules, etc. This task is expected to be easier in the membership space because all features are confined to the interval [0,1], and have meaning within this interval, which makes the distance/similarity measure more meaningful. Moreover, since different areas of the transformed spaces correspond to different concepts, the MembershipMap can be "explored" in a pre-data mining phase to uncover useful underlying information about the data. In this paper, we focus on exploring the MembershipMaps to uncover: seed points (can be used for more effective sampling);

noise/outliers (for data cleaning and outlier identification), and boundary points (which, when removed from the data set, can make subsequent data mining much easier) for the case of unlabeled data in a completely unsupervised way. These identified points of interest could provide additional information to subsequent learning algorithms.

5.1 Identifying Seed Points

In applications involving huge amounts of data,"*Seed points*" identification may be needed to reduce the complexity of data-driven algorithms. This process can be extremely difficult, if not impossible, for data with uneven, noisy, heterogeneous distributions. In addition to sampling, seed points can offer excellent initialization for techniques that are sensitive to the initial parameters such as clustering.

Using the PossibilisticMap, seed point identification is fast and trivial. In particular, seed points can be identified as points that have strong (close to 1) possibilistic labels in at least one cluster of each attribute in the original space. Formally, we use the possibilistic MembershipMap to define the *typicality* of each data point as follows:

Definition 1. *The degree of typicality of a data point* $\mathbf{x}_j$*,* $T_{\mathbf{x}_j}$*, is defined as*

$$T_{\mathbf{x}_j} = \min_{k=1,\cdots,n} \left\{ \underbrace{\max_{l=1,\cdots,C_k} \left\{ u_{lj}^{(k)} \right\}}_{\text{Best Possibilistic label in Dim k}} \right\} \tag{5}$$

where $u_{lj}^{(k)}$ *is the possibilistic label of* $\mathbf{x}_j$ *in the* l^{th} *cluster of dimension* k*, and* C_k *is the number of clusters along dimension* k*. Seed points are defined as points with high values of* $T_{\mathbf{x}_j}$ *(max of 1).*

5.2 Identifying Noise Points and Outliers

Noise and outlier detection is a challenging problem that has its own niche of research in the fields of data mining and statistics. Experts agree that outlier detection cannot succeed using simple statistical routines, or using clustering alone. In fact, when data has a large, unknown proportion of outliers, most learning algorithms break down and cannot yield any useful solution. Using the PossibilisticMap, noise and outliers can be identified as those points located near the origin. In other words, they correspond to points that have low possibilistic membership degrees in all clusters of all attributes. Formally, we define noise points as follows:

Definition 2. *Let* $\mathbf{x}_j^P = \left[\left\{ u_{lj}^{(k)} \right\}, l=1,\cdots,C_k;\ k=1,\cdots,n \right]$ *be the coordinates of point* $\mathbf{x}_j$ *in the PossibilisticMap.* $\mathbf{x}_j$ *is a noise point if* $\|\mathbf{x}_j^P\|$ *is small.*

5.3 Identifying Boundary Points

The process of boundary point identification is paramount in many classification techniques. It can be used to reduce the training data to (or emphasize) a set of boundary points to achieve higher focus in learning class boundaries. Using the FuzzyMap, boundary points can be identified as points that do not have a strong commitment to any cluster of a given attribute. In other words, these are points that are shared by many concepts and have ambiguous "multi-concept" interpretation. Formally, we use the FuzzyMap to define the *purity* of each data point as follows:

Definition 3. *The degree of purity of a data point* $\mathbf{x}_j$*,* $P_{\mathbf{x}_j}$*, is defined as*

$$P_{\mathbf{x}_j} = \min_{k=1,\cdots,n} \left\{ \underbrace{\max_{l=1,\cdots,C_k} \left\{ u_{lj}^{(k)} \right\}}_{\text{Best Fuzzy label in Dim } k} \right\} \tag{6}$$

where $u_{lj}^{(k)}$ *is the fuzzy label of* $\mathbf{x}_j$ *in the* l^{th} *cluster of dimension* k*, and* C_k *is the number of clusters along dimension* k*. Boundary points are defined as points with low purity values.*

In high-dimensional feature spaces, and when clusters are not expected to be well-defined along all features, definitions 1, 2, and 3 may be too restrictive. In this case, the min operators in equations (5) and (6) may be replaced by a more relaxed ordered statistics operator. Similarly, in definition 2, we replace $\|\mathbf{x}_j^P\|$ by the magnitude using only a subset of the attributes. The choice of such aggregation operators and the subset size depends on the subsequent learning algorithms, their sensitivity to noise, the cost of missing noise points, etc.

5.4 Illustrative Example

Fig.4 displays a simple data set with 4 clusters. First, the projection of the data along each dimension is partitioned into 2 clusters. Next, possibilistic labels were assigned to each $\mathbf{x}_j$ in all 4 clusters using equation (4). Each point $\mathbf{x}_j$ is mapped to $\mathbf{x}_j^P = \left[u_{1j}^{(1)}, u_{2j}^{(1)}, u_{1j}^{(2)}, u_{2j}^{(2)}\right]$, where $u_{kj}^{(i)}$ is the possibilistic label of point $\mathbf{x}_j$ in the kth cluster of attribute i. In Fig. 4(b), we have quantized $T_{\mathbf{x}_j}$ into 4 levels: $T_{\mathbf{x}_j} \geq 0.95$ (black squares); $0.80 \leq T_{\mathbf{x}_j} < 0.95$; $0.50 \leq T_{\mathbf{x}_j} < 0.80$; and $0.25 \leq T_{\mathbf{x}_j} < 0.50$ (squares with shade that gets lighter). As it can be seen, points located at the core of each cluster have the highest degree of typicality and could be treated as "seeds" for the entire data set. The degree of typicality decreases as we move away from the clusters. Points near the origin, i.e., $\|\mathbf{x}_j^P\|$ is small (<0.25), are identified and mapped back to the original space. These points are displayed using the "+" symbol in Fig.4(b). As it can be seen, these are generally noise points. We point in particular to noise points located in between clusters. These are challenging to detect using most conventional outlier-detection techniques.

To identify boundary points, fuzzy labels were assigned to each $\mathbf{x}_j$ in all 4 clusters using eq. (3), and each $\mathbf{x}_j$ is mapped to $\mathbf{x}_j^S = \left[u_{1j}^{(1)}, u_{2j}^{(1)}, u_{1j}^{(2)}, u_{2j}^{(2)}\right]$, where u_{ji}^{k} is the fuzzy label of point $\mathbf{x}_j$ in the kth cluster of attribute i. Fig. 4(c) displays the set of points $\left\{\mathbf{x}_j \mid P_{\mathbf{x}_j} < 0.75\right\}$ as small squares. As it can be seen, these are mainly the cluster boundaries.

In the above example, the thresholds (0.25, 0.5, 0.80, and 0.95) were used to illustrate the typicality and purity degrees graphically. In the experimental section, we do not

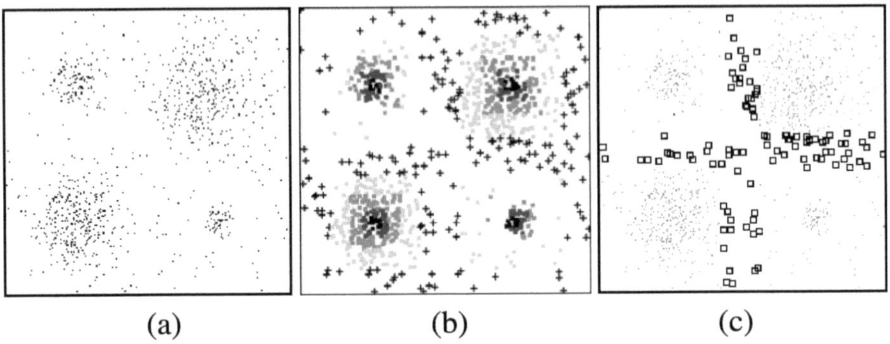

Fig. 4. Identifying regions of interest by exploring the MembershipMaps. (a) Data set. (b) Points with different degrees of typicality: darker square have higher degrees, and '+' signs represent identified noise points. (c) Squares indicate points with low purity.

use such thresholds. We simply sort these quantities and identify points in the top (or bottom) percentile.

5.5 Discussion

The example shown in Figure 4 is over-simplified as it uses a simple data that consists of 4 well-separated clusters, and the projection on each dimension results in two well-defined clusters. As a result, the dimension of the MembershipMap was only 4, and regions of interest were neatly identified. In a more realistic scenario, the data would be more complex with more dimensions. Moreover, it will not yield well-defined clusters when projected on the different attributes. In this case, more clusters would be needed for each dimension, thus, increasing the dimension of the MembershipMap by a factor larger than 2. Moreover, the identified regions of interest will not be very concise and accurate. For instance, in Fig. 4(b), if one selects only 1% of the data points as seeds, points would be sampled from all 4 clusters. In the more realistic scenario, one might need 5 or even 10% seed points to have a high probability of sampling from all the clusters. Several examples involving real and high-dimensional data would be presented and discussed in section 7.

The richness of the discovered semantic information comes at the cost of added dimensionality. In fact, the MembershipMap can significantly increase the dimensionality of the feature space, especially if the data has a large number of clusters. However, the cost of increasing the dimensionality can be offset by the significant benefits that can be reaped by putting this knowledge to use to facilitate data mining tasks. For example, we have shown how the MembershipMap can easily identify seed, border, and noise points. If taken into account, these samples can be used to obtain a significantly reduced data set by sampling, noise cleaning, and removing border points to improve class/cluster separation. This cleaning can facilitate most subsequent supervised and unsupervised learning tasks. It is also possible, after the cleaning, to go back to the original feature space. Thus, benefiting both from lower dimensionality and lower data cardinality.

The cost of the high dimensional MembershipMap can also be offset by the following additional benefits: **(i)** all attribute values are mapped to the interval [0,1]; **(ii)** categorical attributes can be mapped to numerical labels, thus, data with mixed attribute types can be mapped to numerical labels within [0,1]. As a results, a larger set of data mining algorithms can be investigated; and **(iii)** Distance/similarity values can be easily interpreted and thresholded in the new space.

It should be noted that even though the attributes are treated independently, they do get combined at a later stage. Hence, information such as correlations is not lost in the MembershipMap. In fact, treating the attributes independently in the initial (1-D clustering) is analogous to traditional feature scaling methods such as min-max or z-score normalization, except that the MembershipMap approach does not destroy intricate properties of an attribute's distribution using a single global transformation. In a sense, our approach applies a specialized local transformation to each subcluster along an attribute, and is thus non-linear.

The projection and clustering steps in the MembershipMap approach are not restricted to the original individual attributes. In fact, our approach can be combined with other feature reduction techniques. For instance, Principal Component Analysis (PCA) can be used to first project data on a subset of Eigenvectors, and then the MembershipMap can be applied on the data projected on this new set of features in the same way. In fact, one can even project the data on selected 2-D or 3-D subsets of the attributes, and then apply the MembershipMap.

It is also worth mentioning that the MembershipMap approach does not require a "perfect" clustering along each attribute. While crisp labeling may require well-defined clusters, fuzzy labeling has the natural advantage of assigning point membership levels based on the distance of a point from a cluster relative to the point's distances to all other clusters. In fact, regardless of the distribution of the samples, if we select one cluster per attribute that is centerd at the origin, the resulting fuzzy (and possibilistic) mapping would amount to a simple scaling of the original data. Thus, if the original data does not form one compact cluster along each dimension, we expect our multi-cluster projection to yield a richer description than simple scaling.

6 Data Labeling for Semi-Supervised Learning

In supervised learning, each sample is assigned a label that indicates to which of the C classes it belongs. The goal is to learn the classes' distributions and to learn their boundaries. In unsupervised learning (or clustering), the data is not labeled. The clustering algorithm is expected to partition the data, label it, and identify the properties of each cluster. Supervised learning is relatively simpler than unsupervised learning. This is because the labeled data can be used for training, and the problem can be decomposed into C simpler sub-problems, where data from each class can be treated independently in an initial step. Unfortunately, the process of data labeling can be labor intensive, and may even be impossible. A compromise between the two learning paradigms is semi-supervised learning, where only a small fraction of the data is labeled, and used to guide the clustering process. The Semi-Supervised Fuzzy C-Means (SS-FCM)[41] is an example of such algorithms that is based on the FCM.

Let $\mathcal{X}=\mathcal{X}^d \cup \mathcal{X}^u$, where $\mathcal{X}^d$ is a design set of labeled samples and $\mathcal{X}^u$ is the set of unlabeled samples. In general, $N_d=|\mathcal{X}^d|$ is much smaller than $N_u=|\mathcal{X}^u|$. Let b_k be a pointer parameter that indicates if sample x_k is labeled, i.e., $b_k=1$ if $x_k \in \mathcal{X}^d$, and 0 otherwise. In addition, let the matrix $\mathbf{F}=\left[f_{ik}\right]_{c \times n}$ include the fuzzy labels of $x_j \in \mathcal{X}^d$ and zero vectors for $x_j \in \mathcal{X}^u$. The SS-FCM algorithm[41] minimizes

$$J_S = \sum_{i=1}^{C}\sum_{j=1}^{N}(u_{ij})^m d_{ij}^2 \; + \; \alpha \sum_{i=1}^{C}\sum_{j=1}^{N}(u_{ij} - b_j f_{ij})^m d_{ij}^2 \tag{7}$$

subject to the constraint in (2.2). Minimization of J_S with respect to $\mathbf{U}$ yields

$$u_{ij} = \frac{1}{1+\alpha^{1/(m-1)}} \left\{ \frac{1+\alpha^{1/(m-1)}\left[1 - b_j \sum_{k=1}^{c} f_{ik}\right]}{\sum_{k=1}^{c}\left(d_{ij}/d_{kj}\right)^{2/(m-1)}} + \alpha^{1/(m-1)} b_j f_{ij} \right\}. \tag{8}$$

We should note here that the labeled samples have fuzzy labels ($\in [0,1]$) to indicate degrees of belongingness to multiple clusters. Moreover, during the clustering process, the algorithm generates a new partition for all $x_j \in \mathcal{X}$. Thus, after convergence, the supervising labels are replaced by the computed labels. The constant α in (7) controls the importance of the labeled samples during the optimization and thus, the degree to which these labels are allowed to deviate. Therefore, α should be set depending on our confidence in the labeled samples.

In the following, we propose using the MembershipMaps to extract and label candidate samples for the SS-FCM in a completely unsupervised way. Our approach involves the following simple steps:

1. Using the *PossibilisticMap* and definition 1, identify few seed points. Let $S = \{\mathbf{x}_j | T_{\mathbf{x}_j} > \theta_s\}$ be the set of identified seed points. We select θ_s such that $|\theta_s| \approx 5\%|\mathcal{X}|$.
2. Use CA [37] or SOON [38] (or another unsupervised clustering algorithm) to partition S into K clusters (K is determined by the clustering algorithm), and assign a distinct label to each of the K seed cluster: $C_1, C_2, \cdots, C_K$
3. Compute the fuzzy membership degree of each seed point in each of the K clusters using eq. (3);
4. Using the *FuzzyMap* and definition 3, identify few boundary points. Let $B = \{\mathbf{x}_j | P_{\mathbf{x}_j} > \theta_B\}$ be the set of identified boundary points. We select θ_B such that $|\theta_B| \approx 5\%|\mathcal{X}|$.
5. Compute the fuzzy membership degree of each boundary point in each of the K clusters using eq. (3);
6. The identified seed and boundary points constitute the labeled subset for the semi-supervised algorithm, and their fuzzy membership values would be treated as fuzzy labels.

7 Experimental Results

7.1 Data Sets

We illustrate the performance of the proposed transformation using benchmark data sets[1]. These are the Iris, Wisconsin Breast Cancer, and the Heart Disease. The Iris data has 4 features and 50 samples in each of the 3 classes. The Wisconsin Breast Cancer data has 9 features and 2 classes, namely, benign and malignant. These 2 classes have 444 and 239 samples respectively. The Heart Disease data has 13 features and 2 classes. The first class, absence of heart disease, has 150 samples, and the second class, presence of heart disease, has 120 samples. The satellite data has 35 features, 6 classes, and a total of 6,435 samples. This data is partitioned into a training set (4,435) and a testing set (2,000).

We should emphasize here that the membership generation is completely unsupervised. That is, even though the above data sets are actually labeled, the class labels were used only to validate the information extracted from the membership maps, and to evaluate and compare the results.

7.2 Membership Generation

The first step in the MembershipMap transformation consists of a quick and rough segmentation of each feature. This is a relatively easy task that can rely on histogram thresholding or clustering 1-D data. The results reported in this paper were obtained using the Self-Organizing Oscillators Network (SOON) clustering algorithm [38]. SOON has the advantage of clustering noisy data, and finding the optimal number of clusters in an unsupervised way. The Iris maps have 15 dimensions, the Wisconsin Breast Cancer maps have 50 dimensions, and the Heart Disease maps have 39 dimensions. The actual number of clusters per attribute for each data is shown in Table 3. The Satellite data maps t0 107 dimensions. We have also tried other simple clustering algorithms such as the K-Means, where the number of clusters is specified after inspecting the 1-D projected data. The results were comparable. Next, fuzzy and possibilistic labels were assigned to each sample using equations (3) and (4) with m=1.5, η=2, and using the Mahalanobis distance. For the 1-D case, this distance reduces to $d_{ij} = \frac{\mathbf{x}_j - \mu_i}{\sigma_i}$, where μ_i and σ_i are the mean and standard deviation of the ith cluster of the attribute under consideration.

Finally, the fuzzy (possibilistic) labels were aggregated to create the FuzzyMap (PossibilisticMap). Unlike the illustrative example shown in Fig. 4, the projection of these real data sets along each attribute does not provide well-defined clusters. In fact, since fuzzy and possibilistic labeling can tolerate uncertainties and vagueness, there is no need for accurate sub-concept extraction with the correct number of clusters. Moreover, the way we explore these maps (e.g., semi-supervised labeling in Section 6) does not require accurate identification of the points of interest.

Typically, the membership mapping increases the dimensionality considerably. For the above data sets, the new dimensions are still manageable, and thus, we can apply

[1] Available at "http://www.ics.uci.edu/ mlearn/MLRepository.html"

Table 3. Number of clusters per attribute

Data Set	No. of Clusters for each attribute
Iris	4 3 4 4
Breast Cancer	6 6 6 6 5 6 5 6 4
Heart Disease	4 2 4 4 2 2 2 4 2 3 3 4 3

clustering (or classification) to the transformed spaces (as in section 7.4). However, in other applications, the original dimension may be large, and thus, the dimension of the maps would grow even larger. In this case, the membership space would be explored by identifying points of interest (as in section 7.5), and the actual learning would be applied to the original feature space.

7.3 Identifying Regions of Interest

To identify seed points, noise and outliers for the Iris data, we compute the typicality of each sample. Points with high typicality would correspond to seed points, while points with low typicality would correspond to noise. Since the Iris data has 4 dimensions, the identified points could not be visualized as in Fig. 4. Instead, we rely on the ground truth to validate the results. We use the class labels and compute the centroid of each class. Then, we compute the distance from each point to the centroid of its class. Theoretically, points with small distances would correspond to seeds, and points with large distances would correspond to noise. Fig. 5(a) displays a scatter plot of the sample typicality versus their distance to the true class centroid. The distribution clearly suggests a negative correlation. In fact, the lower and upper bounds of the correlation coefficient for a 95% confidence interval are -0.596 and -0.349 respectively. This means that typicality, which was extracted only from the PossibilisticMap *without any knowledge* about the true class labels, is higher for those samples located near the true class centroid.

To validate the degree of purity, we use definition 3 to estimate the purity of each sample. We also use the ground truth and compute the membership of each sample using eq.(3), where the distances are computed with respect to the classes' centroids (we use this to approximate the ground truth purity of each sample). Fig. 6(b) displays a scatter plot of the sample purity versus their membership in the true class. The distribution shows that all samples from class 1 have high purity, and the few samples with low purity belong to either class 2 or 3. This information is consistent with the known distribution of the Iris data where class 1 is well-separated while classes 2 and 3 overlap. Thus, the purity computed using the FuzzyMap without any knowledge about the true class labels can estimate the degree of sharing among the multiple classes, i.e, boundary points.

7.4 Clustering the Membership Maps

This experiment illustrates the advantage of using the Membership Maps to improve clustering. We apply the Fuzzy C-Means to cluster the different data sets in the original,

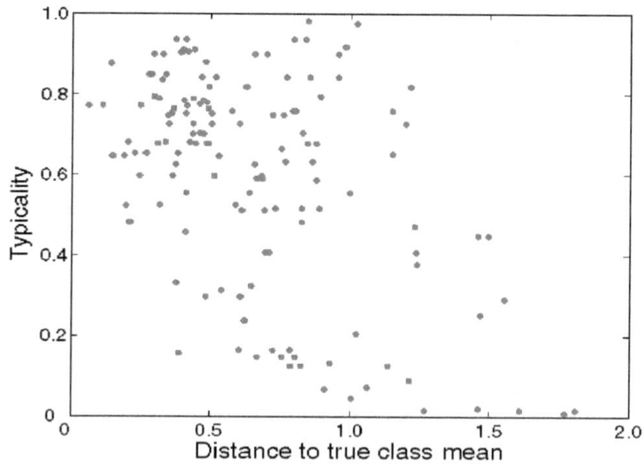

Fig. 5. Validating the typicality values of the Iris data samples extracted using the PossibilisticMap

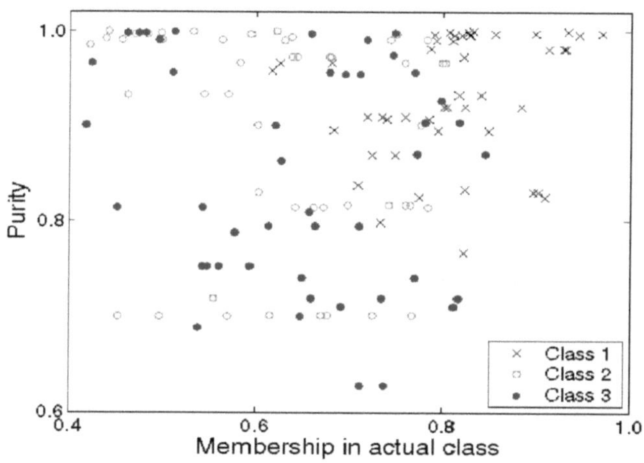

Fig. 6. Validating the purity values of the Iris data samples extarcted using the fuzzy MembershipMap

FuzzyMap, and PossibilisticMap spaces. For all cases, we fix the number of clusters to the actual number of classes, and use the same random points as initial centers.

To assess the validity of each partition we use the true class labels to measure the purity of the clusters. The results of the average of 10 runs with different initializations are shown in Table 4. For each confusion matrix, the column refers to the cluster index while the row refers to the true class index. As it can be seen, both maps improve the overall clustering results. For the breast cancer data, the overall number of misclassification is reduced from 28 to 24 (FuzzyMap) and 25 (PossibilisticMap). This

Table 4. Clustering Results

Data Set	*Original Space*	*FuzzyMap*	*Poss.Map*
Iris	50 0 0	50 0 0	50 0 0
	0 46 4	0 47 3	0 47 3
	0 14 36	0 4 46	0 5 45
Breast Cancer	435 9	423 21	422 22
	19 220	3 236	3 236
Heart Disease	117 33	121 29	121 29
	23 97	21 99	21 99

Table 5. Classification Results

	Original Space	FuzzyMap	PossMap
Training	89.29%	88.86%	93.57%
Testing	85.50%	86.30%	87.30%

improvement is not consistent for the 2 classes as the number of misclassifications of class 1 has increased from 9 to 21 (and 22), while the number of misclassifications for class 2 was reduced from 19 to 3. A possible explanation of this behavior may be the existence of few points located in between the two clusters that could be assigned to either class almost arbitrarily.

We should note here that a better partition of these data sets could be obtained using other clustering algorithms, other distances, and/or multiple clusters to represent each class. However, our goal here is to illustrate that the Membership Maps can improve the distribution of the data so that simple clustering algorithms are more reliable.

7.5 Classification Using the Membership Maps

This experiment uses the 36-D satellite data to illustrate the use of the Membership Maps to improve data classification. Only the training data is used to learn the mapping. The test data is mapped using the same cluster parameters obtained for the training data. We train a Back Propagation Neural Networks with 36 input units, 20 hidden units, and 6 output units on the original data and two other similar nets with 107 input units, 20 hidden units, and 6 output units on the Fuzzy and possibilistic maps. Table 5 compares the classification results in the 3 spaces. As it can be seen, the results on the testing data are slightly better in both transformed spaces. The above results could be improved by increasing the number of hidden nodes to compensate for the increased dimensionality in the mapped spaces. Moreover, the classifications rates could be improved if one exploits additional information that could be easily extracted from the possibilistic and fuzzy maps. For instance, by using boundary points for bootstrapping and filtering noise points.

(a) (b) (c)

Fig. 7. Original images

7.6 Application: Color Image Segmentation

In this section, we illustrate the application of the proposed mapping to the problem of image segmentation. Image segmentation is a complex and largely unsolved problem. Other approaches (not based on clustering) or other features may be more suitable for this problem. However, our goal in this application is to (graphically) illustrate the benefits of the extracted points of interest.

We use the membership maps as pre-processing tools to clean and down-sample the data. We also use the mapping to select few samples and label them so that a semi-supervised clustering algorithm could be used for the segmentation. Fig. 7 displays 3 typical images for which we show and compare the segmentation results. Each image is of size 128 by 192 (or 192 by 128). First, to construct the feature space, each pixel in the image is mapped to a 6-dimensional feature vector consisting of 3 color and 3 texture features [42]. Next, to segment the image, a clustering algorithm is used to group pixels (in the feature space) into homogeneous regions. Thus, the segmentation process involves clustering 24,576 points (128$\times$192) in a 6-dimensional feature space.

To extract regions of interest for the purpose of data reduction and cleaning, we first construct the fuzzy and possibilistic maps for the feature vectors of each image as described in section 7.2. Next, fuzzy and possibilistic labels were assigned to each sample using equations (3) and (4) with m=2.0, and η=σ_i (standard deviation of the features of cluster i). Finally, the fuzzy (possibilistic) labels were aggregated to create the FuzzyMap (PossibilisticMap).

For each image, we use the *PossibilisticMap* to identify seed and noise points. Seed points are selected as the top 5% of the points having the highest typicality values (using eq. (5)), while noise points are the bottom 5% (lowest typicality values). To identify boundary points, we use the *FuzzyMap* and select 5% of the points having the lowest purity values (using eq. (6)). Our choice of 5% is arbitrary. We could have used a lower or higher percentage, or we could have thresholded the typicality and purity values. The identified points of interest are shown in Fig. 8 for the first image in Fig. 7. As it can be seen, seed points correspond to typical points from the different regions in the image. Noise points, on the other hand, correspond to edges and small areas with different color and texture features (e.g. small tree branches). Boundary points are harder to interpret in Fig.8(c).

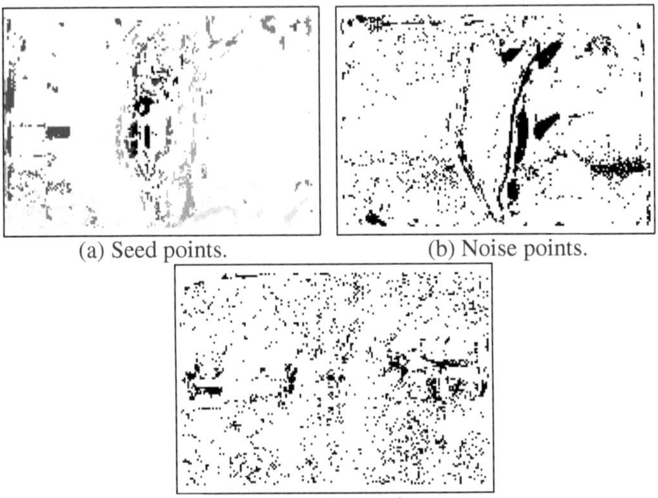

(a) Seed points. (b) Noise points. (c) Boundary points.

Fig. 8. Identified points of interest for the first image in Fig. 7(a)

We should note here that the identified points of interest are labeled with respect to the feature space and do not necessarily correspond to noise or boundary pixels in the actual images. For instance, many of the identified noise points in Fig. 8(b) correspond to edges of objects in the original image. This is caused by the window smoothing during feature extraction. In other words, the few windows superimposed on the objects boundaries will have features that are different from those in the actual objects.

The segmentation results of the images in Fig. 7 using different subsets of the pixels are shown in Fig. 9. For all the cases, we fix the number of clusters to 5, use the Euclidean distance, and fix m to 2.0. Fig. 9(a) displays the segmentation results when all pixels are clustered using the FCM.

To illustrate the sampling feature of the *PossibilisticMap*, we cluster only the identified seed points (1,200 points, compared to clustering 24,576 points) in the original 6-dimensional feature space. Then, all image pixels are assigned to the closest seed cluster. Fig. 9(b) displays these segmentation results. As it can be seen, the results are comparable, even though these segmented images were obtained using only 5% of the CPU time required to cluster all pixels. In fact, for the first two images, segmentation with seeds only yields slightly better results. For instance, in the first image in Fig. 9(a) the tree and bushes were merged, while they were assigned to different clusters in Fig. 9(b). Also, in the second image in Fig. 9(a) the bear was split into two clusters, while it was assigned to a single cluster in Fig. 9(b). The improved performance may be due to the fact that seed points constitute a much cleaner data with compact and well-separated clusters, which makes the FCM less susceptible to local minima. The drawback of clustering seedpoints only is that there is no guarantee that enough seed points would be identified from each region to form a valid cluster. This might explain the reason that the back of the elephant was merged with part of the background in Fig. 9(b).

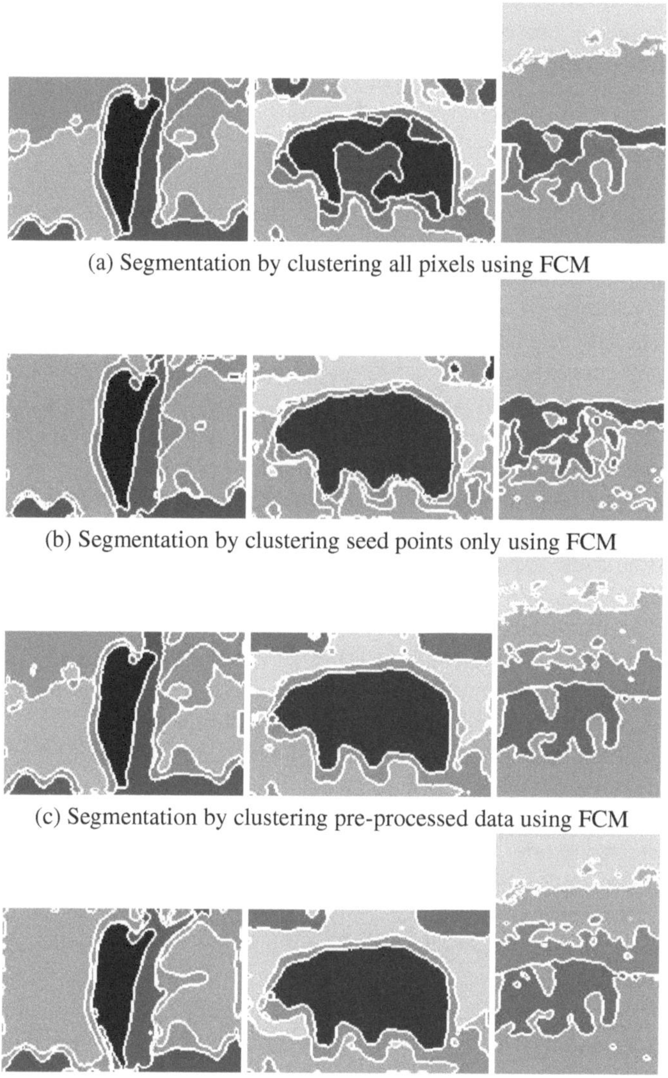

(a) Segmentation by clustering all pixels using FCM

(b) Segmentation by clustering seed points only using FCM

(c) Segmentation by clustering pre-processed data using FCM

(d) Segmentation by clustering pre-processed data using SS-FCM

Fig. 9. Segmentation Results

To illustrate the data cleaning feature of the Membership Maps, we exclude the identified noise and boundary points from the clustering process, and use seed points to initialize the prototypes. The segmentation results are shown in Fig. 9(c). Finally, we illustrate the ability of the MembershipMaps to automate the selection and labeling of representative points for semi-supervised clustering. We follow the steps outlined in Section 6, and use the labeled samples in the Semi-Supervised FCM [41]. The segmentation results are displayed in Fig. 9(d). As it can be seen, data cleaning and using semi-supervised improve the segmentation results. For instance, the back of the elephant is no longer confused with the background. Also, the first image in Fig. 9(d) shows more fine details of the tree branches.

8 Conclusions

We have presented a new mapping that facilitates many data mining tasks. Our approach strives to extract the underlying sub-concepts of each attribute, and uses their orthogonal union to define a new space. The sub-concepts of each attribute can be easily identified using simple 1-D clustering or histogram thresholding. Moreover, since fuzzy and possibilistic labeling can tolerate uncertainties and vagueness, there is no need for accurate sub-concept extraction. In addition to improving the performance of clustering and classification algorithms by taking advantage of the richer information content, the MembershipMaps could be used to formulate simple queries to extract special regions of interest, such as noise, outliers, boundary, and seed points.

In this paper, we have focused on the construction of the MembershipMaps and identifying points of interest using these maps. We have shown that the identified points of interest could be used to pre-process the data to improve the efficiency and effectiveness of data clustering algorithms. We have also shown that identified points of interest could be labeled, using simple rules, so that hard unsupervised learning problems could be converted to simpler semi-supervised problems. Other potential applications of the proposed mapping include converting categorical attributes into a numerical space, feature extraction, and applying other data mining algorithms.

There is a natural trade-off between dimensionality and information gain. Increased dimensionality of the MembershipMap can be offset by the quality of hidden knowledge that is uncovered. In fact, the increased dimension can be avoided by going back to the original feature space after data reduction and cleaning. Thus, subsequent learning algorithms could benefit both from the lower dimensionality of the original feature space, and the lower data cardinality produced by the membership maps. In this case, the high dimensionality of the transformed space is not a major issue as points of interest are extracted using simple queries.

The current implementation of the MembershipMap does not address the case where projections along some of the attributes do not form clusters. Depending on the number of such projections, this case would lead to unnecessary increase of the dimensionality of the map, and may even degrade the performance of the mapping. In future work, we plan to address this issue using two different approaches. The first one would involve examining the validity of each cluster along each attribute. Clusters that do not pass a validity test will not contribute to the construction of the MembershipMap. The second approach would involve combining the mapping with other feature reduction techniques. For instance, Principal Component Analysis can be used to first project data on a subset of Eigenvectors, and then the membership transformation can be applied on the projected data in the same way. In fact, one can even project the data on selected 2-D or 3-D subsets of the attributes, and then apply membership transformations.

Acknowledgments

This work was supported in part by the National Science Foundation Career Award No. IIS-0514319.

References

1. Fayyad, U., Piatetsky-Shapiro, G., Smyth, P., Uthurusamy, R.: Advances in Knowledge Discovery and Data Mining. MIT Press, Cambridge (1996)
2. Famili, A., Shen, W., Weber, R., Simoudis, E.: Data preprocessing and intelligent data analysis. Intelligent Data Analysis 1(1), 3–23 (1997)
3. Han, J., Kamber, M.: Data Mining Concepts and Techniques. Morgan Kaufmann, San Francisco (2001)
4. Pyle, D.: Data Preparation for Data Mining. Morgan Kaufmann, San Francisco (1999)
5. Jolliffe, I.: Principal Component Analysis. Springer, Heidelberg (1986)
6. Shepard, R.N.: The analysis of proximities: multidimensional scaling with an unknown distance function I and II. Psychometrika 27, 125–139, 219–246 (1962)
7. Kohonen, T.: Self-Organization and Associative Memory. Springer, Heidelberg (1989)
8. Sammon, J.W.: A nonlinear mapping for data analysis. IEEE Transactions on Computers 18, 401–409 (1969)
9. Jagadish, H.V.: A retrieval technique for similar shape. In: ACM SIGMOD, pp. 208–217 (1991)
10. Faloutsos, C., Lin, K.-I.: FastMap: A fast algorithm for indexing, data-mining and visualization of traditional and multimedia datasets. In: SIGMOD, pp. 163–174 (1995)
11. Schafer, J.L.: Analysis of Incomplete Multivariate Data. Chapman and Hall, Boca Raton (1997)
12. Wang, X., Barbará, S.D.: Modeling and imputation of large incomplete multidimensional data sets. In: Fourth International Conference on Data Warehousing and Knowledge Discovery, pp. 286–295 (2002)
13. Gray, J., Chaudhuri, S., Bosworth, A., Layman, A., Reichart, D., Venkatrao, M., Pellow, F., Pirahesh, H.: Data cube: A relational aggregation operator generalizing group-by, cross-tab, and sub-totals. J. Data Mining and Knowledge Discovery 1(1), 29–53 (1997)
14. Chauduri, S., Dayal, U.: An overview of data warehousing and OLAP technology. SIGMOD Record 26(1), 65–74 (1997)
15. Almuallim, H., Dietterich, T.G.: Learning with many irrelevant features. In: Ninth National Conf. AI, pp. 547–552 (1991)
16. Dash, M., Liu, H., Yao, J.: Dimensionality reduction for unsupervised data. In: 9th IEEE Int. Conf. on Tools with AI, ICTAI 1997, pp. 532–539 (1997)
17. Frigui, H., Nasraoui, O.: Unsupervised learning of prototypes and attribute weights. Pattern Recognition 37(3), 567–581 (2004)
18. Kivinen, J., Mannila, H.: The power of sampling in knowledge discovery. In: Thirteenth ACM SIGACT-SIGMOD-SIGART Symp. Principles of Database Sys., pp. 77–85 (1994)
19. Dougherty, J., Kohavi, R., Sahami, M.: Supervised and unsupervised discretization of continuous features. In: 12th International Conference on Machine Learning, pp. 194–202 (1995)
20. Ho, K., Scott, P.: Zeta: a global method for discretization of continuous variables. In: 3rd International Conference on Knowledge Discovery and Data Mining (KDD 1997), pp. 191–194. AAAI Press, Menlo Park (1997)
21. Liu, H., Hussain, F., Tan, C., Dash, M.: Discretization: an enabling technique. Journal of Data Mining and Knowledge Discovery 6(4), 393–423 (2002)

22. Barbará, S.D., DuMouchel, W., Faloutsos, C., Haas, P., Hellerstein, J., Ioannidis, Y., Jagadish, H., Johnson, T., Ng, R., Poosala, V., Ross, K., Sevcik, K.: The new jersey data reduction report. Bulletin of the Technical Committee on Data Engineering 20, 3–45 (1997)
23. Duda, R.O., Hart, P.E.: Pattern Classification and Scene Analysis. John Wiley and Sons, Chichester (1973)
24. Kaufman, L., Rousseeuw, P.: Finding Groups in Data: An Introduction to Cluster Analysis. John Wiley and Sons, Chichester (1990)
25. Zimmermann, H.Z.: Fuzzy Set Theory and Its Applications, 4th edn. Kluwer, Dordrecht (2001)
26. Bezdek, J.C.: Pattern Recognition with Fuzzy Objective Function Algorithms. Plenum Press, New York (1981)
27. Dempster, A.P., Laird, N.M., Rubin, D.B.: Maximum likelihood from incomplete data via the EM algorithm. Journal of the Royal Statistical Society Series B 39(1), 1–38 (1977)
28. Frigui, H., Krishnapuram, R.: A robust competitive clustering algorithm with applications in computer vision. IEEE Trans. Patt. Analysis Mach. Intell. 21(5), 450–465 (1999)
29. Krishnapuram, R., Keller, J.: A possibilistic approach to clustering. IEEE Trans. Fuzzy Systems 1(2), 98–110 (1993)
30. Davé, R.N., Krishnapuram, R.: Robust clustering methods: A unified view. IEEE Trans. Fuzzy Systems 5(2), 270–293 (1997)
31. Hampel, F.R., Ronchetti, E.M., Rousseeuw, P.J., Stahel, W.A.: Robust Statistics the Approach Based on Influence Functions. John Wiley & Sons, New York (1986)
32. Pedrycz, W.: Granular Computing: An Emerging Paradigm. Springer, Heidelberg (2001)
33. Yao, Y., Yao, J.: Granular computing as a basis for consistent classification problem. In: Chen, M.-S., Yu, P.S., Liu, B. (eds.) PAKDD 2002. LNCS (LNAI), vol. 2336, pp. 101–106. Springer, Heidelberg (2002)
34. Runkler, T.A., Roychowdhury, S.: Generating decision trees and membership functions by fuzzy clustering. In: Seventh European Congress on Intelligent Techniques and Soft Computing (1999)
35. Ishibuchi, H., Nakashima, T., Murata: Performance evaluation of fuzzy classifier systems for multidimensional pattern classification problems. IEEE Trans. on Systems, Man, and Cybernetics - Part B 29, 601–618 (1999)
36. Klawonn, F., Kruse, R.: Derivation of fuzzy classification rules from multidimensional data. In: Lasker, X.L.G.E. (ed.) Advances in Intelligent Data Analysis, The International Institute for Advanced Studies in Systems Research and Cybernetics, Windsor, Ontario, pp. 90–94 (1995)
37. Frigui, H., Krishnapuram, R.: Clustering by competitive agglomeration. Pattern Recognition 30(7), 1223–1232 (1997)
38. Rhouma, M., Frigui, H.: Self-organization of a population of coupled oscillators with application to clustering. IEEE Trans. Patt. Analysis Mach. Intell. 23(2), 180–195 (2001)
39. Otsu, N.: A threshold selection method from gray level histograms. IEEE Trans. Systems, Man and Cybernetics 9, 62–66 (1979)
40. Wang, Z., Klir, G.: Fuzzy measure theory. Plenum Press, New York (1992)
41. Pedrycz, W., Waletzky, J.: Fuzzy clustering with partial supervision. IEEE Trans. Systems, Man and Cybernetics 27(5), 787–795 (1997)
42. Carson, C., Thomas, M., Belongie, S., Hellerstein, J.M., Malik, J.: Blobworld: A system for region-based image indexing and retrieval. In: Huijsmans, D.P., Smeulders, A.W.M. (eds.) VISUAL 1999. LNCS, vol. 1614, pp. 509–516. Springer, Heidelberg (1999)

Advanced Developments and Applications of the Fuzzy ARTMAP Neural Network in Pattern Classification

Boaz Lerner[1,*] and Hugo Guterman[2]

[1] Department of Industrial Engineering and Management
Ben-Gurion University, Beer-Sheva 84105, Israel
boaz@bgu.ac.il
[2] Department of Electrical and Computer Engineering
Ben-Gurion University, Beer-Sheva 84105, Israel

Abstract. Since its inception in 1992, the fuzzy ARTMAP (FAM) neural network (NN) has attracted researchers' attention as a fast, accurate, off and online pattern classifier. Since then, many studies have explored different issues concerning FAM optimization, training and evaluation, e.g., model sensitivity to parameters, ordering strategy for the presentation of the training patterns, training method and method of predicting the classification accuracy. Other studies have suggested variants to FAM to improve its generalization capability or overcome the prime limitation of the model, which is category proliferation (i.e., model complexity that increases with data complexity). Category proliferation is pronounced in problems that are noisy or contain a large degree of class statistical overlapping. In many investigations, FAM was improved by incorporating elements of optimization theory, Bayes' decision theory, evolutionary learning, and cluster analysis. Due to its appealing characteristics, FAM and its variants have been applied extensively and successfully to real-world classification problems. Numerous applications were reported in, for example, the processing of signals from different sources, images, speech, and text; recognition of speakers, image objects, handwritten, and genetic abnormalities; and medical and fault diagnoses. When compared to other state-of-the-art machine learning classifiers, FAM and its variants showed superior speed and ease of training, and in most cases they delivered comparable classification accuracy.

1 Introduction

Most state-of-the-art pattern classifiers, such as the multilayer perceptron (MLP) neural network (NN) [1] and the support vector machine (SVM) [2], usually require extended training periods and large sample sizes to achieve their ultimate performance level. However, in many real-world applications, the provision of extended training periods and large sample sizes can be fulfilled only partially or sometimes not even at all. Nevertheless, the classifier in these applications is expected to learn rapidly using only a few examples.

* Corresponding author.

L.C. Jain et al. (Eds.): Comp. Intel. Para.: Innov. Applications, SCI 137, pp. 77–107, 2008.
springerlink.com

In addition, the classifier may be expected to learn both in batch (offline) and incrementally (online) to address the different and changing requirements of the domain and provide the system with potential growth. As an illustrative example, take an application for incremental learning based on a robotic vision system that operates under changing environmental conditions and the variable physical characteristics of nonuniform textured objects [3]. We require that models of incremental learning assure plasticity in learning patterns of novel classes while continuing to stably recognize patterns belonging to familiar classes. Also, incremental learning may be beneficial to accommodate changes accumulated not necessarily from the outset but following an initial period for which training was already completed successfully. These changes may not justify re-training the already trained classifier for all the data but rather only modifying classifier parameters based on recent examples. Incremental learning is also advantageous for attaining fast pattern classification at any time. In addition, models of incremental learning improve their accuracies during training and may asymptotically outperform non-incrementally learned models and even reach the ultimate possible accuracy. This is an important advantage of the former models in non-stationary domains and when classifying large databases. Finally, as we are interested in pattern classification, the accomplishment of all theses objectives should not come at the expense of achieving high classification accuracy.

The fuzzy ARTMAP (FAM) [4] is considered one of the leading algorithms for classification. Compared to the MLP, which is learned based on error minimization, FAM performs prototype-based learning. That is, for any given input, MLP minimizes the error between its respective output and the corresponding label, whereas FAM learns by selecting the nearest prototype to the input. FAM excels in fast incremental supervised learning in a non-stationary environment and using few examples. The network enables the learning of new data without forgetting past data, addressing the so-called "plasticity-stability dilemma" [5], which is crucial for incremental learning. Following increase in data complexity, FAM expands its complexity by allocating nodes dynamically without user intervention. Most other NNs, however, have fixed configurations. For example, once the numbers of hidden layers and units in each hidden layer of MLP are determined, they remain unchanged. Furthermore, these numbers are not known beforehand and some preliminary experimentation is required to find the most appropriate values for them. A network that is designed with fewer than the required number of hidden neurons may underfit the data, whereas that having more hidden neurons than required may overfit the data. In both cases, network performance is only sub-optimal. Therefore, the use of a FAM provides design flexibility that is usually missing in other NN models and exempts from excessive preliminary experimentation to determine network configuration. Additional advantages of the FAM algorithm are its dependence on minimal heuristics and parameter settings and its guarantees of short training periods and convergence [4, 6, 7].

FAM and its variants have exemplified themselves as accurate and fast learners in performing various classification tasks, such as automatic target recognition based on radar range profiles [8], speaker-independent vowel recognition [9], online handwritten recognition [10], QRS-wave recognition [11], medical diagnosis of breast cancer and heart disease [12], three-dimensional object understanding and prediction from a series of two-dimensional views [13], classification of noisy signals [14], discrimination of alcoholics from nonalcoholics [15] and recently genetic abnormality diagnosis [16], as well as many other classification tasks such as [17, 18, 19, 20, 21, 22, 23].

The major limitation of FAM is its sensitivity to statistical overlapping between the classes. It is the self-organization nature of the algorithm that while enabling continuous learning of novel patterns also leads to the overfitting of noisy (overlapped) data that are mistakenly considered novel. This sensitivity can lead to uncontrolled growth in the number of categories, also referred to as category proliferation, leading to high computational and memory complexities and possible degradation in classification accuracy [7, 9, 16, 24, 25, 26, 27, 28, 29]. Improving FAM classification accuracy and reducing model category proliferation are the most studied topics in the FAM literature.

We begin this chapter by reviewing the basics of the fuzzy ART and FAM (Section 2). Advanced FAM-based developments that improve classification pitfalls of the original model are reviewed in Section 3. We focus particular attention on modifications and enhancements to reduce category proliferation and improve the generalization capability of FAM. We continue in Section 4 with a review of successful applications of FAM and its variants and complete the chapter in Section 5 by presenting the results of experimental studies comparing FAM to its advanced variants and non-FAM models.

2 Fuzzy ARTMAP Principles and Dynamics

The FAM NN for incremental supervised learning [4] incorporates two fuzzy adaptive resonance theory (ART)[1] modules denoted ART_a and ART_b. These two modules are linked by a map field module associating nodes (categories) from ART_a with nodes in ART_b. The fuzzy ART module [30] performs fast, incremental unsupervised learning by clustering M-dimensional input patterns[2] into categories, each forming a hyperrectangular region in the M-dimensional input space. The jth category is defined by a vector of weights $\boldsymbol{w_j}$ that are initially set at 1 and then monotonically nonincreasing through time. Categorization with the fuzzy ART is performed in three stages: category choice, category match (vigilance test), and learning. In the category choice stage, a choice function is calculated for the current input $\boldsymbol{I}$ and each existing category $\boldsymbol{w}_j$

$$T_j = \frac{||\boldsymbol{I} \wedge \boldsymbol{w}_j||}{\alpha + ||\boldsymbol{w}_j||} \tag{1}$$

[1] ART is a theory of human cognitive information processing developed by Grossberg [5].

[2] Every input $\boldsymbol{I}= (I_1, I_2, ..., I_M)$ to the fuzzy ART is initially complement-coded [4].

where $\wedge$ is the fuzzy AND operation, $(\mathbf{X} \wedge \mathbf{Y})_i = \mathbf{min}(\mathbf{x_i}, \mathbf{y_i})$, $\alpha > \mathbf{0}$ is a choice parameter[3], and the norm is L_1. The chosen category is the one achieving the highest value of the choice function.

When a category J is chosen, a hypothesis test called a vigilance test is performed to measure the category match to the input $\boldsymbol{I}$. If the match function exceeds the vigilance parameter $\rho \in [0, 1]$

$$\frac{||\boldsymbol{I} \wedge \boldsymbol{w}_J||}{||\boldsymbol{I}||} \geq \rho \tag{2}$$

then the chosen category is said to win and learning is performed. Otherwise, the chosen category is removed from the search for the current input. As a result, a new category maximizing the choice function (1) is chosen and the process continues until a chosen category satisfies the vigilance test (2).

The vigilance parameter, the lowering of which provides broader generalization (large categories) and vice versa, controls the level of similarity required between the chosen category and the input to enable learning. If none of the existing categories meets the vigilance test, a new category is formed and learning is performed without a vigilance test. Either way, learning is accomplished by updating the weight vector of the winning (or new) category according to

$$\boldsymbol{w}_J^{new} = \beta(\boldsymbol{I} \wedge \boldsymbol{w}_J^{old}) + (1 - \beta)\boldsymbol{w}_J^{old} \tag{3}$$

where $\beta \in (0, 1]$ is the learning rate and $\beta = 1$ defines fast learning [4].

In pattern recognition tasks, the input $\boldsymbol{I}_a$ to ART_a is the pattern and the input $\boldsymbol{I}_b$ to ART_b is the pattern label. As ART_b inputs are labels, ART_b vigilance parameter ρ_b is configured to one, so each label is clustered by a specific ART_b category. The map field between ART_a and ART_b includes a matrix of weights $\boldsymbol{w}^{ab}$ that maps ART_a categories to ART_b categories. The Jth row vector of $\boldsymbol{w}^{ab}$ denotes the prediction of ART_b categories as a result of the Jth winning category in ART_a. The map field is activated to produce the output

$$\boldsymbol{x}_{ab} = \boldsymbol{y}^b \wedge \boldsymbol{w}_J^{ab} \tag{4}$$

where ART_b output, $\boldsymbol{y}^b$, has boolean coordinates

$$y_k^b = \begin{cases} 1 & \text{if the } k\text{th category wins in } ART_b \\ 0 & \text{otherwise} \end{cases} . \tag{5}$$

$|\boldsymbol{x}_{ab}|$ is the value of the output that predicts the winning ART_b Kth category as a result of the winning ART_a Jth category.

During the training phase, the map field performs a vigilance test similar to the ART_a vigilance test, where if the match function exceeds the map field vigilance parameter $\rho_{ab} \in [0, 1]$

$$\frac{||\boldsymbol{x}_{ab}||}{||\boldsymbol{y}^b||} \geq \rho_{ab} \tag{6}$$

[3] An elaborated examination of the role of the choice parameter can be found in [31].

then resonance and learning occur. This test assures that the prediction of the correct class complies with the label represented by the winning ART_b Kth category. Otherwise, a match tracking procedure is activated for finding a better category in ART_a. In this process, the map field raises the ART_a vigilance parameter ρ_a

$$\rho_a = \frac{||\boldsymbol{I}_a \wedge \boldsymbol{w}_J^a||}{||\boldsymbol{I}_a||} + \delta, \qquad 0 < \delta \ll 1. \tag{7}$$

This ensures that the current Jth category fails the vigilance test in ART_a and is removed from the competition. The search in ART_a proceeds until an ART_a category that predicts the correct ART_b category is chosen, otherwise a new category is created. When the Jth category upholds the map vigilance test (6), its association with the ART_b categories is adapted by the following learning rule

$$\boldsymbol{w}_J^{ab,new} = \beta_{ab}(\boldsymbol{w}_J^{ab,old} \wedge \boldsymbol{y}^b) + (1 - \beta_{ab})\boldsymbol{w}_J^{ab,old} \tag{8}$$

which is activated during resonance in the map field. In fast learning mode ($\beta_{ab} = 1$), the link between the ART_a Jth category and the ART_b Kth category is permanent, i.e., $\boldsymbol{w}_{JK}^{ab} = 1$ for all input presentations. In the test phase, only ART_a is active, so the vigilance test in the map field is avoided. The class prediction is deduced from the map field weights of the winning ART_a category.

3 Advanced FAM-Based Developments

The introduction of the FAM algorithm to the NN and pattern classification communities in 1992, was followed by the development of advanced FAM-based techniques. We divide these developments into two main categories – modifications to the original FAM and new FAM-based algorithms. This division is an attempt to distinguish between (1) new methodologies to train or optimize FAM; (2) simplifications of FAM; and (3) developments in which the alternations do not change FAM substaintially, on the one hand, and developments that are based on new concepts, on the other hand. Nevertheless, the borderline between these two categories is thin and sometimes we may have crossed it unintentially. Additionally, ignoring previous developments that were already described, we focus on and expand recent developments. Note that most of the models described in this section have been compared experimentally to other classifiers in numerous studies, but we defer the description of results to Section 5.

3.1 Modifications to FAM

Different FAM training methods, such as "one-epoch", "until completion" and "with validation", have been studied [16, 27]. In the one-epoch training method, the FAM classifier is trained using a single presentation of all the training patterns. When training until completion, the classifier is repeatedly trained until the training set is predicted perfectly. In the training with validation method, every training epoch (or so) classifier accuracy is evaluated on a validation set and training continues until no further increase in the accuracy is achieved. Training with validation is a popular approach in the machine learning community aimed at reducing overfitting to enhance generalization [1]. This approach also regards the training until completion method as erroneous as it is prone to overfitting. On the other hand, the developers of FAM declared training until completion as the method of choice and as such it is used frequently [4, 16, 27].

The three training methods were compared experimentally on artificial and "real-world" databases [16, 27]. In [27], the accuracy of training with validation was evaluated on a validation set after the presentation of every 100 training patterns. For the artificial databases, training with validation increased accuracy significantly only when class overlapping was "medium" to "high" (the degree of overlapping – medium or high, was determined by the authors [27]). Nevertheless, for all the degrees of overlapping that were tested, the use of a validation set dramatically reduced the number of categories compared to those created by the training until completion or the training for one epoch methods. On the real-world databases, no significant differences in either the accuracy or number of categories between training with validation and training until completion were found. The accuracy of the FAM trained with validation was higher than that of the FAM trained for one epoch but at the expense of having (sometimes substantially) more categories.

A recent alternative methodology for FAM training is particle swarm optimization (PSO) [32], a population-based stochastic optimization technique that belongs to the class of evolutionary algorithms. Each solution in the search space is a particle and the population of particles is called a swarm. PSO co-jointly learns FAM weights, parameters (α (1), ρ (2), β (3) and δ (7); see Section 2), and number of categories, such that the generalization error is minimized.

Usually, researchers overcome the sensitivity of FAM to training pattern presentation order by training an ensemble of classifiers, each with a different order, and then predicting the true classification accuracy by averaging or voting using the ensemble [4, 9, 12, 16, 27]. Dagher et al. [24] suggested a methodology that tries to remedy this dependence of FAM on the presentation order. To that end, they introduced the ordered FAM that uses the max-min clustering algorithm to identify a fixed presentation order. They showed that the ordered FAM achieves generalization performance comparable to that of FAM with the same network size and only small computational overhead. There is no need to train an ensemble of classifiers, however when using the ordered FAM.

Another methodology for using FAM in pattern classification was suggested in [19]. The authors used hypothesis testing and proposed retraining the network with patterns classified with low levels of confidence. They computed the confidence in FAM classification and compared the confidence in classifying each pattern to a threshold. Patterns with high confidences for rejecting both the null and alternative hypotheses were called ambiguous, and the network was retrained to ensure these patterns were classified correctly when encountered again. Using this methodology, FAM accuracy in classifying piping welds at nuclear power plants was doubled after four iterations of retraining.

The simplified FAM (SFAM) of Taghi et al. [33] is a simpler and faster variant of FAM. The SFAM combines the two FAM ART modules into one mechanism that provides the same functionality as the original FAM but using a simpler structure. Kasuba's SFAM model [34] is similar to that of Taghi et al., although the former is slower to train than the latter [33]. Another model that simplifies FAM is the modified and simplified FAM (MSFAM) [35]. Similar to the two SFAM variants of Taghi et al. [33] and Kasuba [34], MSFAM reduces the architectural redundancy of FAM. It also extracts rules from the data that are more comprehensible than those of FAM. Nevertheless, an experimental comparison of the MSFAM model to the SFAM of Kasuba [34], showed almost no difference in classification performance between the two models.

One of the main research targets of the modifications to the FAM and FAM-based new algorithms is the reduction of category proliferation, which in FAM originates mainly from two sources – inadequate representation of the data and sensitivity to class overlapping. Inadequate data representation is the outcome of using the fuzzy set theory "minimum ($\wedge$)" and "maximum ($\vee$)" operators that lead to categories having hyperrectangular class decision boundaries. Although it may suit data distributed uniformly, a hyperrectangle does not fit the most natural (Gaussian) data distribution, which solicits a decision boundary in the form of a hypersphere or hyperellipsoid. As the dimension of the classification problem increases, the ratio between the volumes of a hyperrectangle and a hyperellipsoid (both bounding the data) increases monotonically [9]. That is, as the dimension increases, the hyperrectangle category represents higher volumes in which there are no patterns to support this representation (these are the hyperrectangle "corners"). Patterns residing in these corners but belonging to different classes than that which is associated with the category, are wrongly clustered by the category. In such a case, the match tracking procedure is activated to find a better existing category, and if none is found, then a new category is created that contributes to category proliferation.

The second source of category proliferation is FAM sensitivity to statistical overlapping between classes, which is responsible for misclassifications during FAM training. Each misclassification requires match tracking and raising the vigilance parameter to find a more suitable category for the misclassified pattern. The selected category, then, needs a larger weight vector to beat the new vigilance parameter in the vigilance test (2), and thus a smaller size. Repeated for many patterns of both classes and because small categories cannot represent wide regions, such misclassifications are responsible for the formation of a large number of small categories within the overlapping area. Moreover, if no existing category is found for a misclassified pattern during match tracking, a new category is formed. Either way, the result is category proliferation that is intensified with the degree of class overlapping. In many real-world applications, the result can be the creation of hundreds and thousands of categories, many if not most of which are redundant categories that contribute very little, if at all, to the classification process. Finally, we also note that category proliferation is a sign of overtraining, as increasing numbers of categories are formed to represent larger databases or overlapped classes. In addition to causing increased complexity, category proliferation (overtraining) may also degrade classifier generalization capability.

One of the first attempts to reduce category proliferation was entailed in the distributed ARTMAP (dARTMAP) [36]. Originally credited with the capacity to merge distributed MLP code compression and FAM fast online learning, however, the analysis and experiments done using dARTMAP in [25] and [36] showed no substantial advantage in using this approach.

Perhaps the main technique directed at overcoming category proliferation is pruning. In the FAM algorithm, each category, which corresponds roughly to a rule, has an associated weight vector that can be translated into a verbal or an algorithmic description of the antecedents in the corresponding rule. Since large databases typically lead to an overabundance of rules, the application of a rule-extraction mechanism enables the selection of a small set of highly predictive categories and the description of these categories in a comprehensible form. For example, the cascade ARTMAP [18] incorporates symbolic knowledge into the fuzzy ARTMAP. In addition to the "if-then" symbolic rules improving predictive accuracy and learning efficiency, the inserted rules can be refined and enhanced by the network.

Blume et al. [37] noted several methods for reducing category proliferation. They suggested a pruning method, in which categories that are responsible for the fewest training patterns are temporarily removed. Another method is to limit the number of clusters that can be formed. A third method is to change the choice function (1) to favor large categories over small ones. Allowing categories both to grow and to shrink (as opposed to FAM, which only allows them to grow) is another method for controlling category proliferation. In addition, Blume et al. proposed allowing each category to be associated with more than one class. They tested these methods on several data sets and concluded that no single method or combination of methods is always preferable.

Carpenter and Tan [38] introduced several pruning methods based on category confidence factors, which are scores related to category usage and performance accuracy. By removing low-confidence categories, the algorithm can produce smaller networks. μARTMAP of Sanchez et al. [10] includes a mechanism that avoids raising ART_a vigilance, which is an operation that leads to category proliferation.

Lin and Soo [26] suggested using the minimum description length (MDL) principle [39] to strike a balance between improved accuracy and increased complexity brought by the newly created category. According to the MDL principle, the best model to select is the one that minimizes both the encoding length of the model and that of the data given the model. Lin and Soo computed FAM encoding length as a function of the number of possible values for each category, the number of categories, the dimension of the problem and the number of patterns. They used a greedy algorithm (e.g., best-first search) to find the model minimizing this MDL score from among a variety of models, each having a different number of categories.

3.2 FAM-Based New Algorithms

Two previous FAM-based algorithms of Carpenter and colleagues are ARTMAP-IC [12] and ART-EMAP [13]. ARTMAP-IC encodes inconsistent cases, i.e., instances of identical input having different outcomes, and thereby yields probabilistic predictions. ART-EMAP is an ART-based model that integrates spatial and temporal evidence for dynamic predictive mapping.

Lim and Harrison [40] suggested the probabilistic FAM (PFAM), which is an extension of PROBART [7], in which the weight w_{jk}^{ab} is the frequency of associations between the jth ART_a category and the kth ART_b category. Thus, $w_{jk}^{ab}/|W_j^{ab}|$ is the empirical estimate of the posterior probability $P(C_k|j)$ that ART_a category j is associated with ART_b category k. The strength of the PROBART algorithm is in its data projection and regression functions and hence, it is not discussed further here. However, the PFAM algorithm [40] integrates FAM with the probabilistic NN (PNN) [41] yielding a hybrid algorithm. FAM clusters input patterns into categories and thereby provides a compact representation of the input space to the PNN. The PNN uses this representation to estimate class posterior probabilities, and by selecting the highest posterior estimate or the minimum risk estimate according to Bayes' theory, classification can be made. Lim and Harrison had an earlier, simpler version of PFAM [40], which they called the modified FAM [42]. They introduced two heuristic rules, one for the category formation process and another for the category selection process, which oriented the FAM toward classification. Learning and prediction using the modified FAM were similar to those of PFAM.

Lavoie et al. [43] proposed modifying the FAM choice function by adding an attentional tuning parameter. Using this parameter, the network can learn and use different categories with different degrees of generality, hence enabling mutual representation by both general and specific categories. They suggested replacing the fuzzy ART choice function (1) with

$$T_j = (M - d_j)^2 - (||R_j|| - S + d_j)^2 \quad (9)$$

where $d_j \equiv d(\boldsymbol{w}_j, \boldsymbol{I} \wedge \boldsymbol{w}_j)$ is Kosko's fuzzy Hamming distance – $d(x,y) \equiv \sum_{i=1}^{2M} ||x_i - y_i||$, $||R_j||$ is the size of the M-dimensional hyperrectangle R_j (i.e., the jth category), and S is a parameter equals to the desired category size $0 \leq S \leq M$.

The desired category size S can be set as $S \equiv \kappa M(1 - \rho)$, and thereby a single parameter, κ, controls both vigilance and attentional tuning. Lovoie et al. claimed that by using this choice function and properly determining the vigilance parameter, every category (and not only one as is the case with FAM) can be recalled. Using this modified choice function, then, enables the network to make a trade-off between generalization and discrimination as only the categories of a particular size are chosen.

Another new algorithm is FAMR – fuzzy ARTMAP with relevance factor [44, 45], which estimates the class posterior probability by assigning a relevance factor to each sample-label pair. This factor is proportional to the importance of that pair as learned during training, and it may be user-defined or related to either the misclassification cost or to the closeness of the sample to the decision boundary (weighing outliers low) [45].

Recently, Al-Daraiseh et al. [28] suggested the heuristic genetic FAM (GFAM), an evolved FAM network produced by repeatedly applying genetic operators on an initial population of trained FAM networks. These initial networks have been trained using different values of the vigilance parameter or presentation orders of patterns, i.e., they have different sizes and generalization capabilities. After initialization, genetic operators, such as selection, crossover, and mutation, are applied to the network population. The pth network is evaluated using the fitness score,

$$Fit(p) = PCC(p) - \alpha(N_a(p) - Cat_{min}), \tag{10}$$

where $PCC(p)$ is the correct classification rate of the pth network on a validation set, $N_a(p)$ is the number of categories created by this network, Cat_{min} is the number of classes in the problem, and α is a user-defined parameter that allows trade-off between network accuracy and complexity. Categories are allowed to be deleted as long as in every network there is at least one category for each class. Also, a stopping criterion is determined that is based on a maximum number of iterations and a maximum number of iterations without fitness improvement.

Another algorithm designed to eliminate category proliferation is that of Tan et al. [46]. The authors suggest what they call a "conflict-resolving network" based on FAM and a mechanism of dynamic decay adjustment, i.e., FAMDDA. During learning, FAMDDA heuristically shrinks the width of a category that clusters a pattern if the pattern belongs to a class that is different than that associated with the category. This conflict-resolving mechanism enabled FAMDDA to improve FAM accuracy.

Verzi et al. [47, 48] "boosted" the ARTMAP by using the notion of structural risk minimization [2] to trade-off between training error and model complexity. The so-called boosted ARTMAP (BARTMAP) is forced to make a training error that is bounded by a tolerance value. By allowing an error, BARTMAP encourages a relatively uncomplicated model that can both improve the generalization error and provide a smaller network (number of categories).

Recently, Vigdor and Lerner [29] modified FAM, improving its classification accuracy while simultaneously reducing its category proliferation, using the Bayesian framework. The proposed algorithm, called Bayesian ARTMAP (BA), preserves the advantages of FAM and also enhances its performance by (1) representing a category using a multidimensional Gaussian distribution rather than the conventional hyperrectangular shape, (2) limiting the volume of a selected category, hence allowing the category to grow or shrink, (3) probabilistically associating patterns with categories and categories with classes to perform, respectively, ART and ARTMAP learning, (4) using Bayes' decision theory [49] for learning and inference, and (5) employing the probabilistic association between every category and a class to predict the class. In addition, BA estimates the class posterior probability. Estimating posterior probabilities can be employed for classification by assigning a pattern to the class having the highest posterior probability to generate this pattern. However, being a generative rather than a discriminative model, BA can use the estimated probabilities to incorporate loss into the problem and classify according to the minimum expected loss, to reject patterns which are unlikely to be correctly classified, or to balance between distributions that are different in the training and test sets.

Similar to FAM, BA [29] comprises two hierarchies – the Bayesian ART and the Bayesian ARTMAP. Similar to the fuzzy ART algorithm, the Bayesian ART algorithm includes category choice, vigilance test, and learning. The chosen (winning) category J in the category choice stage is the category maximizing the a posteriori probability to represent the M-dimensional pattern $\boldsymbol{x}$[4]

$$M_j = \hat{P}(w_j|\boldsymbol{x}) = \frac{\hat{p}(\boldsymbol{x}|w_j)\hat{P}(w_j)}{\sum_{l=1}^{N_{cat}} \hat{p}(\boldsymbol{x}|w_l)\hat{P}(w_l)} \ . \tag{11}$$

N_{cat} is the number of categories and $\hat{P}(w_j)$ is the estimated prior probability of the jth category. The likelihood $\hat{p}(\boldsymbol{x}|w_j)$ of w_j with respect to $\boldsymbol{x}$ is estimated using all patterns already associated with the multivariate Gaussian category w_j.

The vigilance test ensures that the chosen category is limited in size. The test restricts the BA category hypervolume S_J to the maximal hypervolume allowed for a category, $S_J \leq S_{MAX}$, where the hypervolume is defined as the determinant of the Gaussian covariance matrix. If the winning category matches this criterion, learning occurs. Otherwise, the category is removed from the competition for this pattern and the Bayesian ART algorithm searches for another category. If all existing categories fail the vigilance test, a new category is formed with a mean vector that constitutes the input pattern and an initial covariance matrix that enables attaining S_{MAX}. If a chosen category matches the maximal hypervolume, then category parameters (mean and covariance matrix) are adjusted according to the patterns that were clustered by the Jth category before introducing the current pattern.

Learning in BA occurs first by mapping the category w_j of the Bayesian ART algorithm to class c_i with probability $P(c_i|w_j)$. Next is match tracking. If the class posterior probability for the winning category $P(c_i|w_J)$ is larger than a threshold, then the Jth category is associated with the ith class. Alternatively, match tracking is activated by lowering the category maximal hypervolume S_{MAX} enough to disqualify the winning Jth Bayesian ART category. As a result, the Bayesian ART algorithm starts searching for a new category with a smaller hypervolume S_J. This procedure continues until a winning category satisfies the threshold or a new category is formed.

Learning by BA is carried out by estimating the joint category-and-class probability $\hat{P}(c_i, w_j)$ using the frequency count N_{ij} for the number of training patterns that belong to the ith class of the C classes and are clustered to the jth category of the N_{cat} categories, i.e.,

[4] In Section 2, we denoted the pattern (input) by $\boldsymbol{I}$.

$$\hat{P}(c_i, w_j) = \frac{N_{ij}}{\sum_{k=1}^{C} \sum_{l=1}^{Ncat} N_{kl}} . \quad (12)$$

Marginalizing this joint probability over the classes, we get the jth category prior probability $\hat{P}(w_j)$. Using Bayes' theorem, the estimate for the class posterior probability given a category is

$$\hat{P}(c_i|w_j) = \frac{\hat{P}(c_i, w_j)}{\hat{P}(w_j)} = \frac{\frac{N_{ij}}{\sum_{k=1}^{C} \sum_{l=1}^{Ncat} N_{kl}}}{\frac{\sum_{k=1}^{C} N_{kj}}{\sum_{k=1}^{C} \sum_{l=1}^{Ncat} N_{kl}}} = \frac{N_{ij}}{\sum_{k=1}^{C} N_{kj}} \quad (13)$$

which is simply the number of patterns from the jth category that are associated with the ith class normalized by the number of patterns clustered by the jth category. When an input pattern belonging to class c is learned by the winning Jth Bayesian ART category, the frequency count is updated $N_{cJ}^{new} = N_{cJ}^{old} + 1$.

BA-based inference corresponds to the association of a category with a class when predicting a test pattern, and is carried out by using all the categories that are associated with a class. That is, the class chosen for a test pattern $\boldsymbol{x}$ is $c_I = \arg\max_i \hat{P}(c_i|\boldsymbol{x})$, where

$$\begin{aligned}\hat{P}(c_i|\boldsymbol{x}) &= \frac{\hat{P}(c_i, \boldsymbol{x})}{\hat{p}(\boldsymbol{x})} = \frac{\sum_{j=1}^{Ncat} \hat{P}(c_i, w_j, \boldsymbol{x})}{\hat{p}(\boldsymbol{x})} \\ &= \frac{\sum_{j=1}^{Ncat} \hat{P}(c_i|w_j, \boldsymbol{x}) \hat{P}(w_j, \boldsymbol{x})}{\hat{p}(\boldsymbol{x})} \\ &= \frac{\sum_{j=1}^{Ncat} \hat{P}(c_i|w_j) \hat{p}(\boldsymbol{x}|w_j) \hat{P}(w_j)}{\sum_{k=1}^{C} \sum_{l=1}^{Ncat} \hat{P}(c_k|w_l) \hat{p}(\boldsymbol{x}|w_l) \hat{P}(w_l)} . \end{aligned} \quad (14)$$

$\hat{P}(c_i|w_j)$, $\hat{P}(w_j)$ and $\hat{p}(\boldsymbol{x}|w_j)$ were defined above and we assume that $\hat{P}(c_i|w_j, \boldsymbol{x}) = \hat{P}(c_i|w_j)$, which means that once Bayesian ART identifies the winning category for the test pattern, only the association between the category and class affects the classification of this pattern.

BA is closely related to the Gaussian ARTMAP (GA) algorithm [9], and both BA and GA are based on FAM. The main differences between the FAM algorithm and the GA/BA models are, respectively: (1) pattern normalization through complement coding versus Mahalanobis distance (the distance from the pattern to the category mean normalized by the category variance [49]); (2) hyperrectangular versus multidimensional Gaussian categories; (3) categories can only grow versus grow or shrink as a consequence of statistical learning; (4) category choice using terms of fuzzy set theory operations versus Bayes' decision theorem. Using Bayes' theorem, both GA and BA favor categories that are either close to the pattern and small (via the likelihood), highly populated (via the prior probability), or both; (5) ART learning by category movement toward the pattern in terms of fuzzy set theory operations versus terms of maximum-likelihood-based sequential updating of parameters (mean, covariance and prior probability); (6) class prediction during inference, based on a single category versus multiple categories associated with the class.

The differences between GA and BA are outlined below, beginning with (1) the multidimensional representation of the Gaussian cluster, which is limited in GA by a diagonal covariance matrix but which is not limited in BA in which any covariance matrix is allowed. (2) Choice function: GA employs a discriminant function whereas BA computes the posterior probability and thus establishes a generative model with the benefits mentioned before. (3) Category match: GA determines how well the winning category matches the pattern using the Mahalanobis distance from the pattern to the category mean. This distance emphasizes the importance, when evaluating a match, of the closeness of the winning category to the input pattern. However, it neglects the role that the category size may have. That is, a large category (the mean of which is not necessarily close to the pattern) has a greater chance of attaining a category match. Since a large category is more likely to represent patterns of different classes, this category tends to cause mismatches during ARTMAP training followed by match tracking to find or establish more suitable categories. The consequence is category proliferation. However, category proliferation is controlled in BA by limiting the size of a category as part of the category match stage. Only a winning category with a limited volume can match category conditions and be learned. (4) ARTMAP learning: GA-based learning is accomplished by mapping the winning category to the pattern class unless this category has already been associated with another class. Learning in BA is achieved by mapping all categories to all classes in probability. (5) Inference: GA assigns a pattern to the class having the highest sum of joint probabilities for the pattern and each category which is associated with the class. BA does the same but using the weighted and normalized sum. Normalization turns the sum of joint probabilities into a posterior probability and thereby enables the calculation of the probability that the test pattern indeed belongs to a class rather than just deciding that it belongs to this class. Weighting the joint probabilities by $\hat{P}(c_i|w_j)$ before the summation in (14) ensures that categories that are strongly associated with the class (as estimated during training) and thereby lead to higher posterior probabilities $\hat{P}(c_i|w_j)$ will influence the selection of the class more than those categories that are marginal to the class. That is, categories representing the class mass of distribution contribute to the posterior $\hat{P}(c_i|\mathbf{x})$ more than categories representing class outliers.

4 Advanced FAM-Based Applications

Along with the advanced FAM-based developments, numerous applications based on FAM and these developments were revealed in the last fifteen years. One of the first such applications was automatic target recognition based on radar range profiles [8]. In a previous application, three-dimensional object recognition based on predictions formed from simulated two-dimensional views of a geometric object observed from various angles was performed by FAM and ART-EMAP [13]. Represented using Gabor-based features, these views were recognized in both noise-free and noisy environments. In another earlier application, Murshed et al. [20] developed a FAM-based offline signature verification system. Williamson [9] presented speaker-independent vowel recognition using the Gaussian ARTMAP (GA) and compared it to FAM. Carpenter et al. [12] exemplified the medical diagnosis of breast cancer and heart disease using ARTMAP-IC.

More recently, Sanchez et al., introducing μARTMAP [10], described an experiment that investigated online handwritten character recognition. Palaniappan et al. [15] discriminated alcoholics from nonalcoholics using a FAM. They extracted spectral power features of a filtered visual evoked potential (VEP) that indicates the effect of alcohol on the nervous system. Parsons and Carpenter [50] reported on the use of the FAM in data mining of geospatial images. They explained their choice of the FAM model because of its computational capabilities in incremental learning, fast stable learning, and visualization.

Aggarwal et al. [21] classified faults that arose in multicircuit transmission systems as caused by mutual coupling between parallel circuits. Dependence of the coupling on electrical variables and fault characterization led to the mutual impact of faults between different circuits, thereby undermining system stability and the continuity of the power supply. In another FAM application applied to fault detection [22], the algorithm was trained to detect faults in the viscous damper bearing of a helicopter drive shaft. The authors examined features based on spectograms, linear prediction coefficients and cepstrum coefficients of time signatures representing "good" and "bad" (faulty) bearings. They eliminated features achieving less than 50% classification accuracy and projected the remaining features onto the space spanned by the eigenvectors corresponding to the largest eigenvalues (i.e., principal component analysis (PCA) [51]). Based on the spectral features, Hush et al. [22] accomplished FAM-based fault detection of "good" and "bad" bearings with a misclassification rate of around 27%. The authors reported that they could further reduce this error when using FAM as a novelty detector, i.e., trained on the "good" signatures to alert when a "bad" signature appears.

Recently, Vigdor and Lerner [16] investigated FAM in both off and online classifications of fluorescence in situ hybridization (FISH) image signals. This classification enables the clinical diagnosis of numerical genetic abnormalities. Real and artifact signals of Down and Patau syndromes were classified using features of size, shape, intensity, and color. The authors evaluated the classification task (detecting the two abnormalities separately or simultaneously), the classifier paradigm (monolithic or hierarchical), the predicting strategy of

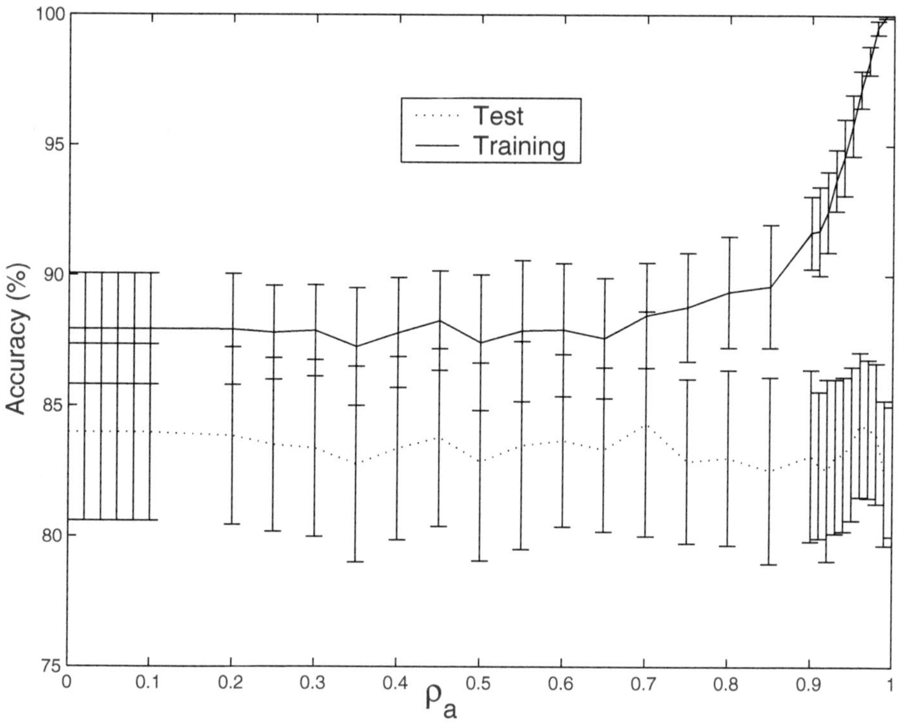

Fig. 1. Mean (± standard deviation) of the training and test accuracies of FAM classifying Down syndrome signals using the averaging strategy for increasing values of the vigilance parameter, ρ_a, and the one-epoch training method (from [16])

the true accuracy (averaging or voting), the training method ("for one-epoch", "with validation" or "until completion") and model sensitivity to the vigilance parameter. Below we detail parts of this evaluation and focus on the differences between the training methods and on incremental learning.

For example, to examine FAM sensitivity to different values of the vigilance parameter ρ_a, Vigdor and Lerner experimented with the Down syndrome signals, averaging strategy and the three training methods for increasing values of ρ_a. Figure 1, Figure 2 and Figure 3 show the training, validation (if applicable), and test classification accuracies for increasing values of ρ_a, for the one-epoch, with validation, and until completion training methods, respectively. In the one-epoch training method and for most of the range of ρ_a (Figure 1), the gap between the training and test accuracies is relatively small (3-4%), but it grows as training continues – 7-8% for training with validation (Figure 2) and $\sim$ 13% for until completion (Figure 3), which is a sign for entering overfitting.

Accuracies for all training methods can be divided roughly into three ranges. In the first range, where $0 \leq \rho_a \leq 0.2$, the accuracies (estimated means and standard deviations) are identical. In the second range, $0.2 < \rho_a < 0.7$, the

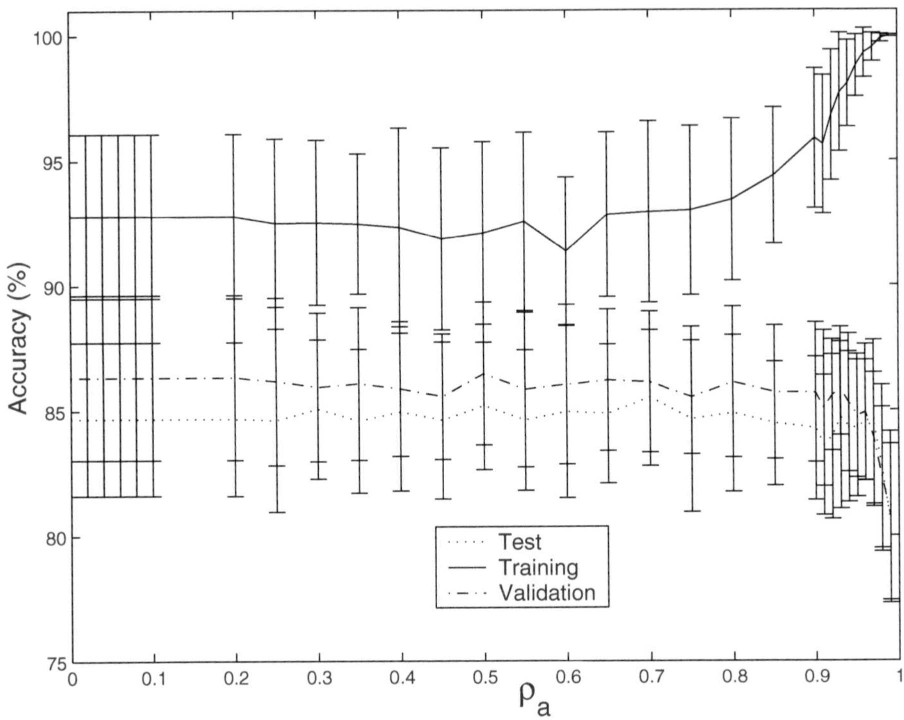

Fig. 2. Mean (± standard deviation) of the training, validation and test accuracies of FAM classifying Down syndrome signals using the averaging strategy for increasing values of the vigilance parameter, ρ_a, and the with validation training method (from [16])

accuracies are similar but not identical. In the third region, $0.7 \leq \rho_a < 1$, the training accuracy rises (usually) monotonically to 100% in the one-epoch and with validation methods and is fixed inherently at (almost) 100% for all values of the vigilance parameter for the until completion method. The test accuracy drops by approximately 4, 3, and 1% for training with validation, until completion and for one-epoch, respectively. Vigilance is very high in this region, so more categories representing only a few signals (and even single signals for vigilance values near 1) are formed. Despite the drop in the FAM test accuracy for the larger vigilance values, this accuracy is high and remarkably stable for a wide range of vigilance values.

Vigdor and Lerner also evaluated FAM while classifying FISH signals of Down syndrome in incremental learning. To that end, they compiled a database comprising 500 Down syndrome signals from both the real and artifact classes, and randomly selected 900 of these signals for training while keeping the remaining 100 signals for the test. In each of 45 training iterations, they continued training the network using ten additional training signals from each class and

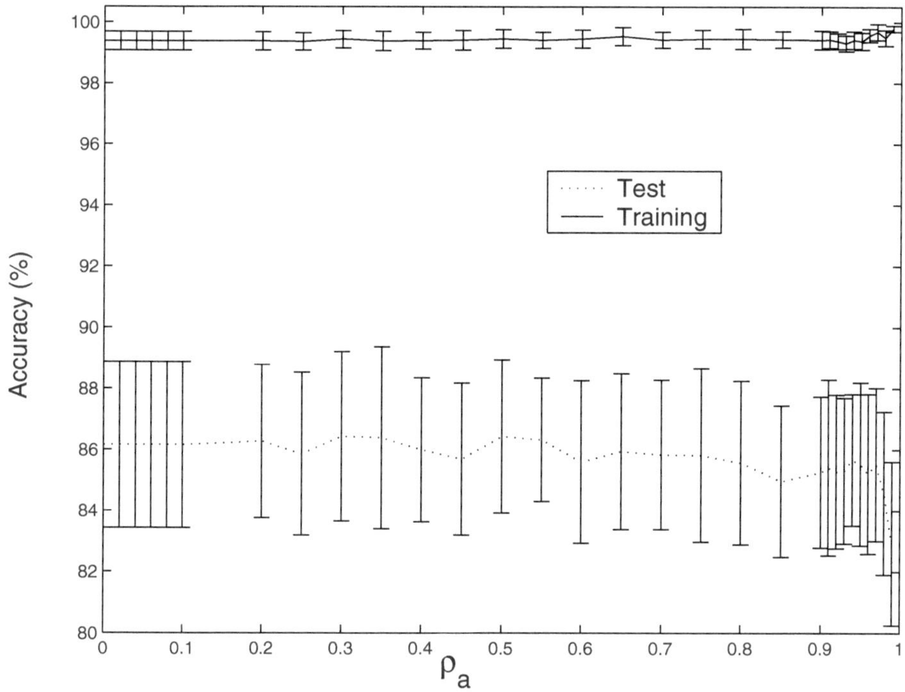

Fig. 3. Mean (± standard deviation) of the training and test accuracies of FAM-classified Down syndrome signals using the averaging strategy for increasing values of the vigilance parameter, ρ_a, and the until completion training method (from [16])

tested the incrementally learned network on the same 100 test signals. In each iteration, they recorded the FAM number of categories and accuracy for both the training and test sets. This procedure was repeated until all 900 signals had been presented to the network (i.e., one training epoch). The procedure continued for several epochs until no change in the training and test accuracies was noticed. The whole experiment was repeated using CV-10 and 5 training pattern orderings, and FAM performance was averaged over all experiments. The number of categories as well as the test and training accuracies are shown in Figure 4 for the first eight epochs. The number of categories rises monotonically and smoothly to 64.4. The training accuracy for the first signals of each epoch is high and it decreases until the end of the epoch as more variants of the signals (noise) participate in the training set. However, from epoch to epoch, the slope of this reduction becomes more moderate until going flat in the seventh epoch, as FAM learns to predict the training set perfectly. The corresponding test accuracy varies between 78.5% and 87.1% during the first epoch and then gradually converges to 85.4% after seven epochs.

To summarize that study, FAM accomplished the classification tasks by requiring only few training epochs. Also, the voting prediction strategy was more

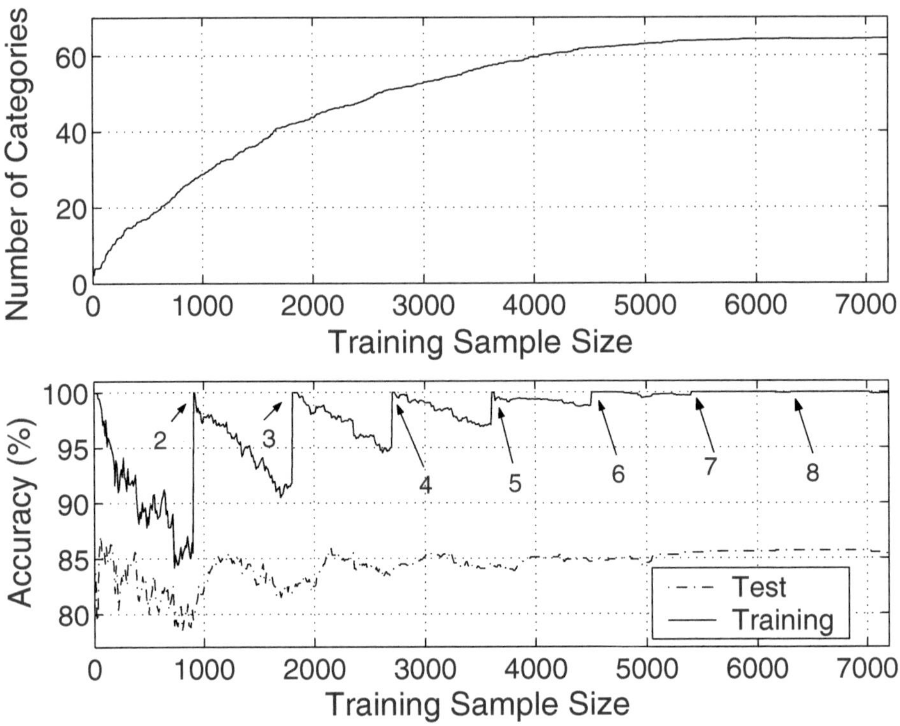

Fig. 4. The number of categories (top) as well as training and test classification accuracies (bottom) of the FAM that learns incrementally the Down syndrome signals. Numbers (2-8) at the bottom figure indicate the start of new training epochs (from [16]).

precise than the averaging strategy. If trained for only one epoch, FAM provided fast, yet stable and accurate learning and was insensitive to model complexity. Early stop of training using a validation set reduced FAM complexity as for other machine learning models but could not improve accuracy beyond that achieved when training was completed. Compared to other machine learning models, FAM did not loose but instead gained accuracy when overtrained, although increasing its complexity (number of categories). Learned incrementally, FAM achieved its ultimate accuracy fast, obtaining most of its data representation capability and accuracy by using only a few examples. The conclusion of the study was that FAM demonstrated rapid and accurate FISH signal classification both off and online, using different classification paradigms, model parameters, ordering strategies and training methods.

Finally, FAM was tested in the classification of incomplete radar pulse data [23]. The pulses were represented by carrier wave frequency, pulse width, and pulse repetition rate and belonged to fifteen different emitters. FAM classified the pulses accurately; Even when only 0.5% of the training date (around

130 pulses) was used, accuracy on the test set was 91.4%, compared to 99.6% when all data was used.

5 Experimental Evaluation of FAM-Based Algorithms

Results of experimental evaluations of the advanced developments described in Section 3 and the advanced applications of Section 4 are provided in this section. Definitions of the acronyms used are given in those sections.

Carpenter et al. [12] compared ARTMAP-IC to FAM, ART-EMAP, logistic regression [52], k-nearest neighbor (KNN) [49], decision tree C4 [53], and some other models using four databases taken from the UCI repository [54]. The performance of the ARTMAP-IC model was superior to the performances of the other ARTMAP-based models, slightly superior to that of the KNN, and similar to that of the logistic regression.

Sanchez et al., introducing μARTMAP [10], described an experiment of online, handwritten character recognition, in which the suggested model reduced the number of FAM categories in 60% though at the same time it undermined the accuracy in 16%. Palaniappan et al. [15] reported that they cut the optimization stage of their system for discriminating between alcoholics and nonalcoholics from around 42 days when experimenting with the MLP classifier to only 17 minutes when using FAM.

Aggarwal et al. [21] analyzed fault diagnosis by several NN classifiers and found that FAM is fast to train, practical, and accurate when compared to other NNs, such as the MLP and self-organizing map (SOM) [55]. Experiments by Ramuhalli et al. [19] showed that the accuracy of classifying piping welds of nuclear power plants was doubled after retraining the FAM algorithm during four iterations on patterns that were classified with low confidence in each iteration. Charalampidis et al. [14] slightly modified FAM to improve its ability to classify noisy signals. They compared their modified FAM with the original FAM using textured images with different noise sources added, and their modified version performed better than the original FAM at high noise levels and independent of the type of the noise and the size of the network.

The modified FAM [42] was examined in binary classification and using synthetic databases. For continuous-valued data, the modified FAM was more accurate than FAM and close to the Bayes' limit, as long as the problem dimension was low and the data were stationary. For discrete data, the modified FAM advantage was not limited. However, this model created more categories than the original FAM. On two other problems – one of animal identification and the other of DNA promoter recognition – the cascade ARTMAP [18] showed higher accuracy than did FAM. On the latter problem, the cascade ARTMAP also outperformed the KNN classifier, MLP, ID3 decision tree, and some other classifiers.

In another comparison that was made for fault detection [22], FAM was trained to detect faults in the viscous damper bearing of a helicopter drive shaft. The classifier was compared with the linear and quadratic classifiers [51], the *1*-nearest neighbor, and the MLP and the radial basis function (RBF) NNs [1]. The accuracy of FAM was amongst the lowest achieved.

In an application of handwritten character recognition [56], the researchers reported a clear advantage of FAM over MLP. That is, recognition rates of 83% versus 60%, respectively. It is known that researchers comparing their suggested algorithm with other algorithms sometimes fail to do that appropriately. While they optimize their model well to the problem, they neglect to do the same for the other models. We believe that in this case [56], proper MLP optimization would have yielded a level of accuracy similar to that of FAM. Nevertheless, in another publication from the same authors in the same year, they reported a prototype mail sorting machine that was successfully designed and developed based on the same FAM.

In this context, we note that in 2003, Taghi et al. [33] searched the INSPEC database and found 42 papers that ran FAM-MLP comparisons based on various datasets and domains. They explored 17 of these papers and identified 16 common pitfalls that undermine performing a fair and reliable experimental comparison. We encourage every practitioner, especially in the NN field, to study [33] carefully.

BARTMAP [47] was compared with FAM, ART-EMAP [13], GA [9], ARTMAP-IC [12], dARTMAP [36], μARTMAP [10], and other ARTMAP-based models on several synthetic databases [48]. In general, BARTMAP was more accurate and had simpler networks than most alternative FAM-based models. MSFAM [35] was compared to SFAM of Kasuba [34] using two synthetic and two real-world databases with almost identical classification accuracies [35]. However, MSFAM predominated over SFAM with respect to the number of extracted rules explaining the data. Taghi et al. [33] used several databases to compare variants of their SFAM with MLP variants. They reported that SFAM is inferior to MLP with respect to classification accuracy and testing time but superior to MLP regarding training speed and ease of parameter tuning.

By pruning FAM categories using confidence factors, Carpenter and Tan [38] reduced FAM network size by two thirds and slightly improved model accuracy on the UCI Pima Indian diabetes database. In a later paper [57], the same authors checked their method on two other UCI databases and compared it with the performances of algorithms that extract symbolic rules from MLP, KNN, and C4.5 [58]. On the mushroom database, MLP, FAM, and the rule-pruning-based FAM achieved similar accuracies. The rule-pruning-based FAM needed a similar number of categories (rules) as did FAM. When testing the classifiers on the DNA promoter database, MLP and FAM exhibited similar error rates of 7.9% and 7.4%, respectively, while the methods of the rule-pruning-based FAM had errors between 7.0% and 10.4%, and KNN showed an error of 5.5%, similar to that of FAM when using ten voters. Nevertheless, pruning cut the number of categories to between 17% and 34% of the number of FAM categories.

Lim and Harrison [40] compared their PFAM model (integration of PNN and FAM) with FAM, MLP, learning vector quantization (LVQ) [55], Boltzmann machine (BM) [59], and probabilistic NN (PNN) [41] in separating two eight-dimensional overlapping Gaussians as formulated in [60]. Among the classifiers, BM exhibited the lowest error (which was very close to the Bayes' error), FAM and PNN had the highest errors, and those of MLP, LVQ, and PFAM fell in between. When FAM and PFAM were compared in incremental learning, PFAM reached the Bayes' limit, which far exceeded the accuracy achieved by FAM. In addition, PFAM provided probability estimations in addition to pure classification decisions. Both FAM and PFAM created the same number of categories. Lim and Harrison also compared FAM and PFAM with the nearest-neighbor and the classification and regression trees (CART) classifiers on the UCI Waveform database. Again, FAM ranked lowest among the classifiers and PFAM was close to the Bayes' limit.

In an interesting comparison of incremental NN classifiers [3], the researchers compared the performances of the growing neural gas (GNG), the growing cell structures (GCS) and the FAM models to MLP on different databases. During training, GNG, GCS, and FAM perform local adaptations whereas MLP performs global adaptations, which may increase its generalization capability. On the other hand, the latter network is fixed in its configuration while the incremental networks can automatically add new nodes that may enhance their performance. The authors employed four databases and evaluated the difficulty in classifying them using the dispersion and (KNN) confusion matrices that measure the degree of class overlapping. The databases presented different levels of classification problem difficulties. By using values defining different network configurations and learning scenarios, the authors established different data sets from each database, enabling examination of the classifiers in a vast range of environments. The evaluation showed that, on average, (1) FAM required fewer nodes (categories) than all other incremental algorithms, (2) FAM needed the lowest number of epochs to terminate learning, and (3) FAM was generally the least accurate classifier. FAM performed reasonably well on small datasets, on Boolean patterns, or on classes consisting of linear decision boundaries. The authors summarized their comparative investigation by saying that "considering the similar classification performance of MLP, GNG, GCS, the rapid convergence of GNG and GCS and the small dependence on variation of parameters of GNG, the overall ranking of networks in descending order is: GNG, GCS, MLP, and FAM. However, when the dataset shows linear boundaries between classes, FAM and MLP can perform better than GNG and GCS."

Palaniappan et al. [15] discriminated alcoholics from nonalcoholics based on spectral power features of a filtered visual evoked potential (VEP) using FAM and MLP. When they averaged classification accuracies over several classifier configurations, MLP achieved accuracies between 94.3% and 96.1% (depending on the number of VEP channels) compared to accuracies between 86.2% and 90.1% of FAM.

Lin and Soo [26] compared their FAM variant, which prunes redundant categories using the minimum description length (MDL) principle [39], with both the original FAM and the C4.5 decision tree learned using MDL. When applied to the UCI Wisconsin breast cancer database, the classification accuracies of their FAM variant and C4.5 were shown to be identical and higher than that of the original FAM. Their MDL-based FAM needed one third of the categories required by FAM. On the UCI Pima Indian diabetes database, although the MDL-based FAM did not improve FAM accuracy, it did cut the number of categories used by an order of magnitude. However, there are two disadvantages to the suggested MDL variant with respect to the original FAM. First is that the MDL variant requires much longer training periods than the original FAM since it has to examine all training patterns before removing a category. Second, the MDL variant, at least in its current form, is applied only to two-class classification problems.

FAM variant was trained via particle swarm optimization (PSO) [32] and compared to the original FAM using two synthetic databases and the NIST Special Database 19 (SD19) [61]. On the synthetic databases, the classification error of the PSO-trained FAM was lower than that of FAM. On databases extracted from SD19, this error of the PSO-trained FAM was 1/3 – 1/2 that of FAM but these errors exceeded those of *1*-nearest neighbor, MLP, and SVM.

FAMDDA [46] was tested against FAM on UCI Iris, Wisconsin breast cancer, and Statlog image segmentation databases, where it demonstrated an average improvement of around 1% to the FAM classification accuracies for each of these databases. FAMDDA also improved FAM accuracy in monitoring the conditions of a circulating water system in a power generation plant. Twelve-dimensional patterns representing heat and water conditions were classified into four classes by FAMDDA, which succeeded in improving FAM accuracy by about 1%.

A FAM-based analysis was suggested as support for manufacturing technology investment decisions [62]. As such, FAM was used to match project proposals for investment with past, successful projects. Projects that showed a match were kept and based on all such successful matches, the system extracted rules to be used in prioritizing new projects. Features related to cost, risk factors, and prioritization scores (derived from, e.g., technology novelty and intellectual property potential) were used by the system to identify the similarity of a given new project to a successful project from the past. The most promising new projects were then presented for evaluation by a project committee that decided whether to invest in the recommended projects. When the system was trained to classify the prospective projects of a company in the pharmaceutical industry as “good”, “average”, or “reject”, it achieved up to 90% and 92% classification accuracies using the averaging and voting strategies, respectively.

Downs et al. [63] reported on using Kasuba’s SFAM model [34] in three medical applications. The first, which predicts the chances of death or survival for patients admitted to a coronary care unit, was run using the voting strategy. In this case, SFAM exhibited classification accuracies between 95.2% and 99.3% for 3 to 13 voters. In the second application concerned, the diagnosis of cancer from fine needle aspirates of the breast, SFAM outperformed the human expert with respect to accuracy and sensitivity but not in terms of specificity. This result was more evident after the network was pruned, in which redundant categories were removed as part of learning. The third application concerned the diagnosis of acute myocardial infarction (AMI) from information available at an early stage of hospital admission. Despite some modifications to SFAM, however, the physicians beat it on all measures with the exception of sensitivity.

When examined in electric load forecasting [64], FAM showed higher accuracy than MLP. Fault diagnoses of integrated navigation systems was performed by a FAM that located the faults based on gyro and accelerometer outputs [65]. Odor discrimination using a FAM-based electronic nose was reported in [66], where the authors reported that FAM accuracy was superior to that of MLP for the odors of coffee and cow breath and that the FAM training period was typically one order of magnitude faster than that of MLP. FAM-based automatic fuel pellet inspection was studied [67] to reduce manual exposure to radiation and to increase inspection accuracy and speed, but the accuracy of FAM was inferior to that of MLP.

FAM was investigated in the off and online classification of fluorescence in situ hybridization image signals enabling the clinical diagnosis of numerical genetic abnormalities [16]. Real and artifact signals of Down and Patau syndromes were classified using twelve features of size, shape, intensity, and color. FAM accuracy was found to be comparable to the accuracies of MLP and SVM and superior to those of the naive Bayesian and linear classifiers [49]. In addition, FAM required fewer than six training epochs compared to the hundreds of epochs required by MLP to perform the same task [68].

GFAM [28] was evaluated on twelve Gaussian databases differing in the number of classes and degree of class overlapping, four other artificial databases, and eleven databases from the UCI repository. GFAM was compared to variants of FAM, the ellipsoid ARTMAP (EFAM) [69], μARTMAP, and GA. On the artificial databases, GFAM performance – classification accuracy and number of categories – was usually the best among the classifiers. Unfortunately, the authors did not complete, or present, classifier comparison for real-world databases.

The ordered FAM [24] and the FAM were compared on nine UCI databases. The FAM was evaluated using ten different presentation orders. The ordered FAM generalization performance was superior to the average performance of FAM, in certain cases the ordered FAM performed as good as, or better than, the best FAM performance. This was achieved with networks having similar sizes and in almost no additional computational load.

Very recently, Vigdor and Lerner compared Bayesian ARTMAP (BA) performance with respect to classification accuracy, learning curves, number of categories, sensitivity to class overlapping, and risk with the performances of FAM and GA using synthetic and real-world databases [29]. FAM was trained for either one epoch or until completion. We report here on three experiments from that study.

In the first experiment, the authors used one dimensional patterns generated from two classes, each represented by a mixture of two Gaussian components. The sensitivity of the classifiers to increasing degrees of statistical class overlapping was examined by changing the variance of each density component. The degree of overlapping was measured using the accuracy achieved by the Bayes' classifier (which is an upper bound on the classification accuracy). Figure 5 shows that all the classifiers produce good results (accuracy and number of categories)

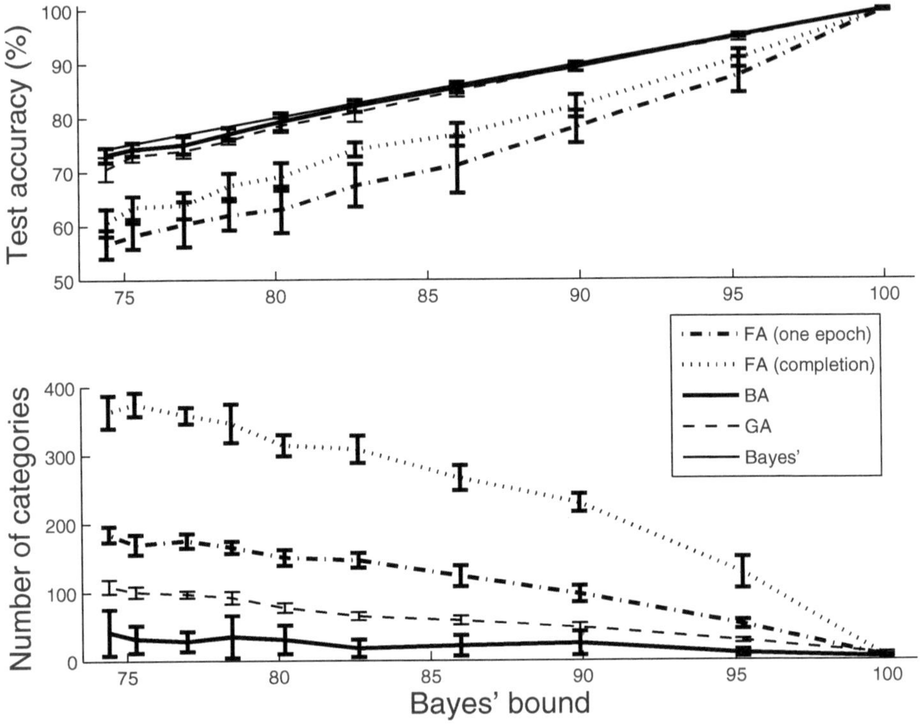

Fig. 5. The test classification accuracy (top) and number of categories (bottom) for the FAM trained for one epoch or until completion, GA and BA for increasing Bayes' bound values (decreasing statistical overlapping). Accuracies are also compared to that of the Bayes' classifier (bound) (from [29]).

for very high (95%-100%) Bayes' bound, i.e., almost no statistical overlapping between the classes. As the Bayes' bound is lowered (i.e., the degree of overlapping is increasing), the average accuracies of BA and GA remain close to the Bayes' bound represented by the line $y = x$. BA and GA accuracies (71.2% and 69.6%, respectively) for the lowest measured Bayes' bound are close to that bound (74.3%) whereas those of the FAM trained for one epoch or until completion are 58.6% or 62.7%, respectively. The advantage of the BA classifier over the other classifiers with respect to the number of categories is even clearer. While the FAM trained until completion, the FAM trained for one epoch, and GA require an average of 363.4, 185, and 108.6 categories, respectively, for the lowest Bayes' bound measured, BA required an average of only 41.7 categories.

In the second experiment of [29], non-Gaussian data of two classes were used. The first class was composed of a mixture of uniform and Rayleigh densities and the second was a mixture of two uniform densities. When measured for increasing sample sizes, the test accuracy of BA was almost insensitive to the sample size, and only slightly lower than the Bayes' bound (85.7%) (Figure 6 and

Table 1. The maximal test accuracy and category growth rate for the FAM trained for one epoch or until completion, GA and BA (non-Gaussian data) (from [29])

Classifier	Test accuracy	Category growth
FA (one epoch)	74.5%	0.03
FA (until completion)	76.7%	0.12
GA	81.0%	0.02
BA	82.9%	0.004

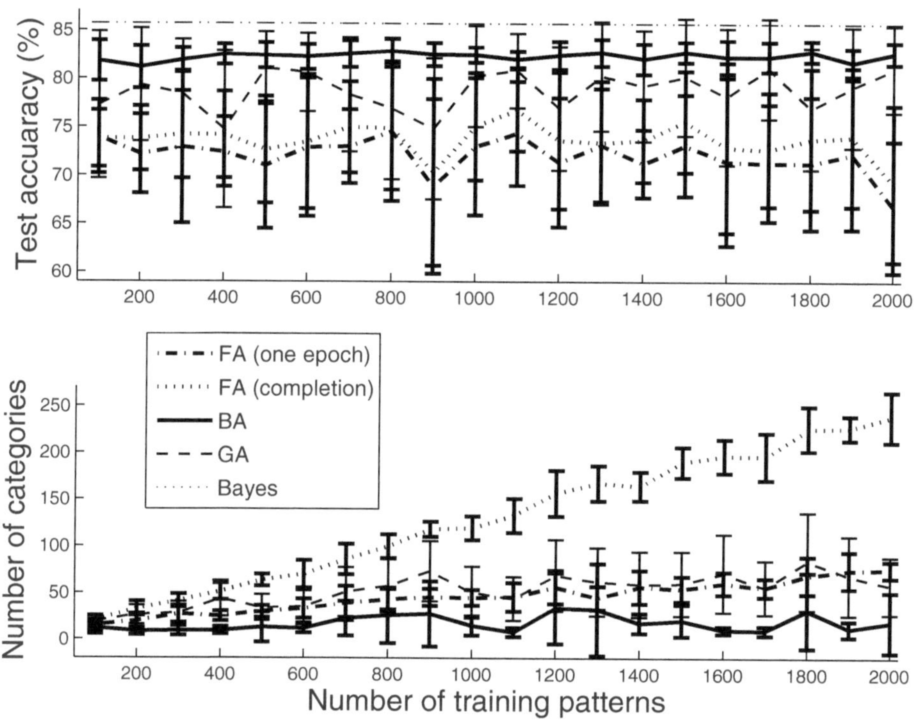

Fig. 6. The test classification accuracy (top) and number of categories (bottom) for the FAM trained for one epoch or until completion, GA and BA for increasing sample size values (non-Gaussian data). Accuracies are also compared to that of the Bayes' classifier (bound) (from [29]).

Table 1). GA accuracy, however was lower than that of BA by about ~ 2-7%, and more sensitive to the sample size. FAM accuracy was lower than that of BA by about 10% for both training modes. Thus, the high test accuracy attained by BA demonstrated its superior ability to approximate non-differentiable and non-Gaussian densities. This superiority was also evident in BA category growth (i.e., the rate of increase in the number of categories as sample size increases), which was lower than those of the other classifiers by at least an order of magnitude

Table 2. The test classification accuracy and variance, and the number of categories for the BA, the FAM trained for one epoch or until completion, the SGE, and the KDE classifiers averaged over twenty real-world UCI databases (from [29])

Classifier	Test accuracy (%)	Test variance (%)	Number of categories
BA	84.8	6.0	54.6
FA (until completion)	79.9	8.3	74.4
FA (one epoch)	78.5	8.3	45.7
Single Gaussian estimation classifier	76.5	5.9	NA
Kernel density estimation classifier	74.8	5.9	NA

(Table 1). In addition, examining the standard deviations in Figure 6 with respect to both accuracy and number of categories, BA was more stable and reliable than the other classifiers.

In the third experiment, Vigdor and Lerner investigated the BA using real-world classification problems from the UCI repository, the United-States Postal Service (USPS) Database [70], and a cytogenetic database [71]. BA was compared to FAM trained for one epoch or until completion, the single Gaussian classifier (SGE) [1], and the kernel density estimation (KDE) classifier [1]. The experiments were performed using CV10 and the averaging strategy with five different data presentation orders. Detailed results are given in [29]. Averaging the test classification accuracy and number of categories over the twenty databases examined show (Table 2) that BA outperforms on average all other evaluated classifiers with respect to accuracy. The second best classifier, the FAM trained until completion, exhibited an average accuracy that was 4.9% lower than that of the BA. In addition, BA was more robust than FAM, as its accuracy variance was smaller (6% compared to 8.3%). Finally, the number of BA categories averaged over all twenty databases was much closer to that of the FAM trained for one epoch than to that of the FAM trained until completion.

In summary, in almost all the experimental comparative studies reported in this section, FAM was the fastest and most easy to train but among the least accurate of the classifiers. Almost all FAM developments, whether modifications to the original model or new algorithms, improved its accuracy and reduced its category proliferation. Nevertheless, these developments did not always provide an advantage with respect to accuracy over other state-of-the-art classifiers.

6 Discussion

Most advanced developments of the FAM aim at enhancing the model classification accuracy, reducing its category proliferation problem, or both. These developments may incorporate elements from different fields, such as optimization theory, evolutionary learning, Bayes' decision theory, or statistical learning theory. Some of the developments simplify the original FAM model and others extend it probabilistically. Other developments use optimization methods, methods

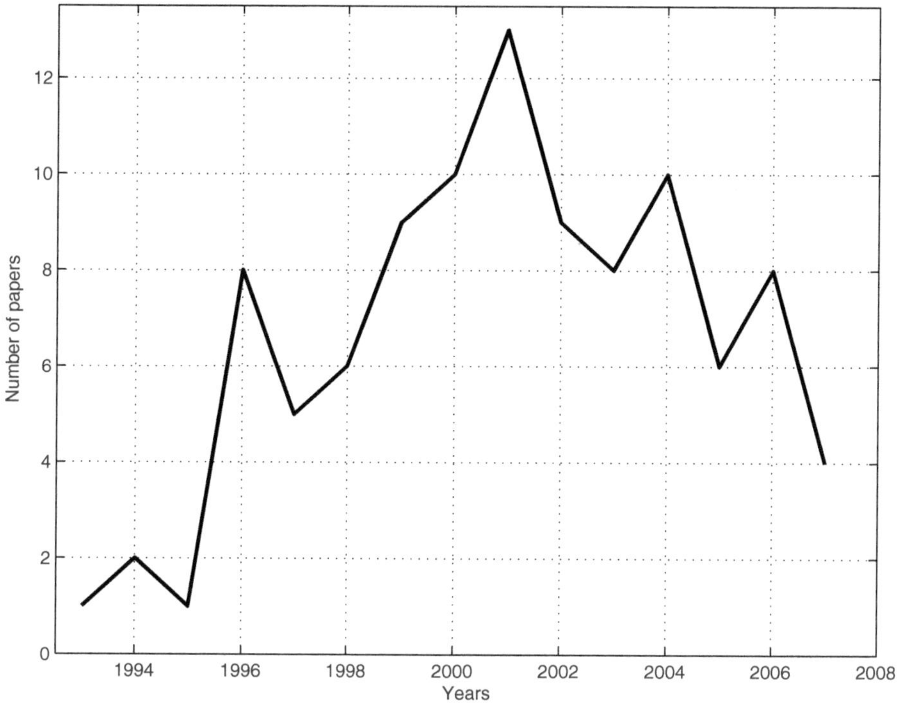

Fig. 7. A histogram for the number of publications investigating the FAM in pattern classification that have been published between 1993 and 2007. Note that the data for 2007 are partial, as the search was made in November of that year.

for rule extraction or pruning, heuristic rules to limit the number or width of categories, and methods to penalize models that are too complex or weigh the importance of patterns to classification. Most advanced developments indeed improve FAM performance – they either reduce category proliferation, improve classification accuracy, or both.

In November, 2007, to understand the current FAM research trend, we searched IEEE *Xplore* for all papers investigating or employing FAM in pattern classification. A histogram (Figure 7) of the number of such publications published between 1993 and 2007 reveals a peak in the number of publications in 2001. In the years after 2001, the graph declines and levels out to around six to seven publications per year. Nevertheless, some of the most important FAM-based developments have occurred since 2001. Furthermore, due to the unique advantages of FAM with respect to the training speed in dual operational modes – off and online, to recent progress in rendering the FAM classifier probabilistic and more accurate, and to the ever-increasing number of FAM-based applications, we expect that the motivation to further improve FAM will remain at least as high as it has been in recent years. If this review intensifies the existing motivation even a bit, then writing it was a worthwhile investment.

Acknowledgment. This work was supported in part by the Paul Ivanier Center for Robotics and Production Management, Ben-Gurion University of the Negev, Beer-Sheva, Israel.

References

1. Bishop, C.M.: Neural networks for pattern recognition. Clarendon Press, Oxford (1995)
2. Vapnik, V.: The nature of statistical learning theory. Springer, New York (1995)
3. Heinke, D., Hamker, F.H.: Comparing neural networks: A benchmark on growing neural gas, growing cell structures, and fuzzy ARTMAP. IEEE Transactions on Neural Networks 9, 1279–1291 (1998)
4. Carpenter, G.A., Grossberg, S., Markuzon, N., Reynolds, J., Rosen, D.: Fuzzy ARTMAP: A neural network architecture for incremental supervised learning of analog multidimensional maps. IEEE Transactions on Neural Networks 3, 698–713 (1992)
5. Grossberg, S.: Adaptive pattern recognition and universal encoding II: Feedback, expectation, olfaction, and illusions. Biological Cybernetics 23, 187–202 (1976)
6. Carpenter, G.A., Grossberg, S., Reynolds, J.H.: ARTMAP: Supervised real-time learning and classification of nonstationary data by a self-organizing neural network. Neural Networks 4, 565–588 (1991)
7. Marriott, S., Harrison, R.F.: A modified fuzzy ARTMAP architecture for the approximation of noisy mappings. Neural Networks 8, 619–641 (1995)
8. Rubin, M.A.: Application of fuzzy ARTMAP and ART-EMAP to automatic target recognition using radar range profiles. Neural Networks 8, 1109–1116 (1995)
9. Williamson, J.R.: Gaussian ARTMAP: A neural network for fast incremental learning of noisy multidimensional maps. Neural Networks 9, 881–897 (1996)
10. Gomez-Sanchez, E., Dimitriadis, Y.A., Cano-Izquierdo, J.M., Lopez-Coronado, J.: μ ARTMAP: Use of mutual information for category reduction in fuzzy ARTMAP. IEEE Transactions on Neural Networks 13, 58–69 (2002)
11. Suzuki, Y.: Self-organizing QRS-wave recognition in ECG using neural networks. IEEE Transactions on Neural Networks 6, 1469–1477 (1995)
12. Carpenter, G.A., Markuzon, N.: ARTMAP-IC and medical diagnosis: Instance counting and inconsistent cases. Neural Networks 11, 323–336 (1998)
13. Carpenter, G.A., Ross, W.D.: ART-EMAP: A neural network architecture for object recognition by evidence accumulation. IEEE Transactions on Neural Networks 6, 805–818 (1995)
14. Charalampidis, D., Kasparis, T., Georgiopoulos, M.: Classification of noisy signals using fuzzy ARTMAP neural networks. IEEE Transactions on Neural Networks 12, 1023–1036 (2001)
15. Palaniappan, R., Raveendran, P., Omatu, S.: VEP optimal channel selection using genetic algorithm for neural network classification of alcoholics. IEEE Transactions on Neural Networks 13, 486–491 (2002)
16. Vigdor, B., Lerner, B.: Accurate and fast off and online fuzzy ARTMAP-based image classification with application to genetic abnormality diagnosis. IEEE Transactions on Neural Networks 17, 1288–1300 (2006)
17. Lakhmi, J.C., Lazzerini, B., Halici, U. (eds.): Innovations in ART neural networks. Studies in Fuzziness and Soft Computing, vol. 43. Springer, Heidelberg (2000)

18. Tan, A.H.: Cascade ARTMAP: Integrating neural computation and symbolic knowledge processing. IEEE Transactions on Neural Networks 8, 237–250 (1997)
19. Ramuhalli, P., Polikar, R., Udpa, L., Udpa, S.S.: Fuzzy ARTMAP network with evolutionary learning. In: Proceedings of the IEEE International Conference on Acoustics, Speech, and Signal Processing (ICASSP 2000), pp. 3466–3469 (2000)
20. Murshed, N.A., Bortolozzi, F., Sabourin, R.: Off-line signature verification using fuzzy ARTMAP neural network. In: Proceedings of the IEEE International Conference on Neural Networks (ICNN 1995), pp. 2179–2184 (1995)
21. Aggarwal, R., Xuan, Q.Y., Johns, T., Li, F., Bennett, A.: A novel approach to fault diagnosis in multicircuit transimission lines using fuzzy ARTMAP neural networks. IEEE Transactions on Neural Networks 10, 1214–1221 (1999)
22. Hush, D.R., Abdallah, C.T., Heileman, G.L., Docampo, D.: Neural networks in fault detection: A case study. In: Proceedings of the Americam Control Conference, pp. 918–921 (1997)
23. Granger, E., Rubin, M.A., Grossberg, S., Lavoie, P.: Classification of incomplete data using the fuzzy ARTMAP neural network. In: Proceedings of the International Joint Conference on Neural Networks (IJCNN 2000), pp. 35–40 (2000)
24. Dagher, I., Georgiopoulos, M., Heilman, G., Bebis, G.: An ordering algorithm for pattern presentation in fuzzy ARTMAP that tends to improve generalization performnace. IEEE Transactions on Neural Networks 10, 768–778 (1999)
25. Hernandez, E.P., Sanchez, E.G., Dimitriadis, Y.A.: Study of distributed learning as a solution of category proliferation in fuzzy ARTMAP based neural systems. Neural Networks 16, 1039–1057 (2003)
26. Lin, T.H., Soo, V.W.: Pruning fuzzy ARTMAP using the minumum description length principle in learning from clinical databases. In: Proceedings of the 9th International Conference on Tools with Artificial Intelligence (ICTAI 1997), pp. 396–403 (1997)
27. Koufakou, A., Georgiopoulos, M., Anagnostopoulos, G., Kasparis, T.: Cross-validation in fuzzy ARTMAP for large databases. Neural Networks 14, 1279–1291 (2001)
28. Al-Daraiseh, A., Kaylani, A., Georgiopoulos, M., Mollaghasemi, M., Wu, A.S., Anagnostopoulos, G.: GFAM: Evolving fuzzy ARTMAP neural networks. Neural Networks 20, 874–892 (2007)
29. Vigdor, B., Lerner, B.: The Bayesian ARTMAP. IEEE Transactions on Neural Networks 18, 1628–1644 (2007)
30. Carpenter, G.A., Grossberg, S., Rosen, D.: Fuzzy ART: Fast stable learning and categorization of analog patterns by an adaptive resonance system. Neural Networks 4, 759–771 (1991)
31. Heilman, G., Bebis, G., Georgiopoulos, M.: Order of search in fuzzy ART and fuzzy ARTMAP: Effect of the choice parameter. Neural Networks 9, 1541–1559 (1996)
32. Granger, E., Henniges, P., Sabourin, R., Oliveira, L.S.: Supervised learning of fuzzy ARTMAP neural networks through particle swarm optimization. Journal of Pattern Recognition Research 1, 27–60 (2007)
33. Taghi, M., Baghmisheh, V., Pavesic, N.: A fast simplified fuzzy ARTMAP network. Neural Processing Letters 17, 273–316 (2003)
34. Kasuba, T.: Simplified fuzzy ARTMAP network. AI Expert 8, 18–25 (1993)
35. Su, M.C., Lu, W.Z., Lee, J., Chen, G.D., Hsieh, C.C.: The MSFAM: A modified fuzzy ARTMAP system. Pattern Analysis and Applications 8, 1–16 (2005)
36. Carpenter, G.A., Milenova, B.L., Noeske, B.W.: Distributed ARTMAP: A neural network for fast distributed supervised learning. Neural Networks 11, 793–813 (1998)

37. Blume, M., Van Blerkom, D.A., Esner, S.C.: Fuzzy ARTMAP modifications for intersecting class distributions. In: Proceedings of the World Congress on Neural Networks (WCNN 1996), pp. 250–255 (1996)
38. Carpenter, G.A., Tan, A.H.: Classification of incomplete data using the fuzzy ARTMAP neural network. In: Proceedings of the World Congress on Neural Networks (WCNN 1993), pp. 501–506 (1993)
39. Rissanen, J.: A universal prior for integers and estimation by minimum description length. Annals of Statistics 11, 416–431 (1983)
40. Lim, C.P., Harrison, R.F.: An incremental adaptive network for online supervized learning and probability estimation. Neural Networks 10, 925–939 (1997)
41. Specht, D.F.: Probabilistic neural networks. Neural Networks 3, 109–118 (1990)
42. Lim, C.P., Harrison, R.F.: Modified fuzzy ARTMAP approaches Bayes optimal classification rates: An empirical demonstration. Neural Networks 10, 755–774 (1997)
43. Lavoie, P., Crespo, J.F., Savaria, Y.: Generalizaion, discrimination, and multiple categorization using adaptive resonance theory. IEEE Transactions on Neural Networks 10, 757–767 (1999)
44. Andonie, R., Sasu, L.: A fuzzy ARTMAP probability estimator with relevance factor. In: Proceedings of the 11th European Symposium on Artificial Neural Networks (ESANN 2003), pp. 367–372 (2003)
45. Andonie, R., Sasu, L.: Fuzzy ARTMAP with input relevances. IEEE Transactions on Neural Networks 17, 929–941 (2006)
46. Tan, S.C., Rao, M.V.C., Lim, C.P.: Fuzzy ARTMAP dynamic decay adjustment: An improved fuzzy ARTMAP with a conflict resolving facility. Applied Soft Computing 8, 543–554 (2008)
47. Verzi, S.J., Heileman, G.L., Georgiopoulos, M., Healy, M.J.: Boosting the performance of ARTMAP. In: Proceedings of the International Joint Conference on Neural Networks (IJCNN 1998), pp. 396–401 (1998)
48. Verzi, S.J., Heileman, G.L., Georgiopoulos, M., Healy, M.J.: Rademacher penalization applied to fuzzy ARTMAP and boosted ARTMAP. In: Proceedings of the International Joint Conference on Neural Networks (IJCNN 2001), pp. 1191–1196 (2001)
49. Duda, R.O., Hart, P.E., Stork, D.G.: Pattern classification, 2nd edn. Wiley, NY (2001)
50. Parsons, O., Carpenter, G.A.: ARTMAP neural networks for information fusion and data mining: Map production and target recognition methodologies. Neural Networks 16, 1075–1089 (2003)
51. Fukunaga, K.: Introduction to statistical pattern recognition, 2nd edn. Academic Press, San Diego (1990)
52. Howell, O.C.: Statistical methods for Psychology. Duxbury Press, Belmont (1992)
53. Quinlan, J.R.: The effect of noise on concept learning. In: Michalski, R.S., Carbonell, J.C., Mitchell, T. (eds.) Machine learning: An artificial intelligence approach, pp. 149–166. Morgan Kaufmann, San Mateo (1986)
54. Blake, C., Merz, C.: UCI repository of machine learning databases (1998), `www.ics.uci.edu/~mlearn/MLRepository.html`
55. Kohonen, T.: Self-organization and associative memory, 2nd edn. Springer, Berlin (1998)
56. Tay, Y.H., Khalid, M.: Comparison of fuzzy ARTMAP and MLP neural networks for hand-written character recognition. In: Proceedings of the IFAC Symposium on AI in Real-Time Control (AIRTC 1997), pp. 363–371 (1997)

57. Carpenter, G.A., Tan, A.H.: Rule extraction: From neural architecture to symbolic representation. Connection Science 7, 3–26 (1995)
58. Quinlan, J.R.: C4.5: Programs for machine learning. Morgan Kaufmann, San Mateo (1993)
59. Akley, D.H., Hinton, G.E., Sejnowski, T.J.: A learning algorithm for Boltzmann machines. Cognitive Science 9, 147–169 (1985)
60. Kohonen, T., Barna, G., Chirsley, R.: Statistical pattern recognition with neural networks: Benchmarking studies. In: Proceedings of the IEEE International Conference on Neural Networks (ICNN 1988), pp. 61–68 (1988)
61. Grother, P.J.: NIST special database 19 – handprinted forms and characters database (1995)
62. Tan, K.H., Lim, C.P., Platts, K., Koay, H.S.: An intelligent decision support system for manufacturing technology investments. International Journal of Production Economics 104, 179–190 (2006)
63. Downs, J., Harrison, R.F., Kennedy, R.L., Cross, S.S.: Application of the fuzzy ARTMAP neural network model to medical pattern classification tasks. Artificial Intelligence in Medicine 8, 403–428 (1996)
64. Lopes, M.L.M., Minussi, C.R., Lotufo, A.D.P.: Electric load forecasting using a fuzzy ART & ARTMAP neural network. Applied Soft Computing 5, 235–244 (2005)
65. Zhang, H.Y., Chan, C.W., Cheung, K.C., Ye, Y.J.: Fuzzy ARTMAP neural network and its application to fault diagnosis of navigation systems. Automatica 37, 1065–1070 (2001)
66. Llobet, E., Hines, E.L., Gardner, J.W., Barlett, P.N., Mottram, T.T.: Fuzzy ARTMAP based electronic nose data analysis. Sensors and Actuators B 61, 183–190 (1999)
67. Keyvan, S., Song, X., Kelly, M.: Nuclear fuel pellet inspection using artificial neural networks. Journal of Nuclear Materials 264, 141–154 (1999)
68. Lerner, B., Lawrence, N.D.: A comparison of state-of-the-art classification techniques with application to cytogenetics. Neural Computing & Applications 10, 39–47 (2001)
69. Anagnostopoulos, G., Georgiopoulos, M.: Ellipsoid ART and ARTMAP for incremental clustering and classification. In: Proceedings of the IEEE International Joint Conference on Neural Networks (IJCNN 2001), pp. 1221–1226 (2001)
70. US postal service handwritten digits recognition corpus, `http://www.cedar.buffalo.edu/Databases/CDROM1/`
71. Lerner, B., Clocksin, W.F., Dhanjal, S., Hultén, M.A., Bishop, C.M.: Feature representation and signal classification in fluorescence in-situ hybridization image analysis. IEEE Transactions on Systems, Man, and Cybernetics A 31, 655–665 (2001)

Large Margin Methods for Structured Output Prediction

Elisa Ricci and Renzo Perfetti

Dept. of Electronic and Information Engineering,
University of Perugia, 06125, Perugia, Italy
{elisa.ricci,renzo.perfetti}@diei.unipg.it

Abstract. Many real-life data problems require effective classification algorithms able to model structural dependencies between multiple labels and to perform classification in a multivariate setting, i.e. such that complex, non-scalar predictions must be produced in correspondence to input vectors. Examples of these tasks range from natural language parsing to speech recognition, machine translation, image segmentation, handwritten character recognition or gene prediction.

Recently many algorithms have been developed in this direction in the machine learning community. They are commonly referred as structured output learning approaches. The main idea behind them is to produce an effective and flexible representation of the data exploiting general dependencies between labels. It has been shown that in many applications structured prediction methods outperform models that do not directly represent correlation between inputs and output labels.

Among the variety of the approaches developed in last few years, in particular large margin methods deserve attention since they have proved to be successful in several tasks. These techniques are based on the smart integration between Support Vector Machines (SVMs) and probabilistic graphical models (PGMs), so they combine the ability to learn in high dimensional feature spaces typical of kernel methods with the algorithmic efficiency and the flexibility in representing data inherited by PGMs.

In this paper we review some of the most recent large margin methods summarizing the main theoretical results, addressing some important computational issues, and presenting the most successful applications. Specifically, we show results in the context of biological sequence alignment and for sequence labeling and parsing in the natural language processing field. We finally discuss some of the main challenges in this new and promising research field.

1 Introduction

In the last few years the increasing of the data available has brought about the proliferation of machine learning and data mining algorithms able to analyze efficiently and automatically a vast amount of complex data. This phenomenon has been observed in several fields such as computational biology, computer vision, natural language processing or financial analysis. Most of the problems arising in these areas involves highly structured data which can be represented by sequences, graphs and images. To deal with this need researchers in machine

L.C. Jain et al. (Eds.): Comp. Intel. Para.: Innov. Applications, SCI 137, pp. 109–132, 2008.
springerlink.com

learning have recently focused most of their efforts in developing algorithms where prediction produces as outputs multiple variables reflecting the structure of complex objects instead of yielding scalar values as in traditional classification and regression algorithms. These algorithms are usually referred as structured output learning approaches.

In the structured output prediction framework usually the observed data are relatively simple while the output actually involves complex structures. Examples of structured output problems are several. One example is the parsing problem in natural language processing: here the input is one sentence, i.e. a sequence of words, and the output is a parse tree. In the sequence alignment problem the input is given by a pair of sequences while the output is the pairwise alignment between them. In gene finding the input is a DNA sequence while the output is another sequence that indicates the presence of genes or other elements which are biologically functional. In protein structure prediction, we are given a sequence of amino acids, while the output is represented by the associated three-dimensional structure.

The learning problem in structured output spaces is treated as an extension of the traditional supervised setting. According to the latter we are given a training set of pairs $(\mathbf{x}^i, y^i)$ of observations $\mathbf{x}^i$ and labels y^i. The goal is to learn a function h such that $h(\mathbf{x}) = y$ on an unseen pair $(\mathbf{x}, y)$ drawn by the same distribution of the data of the training set. In particular in classification problems y takes discrete values, in regression y is a real-valued number. In both cases, however y is a scalar. Learning in structured outputs problems is formulated in the same fashion with the only difference that the outputs are vectors $\mathbf{y}$. The components of the vectors $\mathbf{y}$ belong to a finite set and they are not independent. They are associated with others based on the locations or on same other properties of the corresponding objects in order to capture the correlation between them (e.g. the t-th component of $\mathbf{y}$, y_t, could be dependent on the value of y_{t-1} and y_{t+1}). It is expected that modeling these relationships better performance can be achieved with respect to methods that consider labels independently. The relationships between labels are represented by the formalism of graphical models, ranging from the simple models such as hidden Markov models (HMMs) [20] and Markov random fields, up till to maximum entropy Markov models (MEMMs) [15] and conditional random fields (CRFs) [13].

In particular CRFs have shown very good performance compared to previous approaches in several tasks such as part of speech tagging (POS) and named entity recognition (NER) in natural language processing (NLP) or image segmentation and object recognition in computer vision. CRFs define the conditional distribution $P(\mathbf{y}|\mathbf{x})$ as a function of some features relating input and output variables. For a given input $\mathbf{x}$, the predicted output $\mathbf{y}$ can be found as the one that maximizes the a posteriori probability. Since for each problem an underlying graphical model is considered the prediction corresponds to compute the inference with Viterbi-like algorithms. In the learning phase, instead of modeling the joint distribution of input and output data as in generative models, a discriminative approach is adopted: learning amounts to find a parameter set that give the best possible accuracy on training data in a way such that the empirical risk minimization (ERM) [27] principle can be invoked. In particular in CRFs the empirical risk is defined in terms of an upper bound of a 0/1 loss, so ERM guarantees that the number of uncorrect vectors $\mathbf{y}$ is minimized.

Using this principle but devising different upper bounds or different loss functions other algorithms have been developed in the last few years. In particular, the methods presented in [23, 26] and typically referred as max-margin approaches minimize an hinge loss, so they look for the parameters set such that the maximum margin in the training set is achieved. These approaches can be seen as an adaption of Support Vector Machines [8] to the structured output setting. Crucially they inherit from SVMs the good generalization performance due to the max-margin property and the possibility to model high-order correlations between feature vectors by the use of kernels.

In this paper we provide an overview of the current approaches for max-margin structured classification, including a detailed description of the most important algorithms, and we present some applications in different domains such as NLP and computational biology.

The rest of the paper is organized as follows. In Section 2 we introduce the framework of discriminative learning in structured output spaces together with some notation adopted later on in the paper. In Section 3 we consider the large margin approach for structured prediction problems and we present the main algorithms that have been proposed so far. Section 4 addresses some theoretical issues related to the generalization performance of structured output learning with maximal margin. Some state-of-the-art results obtained in various applications are shown in Section 5. Section 6 concludes this paper with a discussion on the new challenges in this field.

2 Discriminative Models for Structured Output Learning

2.1 Structured Output Problems

Structured output prediction is a classification problem where the objects to be classified can take a finite number of labels (e.g. m labels) and each label assume only a finite number of values (e.g. n values).

Several tasks can be modeled as structured output prediction problems. Typical examples include:

- Sequence labeling: given an input sequence, the task is to reconstruct an associated label sequence of equal length. This problem arises in various applications such as POS and NER in NLP or gene finding in computational biology.
- Sequence parsing: given an input sequence, the goal is to determine the associated parse tree given an underlying grammar. Examples of this problem are syntactic parsing in NLP and RNA structure alignment and prediction in bioinformatics.
- Sequence alignment: given a sequences pair, predict the correct sequence of alignment operations.

- Hierarchical classification: given an object (e.g. a document), the goal is to automatically categorize it into predefined hierarchies or taxonomies (e.g. by topic).
- Matching: given a bipartite graph, find the best possible matching. Examples of this task are word alignment in machine translation and protein structure prediction in computational biology.

Of course this is not an exhaustive list of possible tasks. Recently the structured output framework has been adopted also in other applications, such as document summarization, speech recognition and image segmentation and it is reasonable to believe that this list is intended to grow. In the following Subsection we introduce formally the structured output learning setting.

2.2 Learning with Structured Outputs

We are given a training set $\mathcal{T} = \{(\mathbf{x}^1, \bar{\mathbf{y}}^1)(\mathbf{x}^2, \bar{\mathbf{y}}^2) \dots (\mathbf{x}^\ell, \bar{\mathbf{y}}^\ell)\}$ drawn i.i.d. from some fixed but unknown distribution $P(\mathbf{x}, \mathbf{y})$. We denote with $\mathcal{X}$ the set of structured objects that we want to classify and with $\mathcal{Y} = \{\mathcal{Y}_1 \times \dots \times \mathcal{Y}_m\}$ the set of labels, $\mathcal{Y}_i = \{y_1, y_2 \cdots y_n\}$.

Depending on the applications each pair $(\mathbf{x}, \mathbf{y})$, $\mathbf{x} \in \mathcal{X}$, $\mathbf{y} \in \mathcal{Y}$ can represent different objects. For example, in sequence labeling learning applied to part of speech tagging, $\mathbf{x}$ denotes a sequence of words (i.e. the observed sequence), while $\mathbf{y}$ is the associated sequence of tags (i.e. the hidden sequences). In parse learning each sequence of words represented by $\mathbf{x}$ corresponds to a parse tree $\mathbf{y}$. This concept is illustrated in Fig. 1.

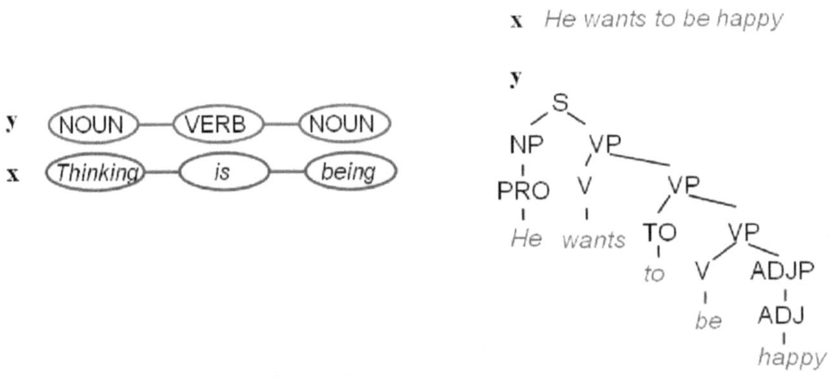

Fig. 1. Example of applications of the structured output framework: (a) sequence labeling learning and (b) parse labeling learning

The goal of structured output learning is to learn a function $h : \mathcal{X} \to \mathcal{Y}$, the so called *prediction function*, such that the optimal hidden variables $\bar{\mathbf{y}}$ can be reconstructed from the observed variables set $\mathbf{x}$ for an unknown data point drawn from the same distribution P. The function h is typically selected from some parametric family $\mathcal{H}$. Usually the linear family is chosen and:

$$h_{\mathbf{w}}(\mathbf{x}) = \arg\max_{\mathbf{y}\in\mathcal{Y}} \mathbf{w}^T \boldsymbol{\phi}(\mathbf{x},\mathbf{y}) \tag{1}$$

where $\mathbf{w} \in \mathbb{R}^d$ is a parameter vector and $\boldsymbol{\phi}(\mathbf{x},\mathbf{y})^T \in \mathbb{R}^d$ is an appropriately defined feature vector. Then the hypothesis space reduces to the set:

$$\mathcal{H} = \{h_{\mathbf{w}} : \mathbf{w} \in \mathbb{R}^d\} \tag{2}$$

In general the computation of $h_{\mathbf{w}}$ is not trivial. In fact the size of the output space $\mathcal{Y}$, i.e. the number of possible assignments $\mathbf{y}$ given an $\mathbf{x}$ is huge. For example in parse labeling it corresponds to the number of all possible parse trees given a sequence of words. To deal with this difficulty, probabilistic graphical models are used.

We define the *scoring function* or *discriminant function* $s_{\mathbf{w}}(\mathbf{x},\mathbf{y})$:

$$s_{\mathbf{w}}(\mathbf{x},\mathbf{y}) = \mathbf{w}^T \boldsymbol{\phi}(\mathbf{x},\mathbf{y}) = \sum_k w_k \phi_k(\mathbf{x},\mathbf{y})$$

For PGMs this function corresponds to the log-probability of the output $\mathbf{y}$ conditioned to an input $\mathbf{x}$. The parameter vector $\mathbf{w}$ contains the logarithms of the clique potentials or conditional probabilities and the vector $\boldsymbol{\phi}(\mathbf{x},\mathbf{y})$ is the associated vector of sufficient statistics. With this formalism the feature vector $\boldsymbol{\phi}(\mathbf{x},\mathbf{y})$ decomposes in terms of k real valued functions $\phi_k(\mathbf{x},\mathbf{y}) : \mathcal{X} \times \mathcal{Y} \to \mathbb{R}$, associated to the edges of the underlying PGM. In other words the nature of the feature map is uniquely defined by the structure of the PGM. However in the structured output framework often one considers more rich and complex feature vectors: features are not necessarily associated to clique potentials or conditional probabilities but any feature that describes the relationship between input and output variables and decomposes over the structure of the PGM is allowed. This means that the score $s_{\mathbf{w}}(\mathbf{x},\mathbf{y})$ looses its interpretation as a log-likelihood function.

Adopting the PGMs formalism, the prediction function $h_{\mathbf{w}}$ can be defined as returning the maximizer of the scoring function:

$$h_{\mathbf{w}}(\mathbf{x}) = \arg\max_{\mathbf{y}\in\mathcal{Y}} s_{\mathbf{w}}(\mathbf{x},\mathbf{y})$$

Crucially for a wide range of PGM structures (e.g., tree-structured networks), despite the huge size of the output space, this maximization can be accomplished efficiently by means of efficient dynamic programming algorithms (e.g. the Viterbi algorithm). For more complex models, approximate inference algorithms can be used.

In generative models the parameters $\mathbf{w}$ (which in PGMs correspond to log conditional probabilities or log-potentials) are chosen during the learning phase from the training set $\mathcal{T}$ with Maximum Likelihood (ML) estimates. However, it is known that this approach is often suboptimal and discriminative approaches so as to maximize the predictive power of the classifier are preferable.

To clarify what is meant by predictive power, analogously to scalar-valued classification problems, loss functions must be introduced. Consider a *loss function* $\mathcal{L}_{\mathbf{W}}(\mathbf{x}, \bar{\mathbf{y}})$ which measures the discrepancy between the prediction $h_{\mathbf{W}}(\mathbf{x})$ and the optimal labels vector $\bar{\mathbf{y}}$. Empirical risk minimization strategies [27] attempt to find the vector $\mathbf{w}$ such that the *empirical risk* for a given training sample $\mathcal{T}$, defined as:

$$E_{\mathcal{T}}\{\mathcal{L}_{\mathbf{W}}(\mathbf{x}, \bar{\mathbf{y}})\} = \frac{1}{\ell}\sum_{i=1}^{\ell} \mathcal{L}_{\mathbf{W}}(\mathbf{x}^i, \bar{\mathbf{y}}^i)$$

is minimized. This will guarantee that the expected loss $E\{\mathcal{L}_{\mathbf{W}}(\mathbf{x}, \bar{\mathbf{y}})\}$ is small as well. Often a regularization parameter is introduced in order to prevent overfitting.

As concerns the loss function, a natural choice is the *0/1 loss*, defined as:

$$\mathcal{L}^{zo}_{\mathbf{W}}(\mathbf{x}, \bar{\mathbf{y}}) = I(h_{\mathbf{W}}(\mathbf{x}) \neq \bar{\mathbf{y}})$$

where $I(\cdot)$ is an indicator function. Basically with this loss the number of incorrect vectors $h_{\mathbf{W}}(\mathbf{x})$ is taken into account. However often for real-life data problem and especially when $|\mathcal{Y}|$ is large other losses can be more informative. A nice alternative loss measures the number of incorrect microlabels. This loss, the so-called *Hamming loss*, is defined as:

$$\mathcal{L}^{hm}_{\mathbf{W}}(\mathbf{x}, \bar{\mathbf{y}}) = \sum_{j=1}^{m} I(h_{\mathbf{W},j}(\mathbf{x}) \neq \bar{y}_j)$$

In general any monotonically non-decreasing function $\mathcal{L}_{\mathbf{W}}$ such that $\mathcal{L}_{\mathbf{W}}(\mathbf{x}, \bar{\mathbf{y}}) = 0$ if $h_{\mathbf{W}}(\mathbf{x}) = \bar{\mathbf{y}}$ and $\mathcal{L}_{\mathbf{W}}(\mathbf{x}, \bar{\mathbf{y}})$ is maximal if $h_{\mathbf{W},j}(\mathbf{x}) \neq \bar{y}_j$, $\forall j = 1, \ldots, m$ can be taken as loss. One can easily note that the 0/1 loss and the Hamming loss satisfy this requirement.

However these losses are discontinuous functions and directly minimizing the empirical risk for any of them is an NP-hard problem. Therefore the common approach is to choose to minimize real valued convex upper bounds on these. All the algorithms developed for structured output learning, such as CRFs and max-margin approaches, adopt this principle. We will further analyze this aspect later on in the paper.

2.3 Sequence Labeling: From HMMs to Chain CRFs

To further clarify the structured output learning framework, we analyze a specific example: the sequence labeling problem. In sequence labeling, given an observed sequence $\mathbf{x}$, the goal is to reconstruct the associated hidden sequence $\mathbf{y}$. This problem arises in several applications ranging from the natural language processing field to the computational biology area and traditionally hidden Markov models [20] are used for this task.

For each pair $(\mathbf{x}, \mathbf{y})$, $\mathbf{x} = (x_1, ..., x_m)$ denotes an observed sequence, $\mathbf{x} \in \mathcal{X}$, and $\mathbf{y} = (y_1, ..., y_m)$, $\mathbf{y} \in \mathcal{Y}$, is the corresponding hidden sequence. Each observed symbol x_i is an element of the observed symbol alphabet Σ_x, and the hidden symbols y_i are elements of Σ_y. Therefore the output space is $\mathcal{Y} = \Sigma_y{}^m$, while $\mathcal{X} = \Sigma_x{}^m$.

HMMs work by computing the joint distribution of observations $\mathbf{x}$ and labels $\mathbf{y}$, $P(\mathbf{x}, \mathbf{y})$. The graphical model representation of an HMMs is shown in Fig. 2.a. Two assumptions are made by the model. The first, called the Markov assumption, states that the current hidden state is dependent only on the previous one. The second assumption states that the output observation at time t is dependent only on the current hidden state and it is independent of previous observations and states. Then an HMM is characterized by two kinds of probability distributions: $P(y_t|y_{t-1})$, expressing the probability that the given state is y_t and the previous state is y_{t-1}, and $P(x_t|y_t)$ of emitting the symbol x_t being in state y_t. The joint probability can be computed with:

$$P(\mathbf{x}, \mathbf{y}) = \prod_{t=1}^{m} P(x_t|y_t) \prod_{t=2}^{m} P(y_t|y_{t-1}) \tag{3}$$

In the decoding phase the hidden state sequence that is most likely to have produced a given observation sequence must be determined, i.e.

$$h(\mathbf{x}) = \arg\max_{\mathbf{y}\in\mathcal{Y}} P(\mathbf{y}|\mathbf{x}) = \arg\max_{\mathbf{y}\in\mathcal{Y}} \frac{P(\mathbf{x}, \mathbf{y})}{P(\mathbf{x})} = \arg\max_{\mathbf{y}\in\mathcal{Y}} P(\mathbf{x}|\mathbf{y})P(\mathbf{y}) \tag{4}$$

Despite the huge size of the output space ($|\Sigma_y|^m$) decoding can be performed efficiently with the Viterbi algorithm. Substituting (3) in (4) and due to the monotonicity of logarithms (4) is equivalent to:

$$\begin{aligned} h(\mathbf{x}) &= \arg\max_{\mathbf{y}\in\mathcal{Y}} \left[\sum_{t=2}^{m} \log P(y_t|y_{t-1}) + \sum_{t=1}^{m} \log P(x_t|y_t) \right] \\ &= \arg\max_{\mathbf{y}\in\mathcal{Y}} \left[\sum_{i,j\in\Sigma_y} T_{ij} \sum_{t=2}^{m} \phi^{ij}(\mathbf{x}, y_t, y_{t-1}) + \sum_{i\in\Sigma_y, o\in\Sigma_x} E_{io} \sum_{t=1}^{m} \phi_t^{io}(\mathbf{x}, y_t) \right] \end{aligned}$$

where we have defined $\phi^{ij}(\mathbf{x}, y_t, y_{t-1}) = I(y_t = j)I(y_{t-1} = i)$ and $\phi_t^{io}(\mathbf{x}, y_t) = I(y_t = i)I(x_t = o)$ and T_{ij} for $i, j \in \Sigma_y$ denotes the logarithm of the transition probability for the transition from state i to j, and E_{io} the logarithm of the emission probability for the emission of $o \in \Sigma_x$ at state $i \in \Sigma_y$.

It is worth noting that the prediction function (4) is in the form (1). In fact we can define the vector $\mathbf{w}$ containing the parameters T_{ij} and E_{io}. Then the corresponding sufficient statistics (i.e. the number of occurrences of each specific transition and emission) $\sum_{t=2}^{m} \phi^{ij}(\mathbf{x}, y_t, y_{t-1})$ and $\sum_{t=1}^{m} \phi_t^{io}(\mathbf{x}, y_t)$ represent the elements of the feature vector $\boldsymbol{\phi}(\mathbf{x}, \mathbf{y})$.

In the learning phase transition and emission probabilities must be determined. They are computed with a ML approach and constraints are imposed to guarantee that the computed parameters represent probabilities. Moreover the conditional independence assumptions are often too restrictive and in most applications rich sets of features must be considered in order to model arbitrary dependencies between hidden variables. To overcome these difficulties discriminative training models such as MEMMs and CRFs have been proposed.

First proposed by Lafferty *et al.* [13], CRFs are undirected graphical models. The graphical model representation for a chain-structured CRF is shown in Fig. 2.b, where we have one state assignment for each observation in the sequence. Specifically the conditional probability $P(\mathbf{y}|\mathbf{x})$ is defined as:

$$P(\mathbf{y}|\mathbf{x}) = \frac{1}{Z(\mathbf{x})} \exp(\sum_k w_k \phi_k(\mathbf{x}, \mathbf{y}))$$

where $Z(\mathbf{x}) = \sum_{\mathbf{y}} \exp(\mathbf{w}^T \boldsymbol{\phi}(\mathbf{x}, \mathbf{y}))$ is a normalization factor also called the partition function. In chain CRFs two kinds of atomic features $\phi_k(\mathbf{x}, \mathbf{y})$ are considered:

- *Label/Observation features:* $\phi_{t'}^{io}(\mathbf{x}, y_t) = I(y_t = i)\psi_o(x_{t'})$, $i \in \Sigma_y, o \in \Sigma_x$ for each state-observation pair $y_t, x_{t'}$. For example in sequence labeling in NER $\phi_{t'}^{io}(\mathbf{x}, y_t) = 1$ may indicate that the word $o =$ "*Rome*" occurs in position t and is labeled as $i =$ "*Location*". In general $\psi_o(x_{t'})$ can be any real-valued function indicating any property (e.g. spelling property) of the word $x_{t'}$, not just its occurrence.
- *Label/Label features:* $\phi^{ij}(\mathbf{x}, y_t, y_{t'}) = I(y_t = i)I(y_{t'} = j)$, $i, j \in \Sigma_y$ for each pair of states $y_t, y_{t'}$.

For example one may decide to consider only HMM features, i.e. features corresponding to the sufficient statistics associated to transition and emission probabilities. Then atomic features are only limited to $\phi^{ij}(\mathbf{x}, y_t, y_{t-1}) = I(y_t = i)I(y_{t-1} = j)$ and $\phi_t^{io}(\mathbf{x}, y_t) = I(y_t = i)I(x_t = o)$. In this way for each step of the chain t one can define a vector $\boldsymbol{\phi}_t(\mathbf{x}, y_t, y_{t-1})$ containing all the features $\phi_{t'}^{io}(\mathbf{x}, y_t)$ and $\phi^{ij}(\mathbf{x}, y_t, y_{t-1})$ and the overall feature vector is given by:

$$\boldsymbol{\phi}(\mathbf{x}, \mathbf{y}) = \sum_{t=1}^{m} \boldsymbol{\phi}_t(\mathbf{x}, y_t, y_{t-1})$$

In CRFs the optimal weight vector $\mathbf{w}$ is determined using the negative log-loss of the training data as optimization criteria, i.e. the learning phase amounts to minimize:

$$\hat{\mathcal{R}}_{\mathbf{w}}^{CRF}(\mathcal{T}) = -\frac{1}{\ell} \sum_{i=1}^{\ell} \log P(\bar{\mathbf{y}}^i|\mathbf{x}^i) = -\frac{1}{\ell} \sum_{i=1}^{\ell} [\mathbf{w}^T \boldsymbol{\phi}(\mathbf{x}^i, \bar{\mathbf{y}}^i) - \log Z(\mathbf{x}^i)]$$

The main difficulty to solve this problem is given by the estimation of the logarithm of the partition function. In most of the situations it can be computed by marginalizing over the set of possible output $\mathbf{y}$. In practice this optimization problem can be solved either using iterative scaling methods [13] or more complex and efficient optimization strategies such as BFGS [22] or stochastic meta-descent [28]. Since it is well known that CRFs are prone to overfitting a Gaussian prior is often used for regularization.

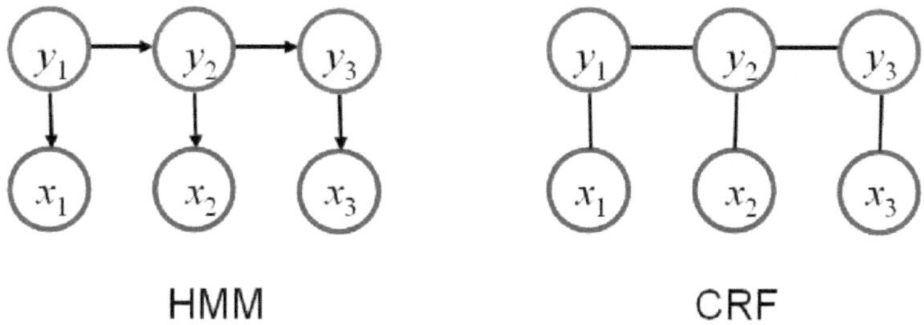

Fig. 2. Graphical models for an hidden Markov model (HMM) and a conditional random field (CRF)

In the last few years CRFs have been extensively applied in many problems, such as information extraction, image processing, parsing and so on [17, 19, 12, 22]. Following the success of CRFs many researchers have developed other methods for structured output learning which basically raise from the idea of utilizing alternative loss functions and constructing more efficient algorithms. For example by the use of the hinge loss max-margin approaches such as the maximum margin Markov networks [23] and hidden Markov support vector machines [1, 26] can be conceived. In the following we provide a detailed description of these models.

3 Large Margin Approaches for Structured Output Learning

The maximal margin approach for structured prediction has been developed as a natural extension of the classifier proposed by Crammer and Singer for multiclass problems [7]. In this Section we first present the algorithm in [7], then we discuss the structured output methods.

3.1 Multiclass SVMs

In multiclass classification a training set $\mathcal{T} = \{(\mathbf{x}^1, y^1), \ldots, (\mathbf{x}^\ell, y^\ell)\}$ is given, with $\mathbf{x}^i \in \mathcal{X}, \mathcal{X} \subseteq \mathbb{R}^n$. The labels y^i are chosen from the set $\mathcal{Y} = \{1, \ldots M\}$, $M \in \mathbb{N}$.

SVMs [8] are inherently binary classifiers but various approaches are possible to apply them to multiclass tasks. One possible approach is to decompose the problem in a set of binary classification tasks and for each of them to train an hyperplane separating the elements of each class from all the other datapoints. This approach is usually referred as *one vs. all.* Another possible strategy, the so called *one vs. one* method, decomposes the multiclass problem into pairwise classifications. In this way, one has to train $M(M-1)/2$ classifiers, separating each class label from another.

However one vs. all and one vs. one classifiers can be suboptimal in some applications because all the binary classifiers are trained independently. Therefore in alternative one can decide to define a global objective function to achieve multiclass classification. This is the basic idea of the algorithm proposed in [7]. The following optimization problem must be solved:

$$\begin{aligned} &\min \frac{1}{2}\|\mathbf{W}\|^2 \\ &\text{s.t. } \mathbf{w}_{y^i}^T \mathbf{x}^i - \max_{r:\ r \neq y^i,\ 1 \leq r \leq M} \mathbf{w}_r^T \mathbf{x}^i \geq 1 \quad \forall 1 \leq i \leq \ell \end{aligned}$$

where $\mathbf{W}$ is a matrix of size $M \times \ell$ over $\mathbb{R}$ and $\mathbf{w}_r$ is the r-th row of $\mathbf{W}$. In practice for each training point the goal is to discriminate as much as possible the correct label from the second best incorrect label. To achieve this task a kind of margin for the multiclass case is defined and one looks for the parameters $\mathbf{W}$ such that this margin is maximized. In the prediction phase the class is assigned by solving:

$$h(\mathbf{x}) = \arg \max_{1 \leq r \leq M} (\mathbf{w}_r^T \mathbf{x})$$

3.2 Maximizing the Margin in Structured Output

The maximum margin approach for structured output learning extends the Crammer and Singer method for multiclass SVMs. In [7] labels are assigned to objects independently in a way such as to guarantee the maximal possible separation between the correct label and its best runner-up. On the other hand max-margin approaches for structured output learning consider labels jointly by defining an appropriate feature map $\boldsymbol{\phi}(\mathbf{x}, \mathbf{y})$. Then the goal is to look for the parameter vector $\mathbf{w}$ such that the difference between the score of the optimal pairs $(\mathbf{x}^i, \bar{\mathbf{y}}^i)$ and the score of the most competitive incorrect ones is maximized. This amounts to solving the following optimization problem:

$$\begin{aligned} &\min_{\mathbf{w}} \frac{1}{2}\|\mathbf{w}\|^2 \\ &\text{s.t. } \ \mathbf{w}^T \boldsymbol{\Delta\phi}(\mathbf{x}^i, \mathbf{y}_j^i) \geq 1 \quad \forall \mathbf{y}_j^i \neq \bar{\mathbf{y}}^i \end{aligned}$$

where we have defined $\boldsymbol{\Delta\phi}(\mathbf{x}^i, \mathbf{y}_j^i) = \boldsymbol{\phi}(\mathbf{x}^i, \bar{\mathbf{y}}^i) - \boldsymbol{\phi}(\mathbf{x}^i, \mathbf{y}_j^i)$.

Since in practical real-life data problems data are not linearly separable non-negative slack variables ξ_i (one for each training pair) are introduced, to allow for constraints to be violated. The resulting optimization problem is:

$$\begin{aligned} \text{SVM}_1: \quad &\min_{\mathbf{w}} \frac{1}{2}\|\mathbf{w}\|^2 + \frac{C}{\ell}\sum_{i=1}^{\ell} \xi_i \\ &\text{s.t. } \ \mathbf{w}^T \boldsymbol{\Delta\phi}(\mathbf{x}^i, \mathbf{y}_j^i) \geq 1 - \xi_i \quad \forall \mathbf{y}_j^i \neq \bar{\mathbf{y}}^i \\ &\xi_i \geq 0 \quad \forall i \end{aligned} \tag{5}$$

where C is a user-defined regularization parameter.

Solving the optimization problem (5) an upper bound on the 0/1 loss is minimized. However especially for problems where $|\mathcal{Y}|$ is large it is more useful to enforce large margins for examples with high loss and smaller ones for small losses. To this aim a possible approach consists in rescaling each slack variable ξ_i (representing the training error) by the loss of the associated linear constraint. In this way pseudo-examples with high loss are penalized more than those with small loss. The resulting optimization problem is:

$$\begin{aligned} \mathrm{SVM}_1^{\lambda_s}: \quad & \min_{\mathbf{w}} \frac{1}{2}\|\mathbf{w}\|^2 + \frac{C}{\ell}\sum_{i=1}^{\ell}\xi_i \\ & \text{s.t.} \quad \mathbf{w}^T \boldsymbol{\Delta\phi}(\mathbf{x}^i, \mathbf{y}_j^i) \geq 1 - \frac{\xi_i}{\lambda(\mathbf{y}_j^i)} \quad \forall \mathbf{y}_j^i \neq \bar{\mathbf{y}}^i \\ & \qquad \xi_i \geq 0 \quad \forall i \end{aligned} \tag{6}$$

where $\lambda(\mathbf{y}) = \mathcal{L}_{\mathbf{W}}(\mathbf{x}, \mathbf{y})$ denotes the loss function.

In structured output learning with max-margin is convenient for algorithmic efficiency to define a loss that decomposes over the structure of the graph. The most common choice is the Hamming loss $\mathcal{L}_{\mathbf{W}}^{hm}(\mathbf{x}, \mathbf{y})$ [23]. Other decomposable losses have been used for example in [21] in the context of hierarchical classification: here the loss is defined in a way such that in the hierarchy predicting the parent microlabel correctly is typically more important than predicting the child correctly.

The optimization problem $\mathrm{SVM}_1^{\lambda_s}$ has been rarely used in practice. In fact in the vast majority of applications a different formulation is adopted:

$$\begin{aligned} \mathrm{SVM}_1^{\lambda_m}: \quad & \min_{\mathbf{w}} \frac{1}{2}\|\mathbf{w}\|^2 + \frac{C}{\ell}\sum_{i=1}^{\ell}\xi_i \\ & \text{s.t.} \quad \mathbf{w}^T \boldsymbol{\Delta\phi}(\mathbf{x}^i, \mathbf{y}_j^i) \geq \lambda(\mathbf{y}_j^i) - \xi_i \quad \forall \mathbf{y}_j^i \neq \bar{\mathbf{y}}^i \\ & \qquad \xi_i \geq 0 \quad \forall i \end{aligned} \tag{7}$$

In practice in $\mathrm{SVM}_1^{\lambda_m}$ the margin is also decomposable over the edges of the graph which is beneficial for algorithmic efficiency especially for large scale problem. A drawback of this approach is the fact that some capacity of the learning machine is wasted in imposing large margins of high-loss examples. We will illustrate this concept more in detail later on in the paper.

As for classical SVMs [8] a dual formulation for problem (5) can be derived which is greatly advantageous in problems with a large number of features such as for example NLP applications. Dual variables $\alpha_{\mathbf{x}}(\mathbf{y})$ can be introduced and the resulting optimization problem is:

$$\begin{aligned} & \max_\alpha \sum_{i,j} \alpha_{\mathbf{x}^i}(\mathbf{y}_j^i) - \frac{1}{2}\sum_{i,j}\sum_{i',j'} \alpha_{\mathbf{x}^i}(\mathbf{y}_j^i)\alpha_{\mathbf{x}^{i'}}(\mathbf{y}_{j'}^{i'}) \boldsymbol{\Delta\phi}(\mathbf{x}^i, \mathbf{y}_j^i) \cdot \boldsymbol{\Delta\phi}(\mathbf{x}^{i'}, \mathbf{y}_{j'}^{i'}) \\ & \text{s.t.} \quad \ell \sum_{i,j} \alpha_{\mathbf{x}^i}(\mathbf{y}_j^i) \leq C \quad \forall\, i, \quad \alpha_{\mathbf{x}^i}(\mathbf{y}_j^i) \geq 0 \quad \forall\, i, j \end{aligned} \tag{8}$$

The dual optimization problems $\mathrm{SVM}_1^{\lambda_s}$ and $\mathrm{SVM}_1^{\lambda_m}$ follow with minor modification from (8). In $\mathrm{SVM}_1^{\lambda_s}$ the box constraints have the Lagrange multipliers rescaled dividing by the associated loss terms, while for $\mathrm{SVM}_1^{\lambda_m}$ the dual problem is the same as (8) but with the linear part in the objective function given by $\sum_{i,j} \alpha_{\mathbf{x}^i}(\mathbf{y}_j^i)\lambda(\mathbf{y}_j^i)$.

Once the optimization problem is solved the primal solution can be obtained from the dual:

$$\mathbf{w} = \sum_{i,j} \alpha_{\mathbf{x}^i}(\mathbf{y}_j^i)\boldsymbol{\Delta\phi}(\mathbf{x}^i, \mathbf{y}_j^i)$$

and the prediction for a new data point is realized computing (1).

With the dual formulation, analogously to classical SVMs, the use of kernels is allowed. In fact both the optimization problem (8) and the decision function (1) can be expressed in terms of dot products of the feature vectors. Then the main advantage is that with the kernel trick high dimensional feature spaces are considered and high order correlation between labels can be taken into account.

However the optimization problem (8) is not practically useful since it relies on an exponential number of variables and constraints. To deal with this difficulty various strategies have been proposed which basically can be classified into three main approaches:

- *Marginal variables method:* in [23] Taskar *et al.* observe that (8) can be reformulated in terms of marginal variables and the objective function depends only on a polynomial number of those. To solve it they develop a coordinate descent method analogous to the sequential minimal optimization (SMO) [18] for SVMs.
- *Iterative method:* in [1, 26] an iterative approach is proposed. A working set of dual variables is incrementally constructed adding at each step the dual variable corresponding to the most violated constraint in the primal. The worst violator is computed solving the inference problem in the underlying graph. The algorithm is shown to converge after adding at most a polynomial number of variables.
- *Min-max method:* in [25] the inference problem is embedded into the primal problem (7) resulting in a convex-concave saddle-point problem. To solve it an approach based on the dual extra-gradient method [16] is proposed.

In the following we briefly review these approaches.

3.3 Marginal Variables Method

In [23] maximum margin Markov networks (M^3N) are introduced. Taskar *et al.* consider the optimization problem $\mathrm{SVM}_1^{\lambda_m}$ in its dual form:

$$\begin{aligned} &\max_\alpha \sum_{i,j} \alpha_{\mathbf{x}^i}(\mathbf{y}_j^i)\lambda(\mathbf{y}_j^i) - \frac{1}{2}\sum_{i,j}\sum_{i',j'} \alpha_{\mathbf{x}^i}(\mathbf{y}_j^i)\alpha_{\mathbf{x}^{i'}}(\mathbf{y}_{j'}^{i'})\boldsymbol{\Delta\phi}(\mathbf{x}^i, \mathbf{y}_j^i) \cdot \boldsymbol{\Delta\phi}(\mathbf{x}^{i'}, \mathbf{y}_{j'}^{i'}) \\ &\text{s.t.} \quad \ell \sum_{i,j} \alpha_{\mathbf{x}^i}(\mathbf{y}_j^i) = C \quad \forall\, i, \quad \alpha_{\mathbf{x}^i}(\mathbf{y}_j^i) \geq 0 \quad \forall\, i,j \end{aligned}$$

This problem is solved by defining a set of marginal variables. The main idea behind this approach is that "the dual variables $\alpha_{\mathbf{x}}(\mathbf{y}_j)$ can be interpreted as a density function over $\mathbf{y}$ conditional on $\mathbf{x}$, as $\sum_j \alpha_{\mathbf{x}}(\mathbf{y}_j) = C$ and $\alpha_{\mathbf{x}}(\mathbf{y}_j) \geq 0$" [23] and the objective function is a function of the expected values of the losses and of the feature vectors. Here, for sake of simplicity of the notation, this concept is illustrated by taking a chain CRF in a way such that the features vectors decompose over the chain of the graphs as follows:

$$\boldsymbol{\Delta}\boldsymbol{\phi}(\mathbf{x}^i, \mathbf{y}^i) = \sum_{t=1}^{m} \boldsymbol{\Delta}\boldsymbol{\phi}_t(\mathbf{x}^i, y_t^i, y_{t-1}^i)$$

Moreover a decomposable loss function (e.g. the Hamming loss) is taken into account:

$$\lambda(\mathbf{y}^i) = \sum_{t=1}^{m} \lambda_t(y_t^i)$$

One can define the marginal variables:

$$\mu_{\mathbf{x}^i}(y_t^i) = \sum_{j:\ \mathbf{y}_j^i \sim y_t^i} \alpha_{\mathbf{x}^i}(\mathbf{y}_j^i)$$

and:

$$\mu_{\mathbf{x}^i}(y_t^i, y_{t-1}^i) = \sum_{j:\ \mathbf{y}_j^i \sim [y_t^i, y_{t-1}^i]} \alpha_{\mathbf{x}^i}(\mathbf{y}_j^i).$$

where the sums are taken over all j such that the vectors $\mathbf{y}_j^i$ contains y_t^i or (y_t^i, y_{t-1}^i) respectively. Then the dual optimization problem can be rewritten in terms of marginal variables as:

$$\max_{\mu_{\mathbf{x}^i}(y_t^i), \mu_{\mathbf{x}^i}(y_t^i, y_{t-1}^i)} \sum_i \sum_t \sum_{y_t^i} \mu_{\mathbf{x}^i}(y_t^i) \lambda_t(y_t^i) \tag{9}$$

$$-\frac{1}{2} \sum_{i,i'} \sum_{t,t'} \sum_{(y_t^i, y_s^i), (y_{t'}^{i'}, y_{s'}^{i'})} \mu_{\mathbf{x}^i}(y_t^i, y_s^i) \mu_{\mathbf{x}^{i'}}(y_{t'}^{i'}, y_{s'}^{i'}) \boldsymbol{\Delta}\boldsymbol{\phi}_t(\mathbf{x}^i, y_t^i, y_s^i) \cdot \boldsymbol{\Delta}\boldsymbol{\phi}_{t'}(\mathbf{x}^{i'}, y_{t'}^{i'}, y_{s'}^{i'})$$

$$\text{s.t. } \ell \sum_{y_t^i} \mu_{\mathbf{x}^i}(y_t^i) = C \quad \forall\, i, \quad \mu_{\mathbf{x}^i}(y_t^i, y_s^i) \geq 0 \quad \forall\, i, j$$

$$\sum_{y_t^i} \mu_{\mathbf{x}^i}(y_t^i, y_s^i) = \mu_{\mathbf{x}^i}(y_t^i)$$

where $s = t - 1$ and $s' = t' - 1$. The last constraint is meant to enforce the consistency condition, i.e. to guarantee that the joint distribution can be always defined. In practice we simply impose that the sum of the pairwise marginals should be equal to the corresponding singleton marginal. This final optimization problem is equivalent to the original one and can be solved efficiently since it requires only a polynomial number of variables.

In [23] an algorithm which is an extension of the classical SMO [18] is presented. Basically the QP problem is decomposed in terms of smaller subproblems. Specifically at each step the smallest possible subproblem (i.e. given by just two samples $\mathbf{y}_j$, $\mathbf{y}_{j'}$) is taken into account, so the solution can be computed analytically. At each step the working set is chosen by considering the variables that violates the Karush-Kuhn-Tucker conditions [9] and is determined by computing the inference of the graph.

It is worth noting that this approach applies to any arbitrary graph structure. In particular if the graph is triangulated, then clique marginals can be defined together with appropriate consistency conditions. If the graph is non-triangulated, then this approach can still be possible but it is highly impractical since the number of constraints is exponential in the size of the cliques. For further details on the topic we remind to the original paper [23].

Based on the idea of using marginal variables other methods have been proposed to improve the optimization phase in M^3Ns and to allow their use in problems with medium sized data sets. Two relevant examples are the conditional-gradient method developed in [21] for hierarchical document classification and the exponentiated gradient algorithm proposed by Bartlett *et al.* [3]. Despite their increased efficiency with respect to M^3Ns, all these approaches rely on dynamic programming and their use is arduous in large scale applications.

3.4 Iterative Method

An alternative approach with respect to M^3Ns is proposed in [1] and further developed in [26]. This algorithm is usually referred as SVMISO (Support Vector Machines in Interdependent and Structured Output spaces).

In [26] a common iterative algorithm is proposed to solve the problems SVM_1, $\text{SVM}_1^{\lambda_s}$ and $\text{SVM}_1^{\lambda_m}$. More specifically to allow the use of kernels the associated dual problems are considered. As we observed before solving these problems is an hard task since they rely on an exponential number of variables (i.e. the Lagrange multipliers). In [1] the sparsity of the SVMs is exploited. Basically the crucial observation of Altun *et al.* is that one may expect that only a very small fraction of the Lagrange multipliers (corresponding to support vectors) will be different from zero in the solution. Moreover the Lagrange multipliers corresponding to one training sample $\mathbf{x}^i$ are independent of those of another training instance $\mathbf{x}^j$.

To solve the dual problems an incremental algorithm derived by [7] and based on a working set selection approach is used and it is schematically presented in Algorithm 1. Basically the following strategy is adopted. At each step the vector $\hat{\mathbf{y}}$ (which corresponds to the best score for the current set of parameters $\mathbf{w}$) is determined with a Viterbi-like algorithm. Then a constraint is imposed to ensure that the difference between the score of the optimal $\bar{\mathbf{y}}$ and the score associated to $\hat{\mathbf{y}}$ is maximized and added to the current set of constraints. Then, subject to these constraints, the objective function of (5), (6) or (7) is minimized. The algorithm terminates when no constraints are violated. Importantly in [26] it is shown that this strategy guarantees that, fixed a precision ϵ, the parameter vector determined considering only a polynomial number of constraints is such that all the exponentially many constraints are satisfied by ϵ.

For the optimization problem (5) the computation of $\hat{\mathbf{y}}$ amounts to solve the inference problem (1). Determining the worst margin violator is slightly more complex for problems (6) and (7) since one must be able to compute a loss-augmented search. More specifically for $\mathrm{SVM}_1^{\lambda_s}$ the loss-augmented search is:

$$\arg\max_{\mathbf{y}\in\mathcal{Y}} [\mathbf{w}^T\boldsymbol{\phi}(\mathbf{x},\mathbf{y})]\lambda(\mathbf{y}) \tag{10}$$

while for $\mathrm{SVM}_1^{\lambda_m}$ the inference consists in solving:

$$\arg\max_{\mathbf{y}\in\mathcal{Y}} \{\mathbf{w}^T\boldsymbol{\phi}(\mathbf{x},\mathbf{y}) + \lambda(\mathbf{y})\} \tag{11}$$

As argued in [26] and as we observed before, rescaling slacks as in (6) leads to a more appropriate optimization problem with respect to the scaled margin formulation (7). However with the latter the loss-augmented search is quite easy if the loss is decomposable over the structure of the graph since the atomic contributes to the loss are considered additively. On the other hand solving (10) is a more complex task, even intractable in most applications, being the loss a multiplicative term. Then in general considering (11) and decomposable losses as the Hamming loss the computation of the *argmax* is noticeably simplified but it is worth noting that, opposite to M^3Ns, the algorithm in [26] is not limited to these cases.

3.5 Min-Max Method

The last approach we present is proposed in [25]. Taskar *at al.* consider the optimization problem $\mathrm{SVM}_1^{\lambda_m}$ and rewrite the objective function in the form:

$$\min_{\mathbf{w}} \frac{1}{2}\|\mathbf{w}\|^2 - C\sum_{i=1}^{\ell}\{\mathbf{w}^T\boldsymbol{\phi}(\mathbf{x}^i,\bar{\mathbf{y}}^i) - \max_{\mathbf{y}^i\in\mathcal{Y}_i}[\mathbf{w}^T\boldsymbol{\phi}(\mathbf{x}^i,\mathbf{y}^i_j) + \lambda(\mathbf{y}^i_j)]\} \tag{12}$$

In practice the final optimization problem is obtained embedding the inference procedure into the QP problem resulting in a min-max problem. To solve it efficiently one must be able to deal with the loss-augmented inference problem:

$$\max_{\mathbf{y}^i\in\mathcal{Y}_i} [\mathbf{w}^T\boldsymbol{\phi}(\mathbf{x}^i,\mathbf{y}^i) + \lambda(\mathbf{y}^i)] \tag{13}$$

To this aim in [25] a loss function that is decomposable over the structure of the graph is always used, such as the Hamming loss. Then, with this assumption, the basic observation of Taskar *at al.* is that the loss augmented inference problem can be solved through its equivalent integer linear programming problem instead of using a dynamic programming approach. For example for sequence labeling learning the solution of the Viterbi algorithm is in fact the output of a

Algorithm 1. Incremental algorithm to solve SVM_1, $\text{SVM}_1^{\lambda_s}$, $\text{SVM}_1^{\lambda_m}$

```
Input: $\mathcal{T} = \{(\mathbf{x}^1, \bar{\mathbf{y}}^1), (\mathbf{x}^2, \bar{\mathbf{y}}^2), \ldots, (\mathbf{x}^\ell, \bar{\mathbf{y}}^\ell)\}$, $C$, the required accuracy $\epsilon$.

$\mathcal{S}_i \leftarrow \emptyset$ for all $i = 1 \ldots \ell$
Repeat
    for $i = 1 \ldots \ell$ do
        Set up cost function
            $\text{SVM}_1$: $H(\mathbf{y}) = 1 - \mathbf{w}^T \boldsymbol{\Delta}\phi(\mathbf{x}, \mathbf{y})$
            $\text{SVM}_1^{\lambda_s}$: $H(\mathbf{y}) = 1 - [\mathbf{w}^T \boldsymbol{\Delta}\phi(\mathbf{x}, \mathbf{y})]\lambda(\mathbf{y})$
            $\text{SVM}_1^{\lambda_m}$: $H(\mathbf{y}) = \lambda(\mathbf{y}) - \mathbf{w}^T \boldsymbol{\Delta}\phi(\mathbf{x}, \mathbf{y})$
            where $\mathbf{w} = \sum_i \sum_{\mathbf{y}_j^i \in \mathcal{S}_j} \alpha_{\mathbf{x}^i}(\mathbf{y}_j^i) \boldsymbol{\Delta}\phi(\mathbf{x}^i, \mathbf{y}_j^i)$
        Compute $\hat{\mathbf{y}}^i := \arg\max_{\mathbf{y}_j^i} H(\mathbf{y}_j^i)$
        Compute $\xi_i = \max\{0, \arg\max_{\mathbf{y}_j^i \in \mathcal{S}_i} H(\mathbf{y}_j^i)\}$
        if $H(\mathbf{y}_j^i) \geq \xi_i + \epsilon$
            $\mathcal{S}_i \leftarrow \mathcal{S}_i \cup \{\hat{\mathbf{y}}^i\}$
            $\boldsymbol{\alpha}_{\mathcal{S}} \leftarrow$ optimize dual over $\mathcal{S} \cup \mathcal{S}_i$
        endif
    endfor
until no $S_i$ have changed during iterations.
```

shortest path problem. More importantly in most situations the integer linear programming problem can be solved considering the associated continuous variables problem since it is guaranteed to have an integral solution [29]. Then (13) is equivalent to the following:

$$\max_{\mathbf{z}^i \in \mathcal{Z}_i} [\mathbf{w}^T \boldsymbol{\Phi}_i + \mathbf{c}_i^T]\mathbf{z}^i \tag{14}$$

where $\boldsymbol{\Phi}_i$ and $\mathbf{c}_i$ are appropriate matrices derived from the features vectors and from the loss and we have defined the set $\mathcal{Z}_i = \{\mathbf{z}^i : \mathbf{A}\mathbf{z}^i \leq \mathbf{b}, \quad \mathbf{z}^i \geq 0\}$. For example for the shortest path the constraints in the set $\mathcal{Z}_i$ basically ensure that for each node in the graph the number of inward edges should be the same as the number of the outward ones.

Then, the overall estimation problem (12) can be formulated as a convex-concave saddle-point problem:

$$\min_{\mathbf{w}} \max_{\mathbf{z} \in \mathcal{Z}} L(\mathbf{w}, \mathbf{z}) = \frac{1}{2}\|\mathbf{w}\|^2 + C \sum_{i=1}^{\ell} [\mathbf{w}^T \boldsymbol{\Phi}_i \mathbf{z}^i + \mathbf{c}_i^T \mathbf{z}^i - \mathbf{w}^T \phi(\mathbf{x}^i, \bar{\mathbf{y}}^i)]$$

with $\mathcal{Z} = \{\mathcal{Z}_1 \times \mathcal{Z}_2 \times \ldots \times \mathcal{Z}_\ell\}$.

To solve it the dual extra-gradient method [16] is proposed. The gradients of $L(\mathbf{w}, \mathbf{z})$ with respect to $\mathbf{w}$ and $\mathbf{z}$ are considered:

$$\nabla_{\mathbf{w}} L(\mathbf{w}, \mathbf{z}) = \|\mathbf{w}\| + C \sum_{i=1}^{\ell} [\boldsymbol{\Phi}_i \mathbf{z}^i - \phi(\mathbf{x}^i, \bar{\mathbf{y}}^i)]$$

$$\nabla_{\mathbf{z}^i} L(\mathbf{w}, \mathbf{z}) = \boldsymbol{\Phi}_i \mathbf{w} + \mathbf{c}_i$$

together with the cumulative gradient $\mathbf{g}$ which is initialized to zero value ($\mathbf{g}^0 = 0$). Then the extragradient strategy is applied which basically consists in three steps: the prediction step, the correction step and the cumulative gradient update. In the prediction step variables are updated with the following:

$$\mathbf{w}' = \pi_{\mathbb{R}}(\mathbf{w}^t + \beta \mathbf{g}^t_{\mathbf{W}})$$
$$\mathbf{z}^{i'} = \pi_{\mathcal{Z}_i}(\mathbf{z}^{i^t} + \beta \mathbf{g}^t_{\mathbf{Z}^i})$$

where $\pi_{\mathcal{V}}(\mathbf{v}') = \arg\min_{\mathbf{v}\in\mathcal{V}} \|\mathbf{v}' - \mathbf{v}\|$ is the Euclidean projection of a vector $\mathbf{v}$ onto a convex set $\mathcal{V}$ and β is an appropriate chosen step size which is set equal to 1 at the beginning of the optimization and is decreased using an Armijo type rule. In the correction step variables are updated with the following:

$$\mathbf{w}^{t+1} = \pi_{\mathbb{R}}(\mathbf{w}' - \beta \nabla_{\mathbf{W}} L(\mathbf{w}', \mathbf{z}'))$$
$$\mathbf{z}^{i^{t+1}} = \pi_{\mathcal{Z}_i}(\mathbf{z}^{i'} + \beta \nabla_{\mathbf{Z}_i} L(\mathbf{w}', \mathbf{z}'))$$

and for the cumulative gradient the updates are given by:

$$\mathbf{g}^{t+1}_{\mathbf{W}} = \mathbf{g}^t_{\mathbf{W}} - \nabla_{\mathbf{W}} L(\mathbf{w}^t, \mathbf{z}^t)$$
$$\mathbf{g}^{t+1}_{\mathbf{Z}^i} = \mathbf{g}^t_{\mathbf{Z}^i} + \nabla_{\mathbf{Z}^i} L(\mathbf{w}^t, \mathbf{z}^t)$$

Further details about the algorithm can be found in the original paper [25].

This approach is much more efficient than the original SMO and the exponentiated gradient method by Bartlett [3]. In fact both of them rely on dynamic programming in the learning phase so they do not apply to problems with combinatorial structures such as matchings and min-cuts. Examples of such tasks arise in 3D image segmentation and word alignment in translation. On the other hand the min-max method is applicable since matchings and min-cuts as well as the shortest path problems can be casted in the form (14). Moreover the extragradient algorithm provides guarantees of linear convergence and it is simple to implement since it is based on simple gradient and projection calculations. An improvement of this method is the sub-gradient technique proposed in [2]: interestingly this approach can be applied to any loss augmented search problem (not only to those equivalent to linear programs) and extends naturally to the online learning setting.

4 Generalization Bounds

While the analysis of uniform convergence bounds for SVMs is a well established topic, the study of the generalization properties of max-margin approach in structured output learning is considered still an open issue and deserves further study. To the best of our knowledge few results exist on this topic [23, 6] and perhaps the most important is the theorem proposed in [23]. Here a generalization bound for M^3N is given by relating the generalization error to the margin γ of the classifier. It is obtained in terms of covering numbers by an explicit covering construction similar to previous results for classical SVMs [30]. Here we omit details about the derivation but we simply give the statement of the theorem in [23].

The bound is given in terms of an *average per-label loss*:

$$\bar{\mathcal{L}}_{\mathbf{w}}(\mathbf{x}, \bar{\mathbf{y}}) = \frac{1}{m} \lambda^{hm}(\arg\max_{\mathbf{y}} \mathbf{w}^T \phi(\mathbf{x}, \mathbf{y}))$$

where $\lambda^{hm}(\cdot)$ here indicates the Hamming loss and of a *γ-margin per-label loss:*

$$\mathcal{L}^{\gamma}_{\mathbf{w}}(\mathbf{x}, \bar{\mathbf{y}}) = \max_{\mathbf{y}: \mathbf{w}^T \phi(\mathbf{x}, \mathbf{y}) \leq \mathbf{w}^T \phi(\mathbf{x}, \bar{\mathbf{y}}) + \gamma \lambda^{hm}(\mathbf{y});\ \ \forall \mathbf{y}} \frac{1}{m} \lambda^{hm}(\mathbf{y})$$

The expected value of the average per-label loss reflects the generalization error and minimizing it means minimize the number of uncorrect microlabels since the Hamming loss is considered. To state the bound, we also define need to define two other quantities. Let q be the maximum edge degree of the network and R_E be an upper-bound on the 2-norm of the edge-feature vectors. Having defined these two losses we can reformulate the main theorem in [23] with the terminology of the present paper as follows:

Theorem 1. *Given a hypothesis space $\mathcal{H}$ of prediction functions $h_{\boldsymbol{w}}$ as defined in (2), consider a training set $\mathcal{T}$ with size $\ell > 1$ sampled i.i.d. from a fixed but unknown distribution. Then, for any $\delta > 0$, there exists a constant K such that for any $\gamma > 0$ per-label margin the expected value of the per-label loss is bounded by:*

$$E\{\bar{\mathcal{L}}_{\boldsymbol{w}}(\boldsymbol{x}, \bar{\boldsymbol{y}})\} \leq E_{\mathcal{T}}\{\mathcal{L}^{\gamma}_{\boldsymbol{w}}(\boldsymbol{x}, \bar{\boldsymbol{y}})\} + \sqrt{\frac{K}{\ell}\left[\frac{R_E^2 \|\boldsymbol{w}\|^2 q^2}{\gamma^2}[\ln \ell + \ln q + \ln m + \ln n] + \ln \frac{1}{\delta}\right]}$$

with probability at least $1 - \delta$.

In practice the generalization error is bounded by the sum of the expected value of the average γ-margin per-label loss computed on the training set plus a term which is inversely proportional to the margin γ. This sum reflects the trade-off between the training error and the term $\|\mathbf{w}\|/\gamma$ which basically corresponds to the complexity of the classifier. In fact as γ decreases the average γ-margin per-label loss becomes small while the second term increases and viceversa.

This bound, developed specifically for max-margin structured output prediction and Hamming loss, improves the result in [6] which consider a 0/1 loss. In particular this bound is tighter since it depends logarithmically from the number of labels m while in [6] the dependence from m is linear.

5 Experimental Results

In this section we present some experimental results associated to large margin approaches for structured output prediction. We consider three typical applications involving sequences: sequence labeling learning, sequence alignment learning and parse learning.

5.1 Sequence Labeling Learning

In sequence labeling, given a sequence of observations, the associated sequence of labels must be determined. We present two applications of sequence labeling learning: NER and handwritten character recognition.

In [26] the performance of SVMISO for NER problem is presented. The Spanish news wire article corpus of the special session of CoNLL2002 is used as dataset. In particular a small sub-corpus consisting of 300 sentences is considered. The task is to assign 9 different labels to words in the text consisting in non-name and the beginning and the continuation of person names, organizations, locations and miscellaneous names. The feature vector includes the HMM features (i.e. the sufficient statistics associated to transition and emission probabilities) and some features reflecting the spelling properties of the words. The performance of SVMISO with polynomial kernels are compared with those of the averaged structured perceptron (also with polynomial kernel) [5], of CRFs and classical HMMs trained with generative approaches. For SVMISO the 0/1 loss is considered, i.e. the optimization problem (5) is solved.

Table 1 shows the average test error on a 5-fold cross-validation experiment. The results demonstrate that all discriminative approaches outperform the traditional generative model of HMM. Moreover the large margin approach provides further benefits with respect to CRFs and structured perceptron.

Table 1. Test error on the NER task

Method	HMM	CRF	Perceptron	SVMISO
Error	9.36	5.17	5.94	5.08

In [23] handwritten-character recognition is considered as sequence labeling task. The dataset used for experiments [10] consists in about 6100 handwritten words, with average length of 8 characters. Each character of the words is contained in an image with 16 by 8 binary pixels. In this problem the labels correspond to the 26 letters of the alphabeth and the pixels of the images are represented by the vector $\mathbf{x}$. Experiments have been performed with 10 fold crossvalidation. The performance of M^3Ns are compared with logistic regression classifier, multiclass SVMs, and CRFs. For CRFs and M^3Ns approaches only first order dependencies between microlabels y_t and y_{t+1} are considered. The results are shown in Fig. 3. In the logistic regression classifier and multiclass SVMs dependencies between labels are ignored and their accuracy is worse then that of structured output learning approaches (i.e. CRFs and M^3Ns) if linear kernels are considered. However multiclass SVMs with cubic kernel still outperform the structured output approaches. Finally and more importantly if kernels are employed in M^3Ns the lower possible error rate is achieved.

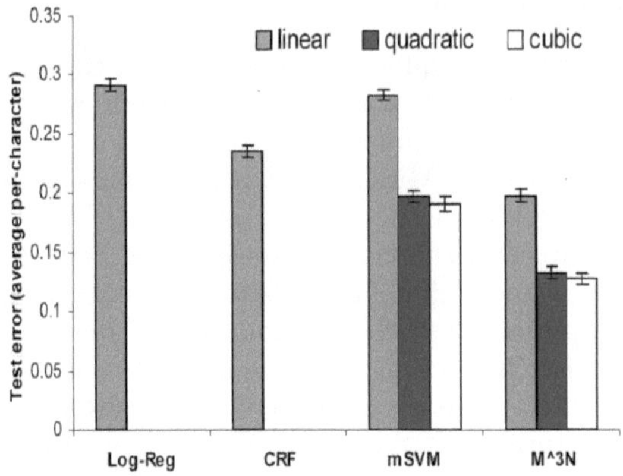

Fig. 3. Error rates for handwritten recognition [23]

5.2 Sequence Alignment Learning

The sequence alignment learning problem consists in learning the optimal substitution matrix and the gap scoring parameters from a collection of pairs of aligned amino acid sequences.

The experiments reported in [26] consider the problem of homology detection and show the better performance of a discriminative large margin approach with respect to a generative model (Table 2). Training sets of different sizes ℓ made up by sequences randomly generated and aligned are considered. In the training set for each native sequence an homologue sequence and the corresponding alignment is given. The task is to find the optimal parameters such that the native is close to its homologue and distant from a set of decoy sequences. The prediction phase is performed with the Smith-Waterman algorithm and the performance is evaluated in terms of error rates (i.e. fraction of times the homolog is not selected).

Table 2 shows the average training and test errors on 10 runs together with the number of constraints required by Algorithm 1. Interestingly it is shown that this number grows sub-linearly with the number of examples. The performance of SVMISO is compared with that obtained with a generative sequence alignment model, where the substitution matrix is computed using Laplace estimates and the gap penalty is chosen such that the best performance on test set is obtained. As expected it is shown that SVMISO outperforms the generative approach especially for datasets of small size.

5.3 Sequence Parse Learning

For sequence parse learning we report the results of the experiments shown in [24] where the Penn English Treebank is used as dataset. Various models have been developed and results are shown in Table 3. All the models have been trained using sections 2-21, while the development set is given by section 22 and section 23 has been used for testing. In the experiment only sentences of length ≤ 15 words are

Table 2. Error rates and number of constraints depending on the number of training examples

	Training error		Test error		
ℓ	Generative	SVMISO	Generative	SVMISO	Const
1	20.0 ± 13.3	0.0 ± 0.0	74.3 ± 2.7	47.0 ± 4.6	7.8 ± 0.3
2	20.0 ± 8.2	0.0 ± 0.0	54.5 ± 3.3	34.3 ± 4.3	13.9 ± 0.8
4	10.0 ± 5.5	2.0 ± 2.0	28.0 ± 2.3	14.4 ± 1.4	31.9 ± 0.9
10	2.0 ± 1.3	0.0 ± 0.0	10.2 ± 0.7	7.1 ± 1.6	58.9 ± 1.2
20	2.5 ± 0.8	1.0 ± 0.7	3.4 ± 0.7	5.2 ± 0.5	95.2 ± 2.3
40	2.0 ± 1.0	1.0 ± 0.4	2.3 ± 0.5	3.0 ± 0.3	157.2 ± 2.4
80	2.8 ± 0.5	2.0 ± 0.5	1.9 ± 0.4	2.8 ± 0.6	252.7 ± 2.1

Table 3. Test error rates on Penn Treebank sentences of length ≤ 15

Model	Precision	Recall	F_1
[11]	88.25	87.73	87.99
[4]	*89.18*	88.20	88.69
Basic	88.08	88.31	88.20
Lexical+Aux	89.14	*89.10*	*89.12*

considered. For other details about the experimental setup we remind the reader to the original paper [24] while here we simply discuss the results.

The first two models are generative approaches previously proposed by Klein and Manning [11] and Collins [4] and they are considered as baseline. The first is an unlexicalized model while Collins used also lexical features therefore superior performance are obtained. Among the discriminative approaches two different max-margin models have been considered. In the first model, referred as "basic", the same features of the model by Klein and Manning [11] are used but weights are estimated with a large margin criterion, so there is a small increase in terms of accuracy (F_1 raised from 87.99 to 88.20). The great improvement in terms of performance ($F_1 = 89.12$) is obtained considering a richer set of features, such as lexical features and other features given by the output of auxiliary classifiers. This model is referred as "Lexical+Aux" in the table. Then again the max-margin approach is advantageous with respect to traditional generative models in terms of accuracy. On the other hand it is worth noting that the training phase is rather expensive in terms of computational cost especially for large training set size as in this case (about 9700 sentences).

6 Discussion and Conclusions

In this paper a survey of max-margin methods for structured output learning problems has been presented. The main learning algorithms such as in [1, 23, 25] have been summarized along with some associated experimental results. From a theoretical point of view we briefly discussed the generalization performance showing the bound in [23].

Max-margin methods for structured output domains have proven their effectiveness in most applications and the reason for their success is twofold. First, opposite to previous generative approaches, the learning problem is treated in a discriminative fashion and the optimal parameters are obtained following the ERM principle and avoiding any modeling assumption. Furthermore, compared with other discriminative methods as CRFs, the max margin property makes them robust to overfitting and the kernel trick allows to handle high dimensional feature spaces and to gain increased performances.

Moreover, with respect to CRFs, the max-margin approaches overcome the bottleneck given by the computation of the partition function which is well known to be a #P-complete task. This allows to apply max-margin algorithms to a wider range of problems with combinatorial structures.

Another favorable property of large margin approaches is the sparsity of the solution which can be exploited to gain algorithmic efficiency. However still the main challenge with large margin methods in structured output domain is represented by the complexity of the learning process since most of the algorithms do not scale to very large training sets which are needed in practical applications as NLP or computational biology problems.

Perhaps other open issues that need further works in the future are the study of efficient inference techniques, a deepen analysis of the generalization performance and the development of methods dealing with hidden variables, imbalanced data and semi-supervised learning.

References

1. Altun, Y., Tsochantaridis, I., Hofmann, T.: Hidden markov support vector machines. In: 20th International Conference on Machine Learning (ICML) (2003)
2. Bagnell, J.A., Ratliff, N., Zinkevich, M.: Maximum margin planning. In: Proceedings of the International Conference on Machine Learning (ICML) (2006)
3. Bartlett, P.L., Collins, M., Taskar, B., McAllester, D.: Exponentiated gradient algorithms for large-margin structured classification. In: Advances in Neural Information Processing Systems (NIPS) (2004)
4. Collins, M.: Head-Driven Statistical Models for Natural Language Parsing. Ph.D. thesis, University of Pennsylvania (1999)
5. Collins, M.: Discriminative training methods for hidden markov models: Theory and experiments with perceptron algorithms. In: Proc. Conference on Empirical Methods in Natural Language Processing (EMNLP) (2002)
6. Collins, M.: Parameter estimation for statistical parsing models: Theory and practice of distribution-free methods. In: IWPT (2001)
7. Crammer, K., Singer, Y.: On the algorithmic implementation of multiclass kernel-based vector machines. Journal of Machine Learning Research 2, 265–292 (2001)
8. Cristianini, N., Shawe-Taylor, J.: An Introduction to Support Vector Machines and other kernel-based learning methods. Cambridge University Press, Cambridge (2000)

9. Fletcher, R.: Practical Methods of Optimization. 2nd edn. John Wiley & Sons, Chichester (1987)
10. Kassel, R.: A Comparison of Approaches to On-line Handwritten Character Recognition. PhD thesis, MIT Spoken Language Systems Group (1995)
11. Klein, D., Manning, C.D.: Accurate unlexicalized parsing. In: ACL, vol. 41, pp. 423–430 (2003)
12. Kumar, S., Hebert, M.: Discriminative random fields: A discriminative framework for contextual interaction in classification. In: IEEE International Conference Computer Vision (ICCV), vol. 2, pp. 1150–1157 (2003)
13. Lafferty, J., Pereira, F., McCallum, A.: Conditional random fields: Probabilistic models for segmenting and labeling sequence data. In: International Conference on Machine Learning (ICML) (2001)
14. Lafferty, J., Zhu, X., Liu, Y.: Kernel conditional random fields: representation and clique selection. In: Proceedings of the Twenty-first International Conference on Machine Learning (ICML) (2004)
15. McCallum, A., Freitag, D., Pereira, F.: Maximum entropy Markov models for information extraction and segmentation. In: Proceedings of the International Conference on Machine Learning (ICML) (2000)
16. Nesterov, Y.: Dual extrapolation and its application for solving variational inequalites and related problems. Technical report, CORE, Catholic University of Louvain (2003)
17. Pinto, D., McCallum, A., Wei, X., Bruce Croft, W.: Table extraction using conditional random fields. In: Proceedings of the 26th ACM SIGIR conference, pp. 235–242 (2003)
18. Platt, J.: Using analytic QP and sparseness to speed training of support vector machines. In: Advances in Neural Information Processing Systems (NIPS), pp. 557–563 (1999)
19. Quattoni, A., Collins, M., Darrel, T.: Conditional random fields for object recognition (2005)
20. Rabiner, L.R.: A Tutorial on Hidden Markov Models and Selected Applications in Speech Recognition. Proceedings of the IEEE 77, 257–286 (1989)
21. Rousu, J., Saunders, C., Szedmak, S., Shawe-Taylor, J.: Learning hierarchical multi-category text classification models. In: Proc. of the International Conference on Machine Learning (ICML), pp. 744–751 (2005)
22. Sha, F., Pereira, F.: Shallow parsing with conditional random fields. In: Proceedings of Human Language Technology Conference / North American Chapter of the Association for Computational Linguistics annual meeting (HLT/NAACL) (2003)
23. Taskar, B., Guestrin, C., Koller, D.: Max-margin markov networks. In: Neural Information Processing Systems (NIPS) (2003)
24. Taskar, B., Klein, D., Collins, M., Koller, D., Manning, C.: Max-margin parsing. In: Proceedings of the Conference on Empirical Methods in Natural Language Processing (EMNLP) (2004)
25. Taskar, B., Jordan, M., Lacoste-Julien, S.: Structured prediction via the extragradient method. In: Neural Information Processing Systems (NIPS) (2005)
26. Tsochantaridis, I., Hofmann, T., Joachims, T., Altun, Y.: Support vector machine learning for interdependent and structured output spaces. In: Proceedings of the International Conference on Machine Learning (ICML) (2004)
27. Vapnik, V.: Statistical Learning Theory. Wiley, Chichester (1998)

28. Vishwanathan, S.V.N., Schraudolph, N.N., Schmidt, M.W., Murphy, K.: Accelerated training of conditional random fields with stochastic meta-descent. In: Proceedings of the International Conference on Machine Learning (ICML) (2006)
29. Wolsey, L.: Integer programming. John Wiley & Sons Inc., Chichester (1998)
30. Zhang, T.: Covering number bounds of certain regularized linear function classes. Journal of Machine Learning Research 2, 527–550 (2002)

Ensemble MLP Classifier Design

Terry Windeatt

Centre for Vision, Speech and Signal Proc., Department of Electronic Engineering,
University of Surrey, Guildford, Surrey, United Kingdom GU2 7XH
t.windeatt@surrey.ac.uk

Abstract. Multi-layer perceptrons (MLP) make powerful classifiers that may provide superior performance compared with other classifiers, but are often criticized for the number of free parameters. Most commonly, parameters are set with the help of either a validation set or cross-validation techniques, but there is no guarantee that a pseudo-test set is representative. Further difficulties with MLPs include long training times and local minima. In this chapter, an ensemble of MLP classifiers is proposed to solve these problems. Parameter selection for optimal performance is performed using measures that correlate well with generalisation error.

1 Introduction

The topic of this chapter concerns solving problems in pattern recognition using a combination of neural network classifiers. Pattern classification involves assignment of an object to one of several pre-specified categories or classes, and is a key component in many data interpretation activities. Here we focus on classifiers that learn from examples, and it is assumed that each example pattern is represented by a set of numbers, which are known as the pattern features. In the case of face recognition (Section 5), these features consist of numbers representing different aspects of facial features. In order to design a learning system it is customary to divide the example patterns into two sets, a training set to design the classifier and a test set, which is subsequently used to predict the performance when previously unseen examples are applied. A problem arises when there are many features and relatively few training examples, and the classifier can learn the training set too well, known as over-fitting so that performance on the test set degrades.

Automating the classification task to achieve optimal performance has been studied in the traditional fields of pattern recognition, machine learning and neural networks as well as newer disciplines such as data fusion, data mining and knowledge discovery. Traditionally, the approach that has been used in the design of pattern classification systems is to experimentally assess the performance of several classifiers with the idea that the best one will be chosen. Ensemble classifiers, also known as Multiple Classifier Systems (MCS), were developed to address the problem of reliably designing a system with improved accuracy. Recognising that each classifier may make different and perhaps complementary errors, the idea is to pool together the results from all classifiers to find a composite system that outperforms any individual (base) classifier. In this way a single complex classifier may be replaced by a set of relatively simple classifiers.

L.C. Jain et al. (Eds.): Comp. Intel. Para.: Innov. Applications, SCI 137, pp. 133–147, 2008.
springerlink.com

Even though an ensemble is less likely to over-fit, there is still the difficulty of tuning individual classifier parameters. Multi-layer perceptrons (MLP) make powerful classifiers that may provide superior performance compared with other classifiers, but are often criticized for the number of free parameters. The common approach to adjusting parameters is to divide the training set into two to produce a validation set. When the number of examples is in short supply, cross-fold validation may be used. For example, in ten-fold cross-validation, the set is randomly split into ten equal parts with nine parts used for training and one part used as a validation set to tune parameters. Training is repeated ten times with a different partition each time, and the results averaged. However, it is known that these approaches to validation are either inappropriate or very time-consuming. Ideally all the training set should be used for training, so that there is no need for validation. However, this requires that over-fitting be detected by looking at performance on only the training set, which is a difficult problem.

It is known that certain conditions need to be satisfied to realise ensemble performance improvement, in particular that the constituent classifiers be not too highly correlated. If each classifier solves a slightly different problem, then composite performance may improve. Various techniques have been devised to reduce correlation by injecting randomness. For example, two techniques that are used in this chapter are Bagging [1] (Section 2), which resamples the training set and Error-Correcting-Output-Coding (ECOC Section 4), which solves multi-class problems by randomly decomposing into two-class problems.

Although it is known that diversity among base classifiers is a necessary condition for improvement in ensemble performance, there is no general agreement about how to quantify the notion of diversity among a set of classifiers. Diversity Measures can be categorised into pair-wise (Section 3) and non-pair-wise, and to apply pair-wise measures to finding overall diversity it is necessary to average over the classifier set. These pair-wise diversity measures are normally computed between pairs of classifiers independent of target labels. As explained in [14], the accuracy-diversity dilemma arises because when base classifiers become very accurate their diversity must decrease, so that it is expected that there will be a trade-off.

A possible way around this dilemma is to incorporate diversity and accuracy to produce a single class separability measure, as suggested in Section 3. The measure is based on a spectral representation that was first proposed for two-class problems in [2], and later developed in the context of Multiple Classifier Systems in [3]. It was shown for two-class problems in [4] that over-fitting of the training set could be detected by observing the separability measure as it varies with base classifier complexity. Since realistic learning problems are in general ill-posed [5], it is known that any attempt to automate the learning task must make some assumptions. The only assumption required here is that a suitable choice be made for the range over which base classifier complexity is varied.

2 MLP Classifiers and Ensembles

There are many text books that contain an introduction to MLP classifiers. Since the topic of neural networks is multi-disciplinary, it is useful to find a text that is written from a stand-point similar to the reader. For students of engineering and computer science, reference [6] is recommended.

A multi-layer rather than a single layer network is required since a single layer perceptron (SLP) can only compute a linear decision boundary, which is not flexible enough for most realistic learning problems. For a problem that is linearly separable, (that is capable of being perfectly separated by linear decision boundary), the perceptron convergence theorem guarantees convergence. In its simplest form, SLP training is based on the simple idea of adding or subtracting a pattern from the current weights when the target and predicted class disagrees, otherwise the weights are unchanged. For a non-linearly separable problem, this simple algorithm can go on cycling indefinitely. The modification known as least mean square (LMS) algorithm uses a mean squared error cost function to overcome this difficulty, but since there is only a single perceptron, the decision boundary is still linear.

An MLP is a universal approximator [6] that typically uses the same squared error function as LMS. However, the main difficulty with the MLP is that the learning algorithm has a complex error surface, which can become stuck in local minima. There does not exist any MLP learning algorithm that is guaranteed to converge, as with SLP. The popular MLP back-propagation algorithm has two phases, the first being a forward pass, which is a forward simulation for the current training pattern and enables the error to be calculated. It is followed by a backward pass, that calculates for each weight in the network how a small change will affect the error function. The derivative calculation is based on the application of the chain rule, and training typically proceeds by changing the weights proportional to the derivative. For the back-propagation equations see reference [6].

Although there are different ways of trying to overcome the local minima problem [6], in this chapter an ensemble of MLPs is proposed. An MLP with random starting weights is a suitable base classifier since randomisation has shown to be beneficial in the MCS context. Problems of local minima and computational slowness may be alleviated by the MCS approach of pooling together the decisions obtained from

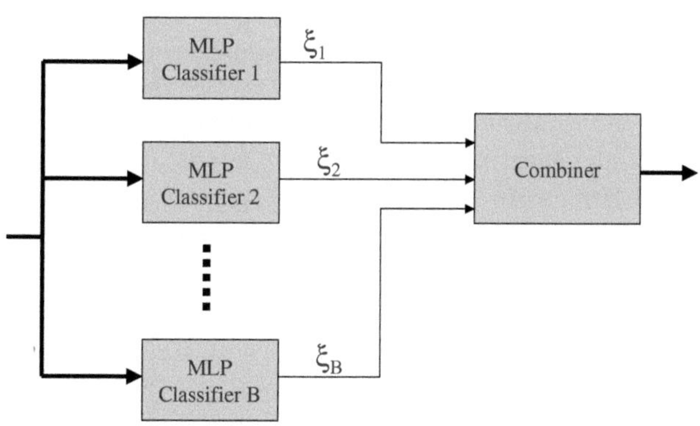

Fig. 1. Ensemble MLP architecture, showing base MLP classifier decisions

locally optimal classifiers. The architecture envisaged is a simple MCS framework in which there are B parallel MLP base classifiers, as shown in figure 1.

Bootstrapping [7] is a popular ensemble technique that provides a beneficial increase in diversity among base classifiers. The implication in bootstrapping is that if μ training patterns are randomly sampled with replacement, $(1-1/\mu))^{\mu} \cong 37\%$ are removed with remaining patterns occurring one or more times. The base classifier out-of-bootstrap (OOB) estimate uses the patterns left out, and should be distinguished from the ensemble OOB. For the ensemble OOB, all training patterns contribute to the estimate, but the only participating classifiers for each pattern are those that have not been used with that pattern for training (that is, approximately thirty-seven percent of classifiers). Note that OOB gives a biased estimate of the absolute value of generalization error [8], but here the estimate of the absolute value is not important. The OOB estimate for the ECOC ensemble is given in Section 4. In Section 5 we demonstrate by example that the OOB estimate can predict the optimal base classifier complexity.

3 Diversity/Accuracy and MCS

Attempts to understand the effectiveness of the MCS (ensemble) framework have prompted the development of various diversity measures with the intention of determining whether they correlate with ensemble accuracy. However, the question of whether the information available from these measures can be used to assist MCS design is open. Most commonly, ensemble parameters are set with the help of either a validation set or cross-validation techniques [9]. In this section, we review the definition of a popular pair-wise diversity measure (equation (6)), and show that diversity and accuracy can be incorporated within a single class separability measure.

Classical class separability measures refer to the ability to predict separation of patterns into classes using original features and rely on a Gaussian assumption [10]. The problem with applying classical class separability measures to the binary mapping associated with MCS (equation ((1))) is that the implicit Gaussian assumption is not appropriate [11]. In [4], [12] a class separability measure is proposed for MCS that is based on a binary feature representation, in which each pattern is represented by its binary ensemble classifier decisions. It is restricted to two-class problems and results in a binary-to-binary mapping.

Let there be μ training patterns with the label ω_m given to each pattern $\boldsymbol{x_m}$ where $m = 1,\dots \mu$. In the MCS framework shown in figure 1, the mth pattern may be represented by the B-dimensional vector formed from the B base classifier decisions given by

$$x_m = (\xi_{m1}, \xi_{m2}, \dots, \xi_{mB}) \qquad \xi_{mi}, \omega_m \in \{0,1\}, \quad i = 1 \dots B \tag{1}$$

In equation (1) $\omega_m = f(\boldsymbol{x_m})$ where f is the unknown binary-to-to binary mapping from classifier decisions to target label. Following [13] the notation in equation (1) is modified so that the classifier decision is 1 if it agrees with the target label and 0 otherwise

$$x_m = (y_{m1}, y_{m2}, \dots, y_{mB}) \qquad y_{mi}, \omega_m = \in \{0,1\}, \; y_{mi}=1 \text{ iff } \xi_{mi} = \omega_m \tag{2}$$

Pair-wise diversity measures, such as Q statistic (equation (6)), Correlation Coefficient, Double Fault and Disagreement measures [13] take no account of class assigned to a pattern. In contrast, class separability [14] is computed between classifier decisions (equation (2)) over pairs of patterns of opposite class, using four counts defined by logical AND ($\wedge$) operator

$$\tilde{N}^{ab}_{mn} = \sum_{j=1}^{B} \psi^{a}_{mj} \wedge \psi^{b}_{nj}, \quad \omega_m \neq \omega_n \quad a,b \in \{0,1\},\ \psi^1 = y, \psi^0 = \overline{y} \tag{3}$$

The *nth* pattern for a two-class problem is assigned

$$\sigma'_n = \frac{1}{\tilde{K}_\sigma}\left(\frac{\tilde{N}_n^{11}}{\sum\limits_{m=1}^{\mu} \tilde{N}_m^{11}} - \frac{\tilde{N}_n^{00}}{\sum\limits_{m=1}^{\mu} \tilde{N}_m^{00}} \right) \tag{4}$$

where
$$\tilde{K}_\sigma = \left(\frac{\tilde{N}_n^{11}}{\sum\limits_{m=1}^{\mu} \tilde{N}_m^{11}} + \frac{\tilde{N}_n^{00}}{\sum\limits_{m=1}^{\mu} \tilde{N}_m^{00}} \right), \tilde{N}_n^{ab} = \sum_{m=1}^{\mu} \tilde{N}_{mn}^{ab}$$

The motivation for σ'_n in equation (4) comes from estimation of the first order spectral coefficients [4] of the binary-to-binary mapping defined in equation (1). Each pattern is compared with all patterns of the other class, and the number of jointly correctly ($\tilde{N}_n^{11}$) and incorrectly ($\tilde{N}_n^{00}$) classified patterns is counted. Note that a classifier that correctly classifies one pattern but incorrectly classifies the other does not contribute. The two terms in equation (4) represent the relative positive and negative evidence that the pattern comes from the target class. A simple example for nine classifier pairs demonstrates the idea.

classifier pair	1 2 3 4 5 6 7 8 9	
	0 0 1 1 1 0 0 1 1	class ω_1
	1 1 1 0 1 0 0 1 1	class ω_2

If 1 indicates correct classification, the contradictory evidence from classifier pair 1, 2, 4 is ignored, evidence from classifiers 3, 5, 8, 9 is for positive correlation with change of class and the remaining classifiers give evidence for negative correlation. The evidence is summed over all pairs, as given in equations (3) and (4), where a pair must contain a pattern from each class.

Furthermore, we sum over patterns with positive coefficient to produce a single number between 0 and 1 that represents the separability of a set of patterns

$$\sigma' = \sum_{n=1}^{\mu} \sigma'_n, \sigma'_n > 0 \tag{5}$$

In our examples in Section 5 we compare the Q diversity measure, as recommended in [13]. Diversity Q_{ij} between *ith* and *jth* classifiers is defined as

$$Q_{ij} = \frac{N_{ij}^{11} N_{ij}^{00} - N_{ij}^{01} N_{ij}^{10}}{N_{ij}^{11} N_{ij}^{00} + N_{ij}^{01} N_{ij}^{10}} \tag{6}$$

where $N_{ij}^{ab} = \sum_{m=1}^{\mu} \psi_{mj}^{a} \wedge \psi_{mj}^{b}$ with a,b, Ψ defined in equation (3). The mean is taken over B base classifiers $Q = \frac{2}{B(B-1)} \sum_{i=1}^{B-1} \sum_{j=i+1}^{B} Q_{ij}$.

4 Error Correcting Output Coding (ECOC) and Multi-class Problems

A single MLP with multiple outputs can handle directly a K-class problem, K > 2. The standard technique is to use a K-dimensional binary target vector that represents each one of K classes using a single binary value at the corresponding position, for example [0,…,0,1,0,…..0]. In practice the target value is often specified as 0.9 and 0.1 rather than 1 and 0, to stop the MLP from saturating at the extremities of the sigmoid activation function, for which the gradient is zero [6]. The reason that a single multi-class MLP is not a suitable candidate for use with a more elaborate coding scheme is that all nodes share in the same training, so errors are far from independent and there is not much benefit to be gained from combining.

There are several motivations for decomposing a multi-class problem into complementary two-class problems. The decomposition means that attention can be focused on developing an effective technique for the two-class MLP classifier, without having to consider explicitly the design of the multi-class case. Also, it is hoped that the parameters of an MLP base classifier run several times may be easier to determine than a single complex MLP classifier run once, and perhaps facilitate faster and more efficient solutions. Finally, solving different two-class sub-problems repeatedly with random perturbation may help to reduce error in the original problem.

The ECOC method [15] is an example of distributed output coding [16], in which a pattern is assigned to the class that is closest to a corresponding code word. Rows of the ECOC matrix act as the code words, and are designed using error-correcting principles to provide some error insensitivity with respect to individual classification errors. The original motivation for encoding multiple classifiers using an error-correcting code was based on the idea of modelling the prediction task as a communication problem, in which class information is transmitted over a channel. Errors introduced into the process arise from various stages of the learning algorithm, including features selected and finite training sample. From error-correcting theory, we know that a matrix designed to have d bits error-correcting capability implies that there is a minimum Hamming Distance $2d+1$ between any pair of code words. Assuming each bit is transmitted independently, it is then possible to correct a received pattern having fewer than d bits in error, by assigning the pattern to the code word closest in Hamming distance. Clearly, from this perspective it is desirable to use a matrix containing code words having high minimum Hamming distance between any pair.

To solve a multi-class problem in the ECOC framework we need a set of codes to decompose the original problem, a suitable two-class base classifier, and a decision-making framework. For a K-class problem, each row of the $K \times B$ binary ECOC matrix Z acts as a code word for each class. Each of the B columns of Z partitions the training data into two 'superclasses' according to the value of the corresponding binary element. Consider the following 6 x B code matrix for a six-class problem.

$$\begin{bmatrix} 0 & 1 & 1 & 0 & \cdots \\ 1 & 0 & 0 & 0 & \cdots \\ 0 & 1 & 0 & 0 & \cdots \\ 1 & 0 & 1 & 1 & \cdots \\ 1 & 1 & 0 & 1 & \cdots \\ 0 & 0 & 1 & 1 & \cdots \end{bmatrix}$$

The first column would naturally be interpreted as putting patterns from class 2, 4, 5 into one 'superclass' and remaining patterns in the second 'superclass'. Columns 2,3,4 ... correspond to different classifier runs and represent different two-class decompositions.Consider the decoding step for a three class problem, represented in figure 2. To classify a pattern $\boldsymbol{x}_m$, it is applied to the B trained base classifiers as shown in figure 2, forming vector $[x_{m1}, x_{m2}, ..., x_{mB}]$ where x_{mj} is the output of the *jth* base classifier. The L^1 norm distance L_i (where i = 1.... K) between output vector and code word for each class is computed

$$L_i = \sum_{j=1}^{b} \left| Z_{ij} - x_{mj} \right| \tag{7}$$

and $\boldsymbol{x}_m$ is assigned to the class ω_m corresponding to closest code word .

To use the ensemble OOB estimate, pattern $\boldsymbol{x}_m$ is classified using only those classifiers that are in the set OOB_m, defined as the set of classifiers for which $\boldsymbol{x}_m$ is OOB. For the OOB estimate, the summation in equation (7) is therefore modified to $\sum_{j \in OOB_m}$. Therefore only a subset of the columns of Z is used in the decoding step in figure 2, so that approximately B/3 columns are used for each pattern.

The topic of designing suitable code matrices has received much attention. In [17] it is shown that any variation in Hamming distance between pairs of code words will reduce the effectiveness of the combining strategy. In [18] it is shown that maximising the minimum Hamming Distance between code words implies minimising upper bounds on generalisation error. In classical coding theory, theorems on error-correcting codes guarantee a reduction in the noise in a communication

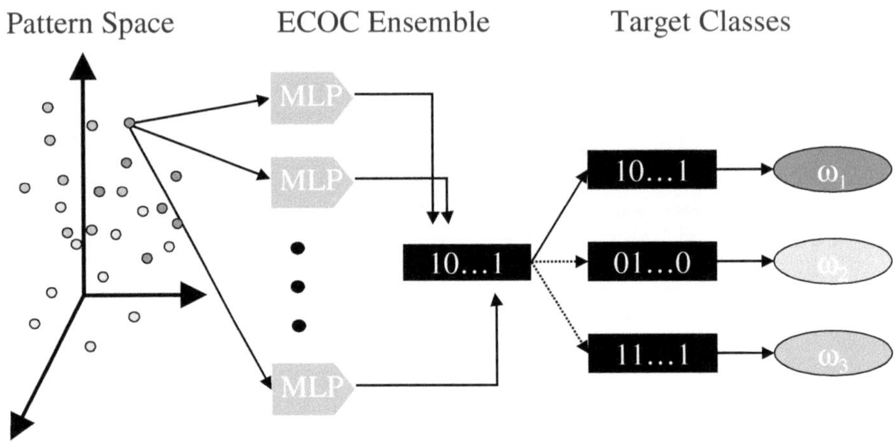

Fig. 2. ECOC Ensemble showing code words for target classes of 3 class problem

channel, but the assumption is that errors are independent. When applied to machine learning the situation is more complex, in that error correlation depends on the data set, base classifier as well as the code matrix Z. In the original ECOC approach [15], heuristics were employed to maximise the distance between the columns of Z to reduce error correlation. Random codes, provided that they are long enough, have frequently been employed with almost as good performance [17]. It would seem to be a matter of individual interpretation whether long random codes may be considered to approximate required error-correcting properties. In this chapter, a random code matrix with near equal split of classes (approximately equal number of 1's in each column) is chosen, as proposed in [19]. Note that some papers prefer to refer to the method as Output Coding, in recognition of the discussion over whether error-correcting properties are significant.

5 Examples

The following examples use natural benchmark data that demonstrate the usefulness of the measures proposed in this chapter. The main purpose of these examples is to demonstrate how well the measures correlate with test error as the number of training epochs of single hidden-layer MLP base classifiers are systematically varied. All the measures are computed on the training data and the datasets use random training/testing split, respecting the class distribution as closely as possible. Experiments are repeated ten times, except for the face database which is repeated twenty times. The split is 50/50 for the ORL face database (described in Section 5.2) and implies five training images per class, while other datasets are split 20/80. The ensemble MLP architecture is shown in figure 1. Each base classifier run uses identical classifier parameters, with variation arising from three sources, (i) random

Table 1. Benchmark Datasets showing numbers of patterns, classes, continuous and discrete features

DATASET	#pat	#class	#con	#dis
diabetes	768	2	8	0
dermatology	366	6	1	33
ecoli	336	8	5	2
glass	214	6	9	0
iris	150	3	4	0
segment	2310	7	19	0
soybean	683	19	0	35
vehicle	846	4	18	0
vowel	990	11	10	1
wave	3000	3	21	0
yeast	1484	10	7	1

initial weights, (ii) bootstrapping, and (iii) for multi-class, the problem decomposition defined by the respective code matrix column.

Natural benchmark problems, selected from [20] and [21] are shown in table 1 with numbers of patterns, classes, continuous and discrete features. For datasets with missing values the scheme suggested in [20] is used. The two-class dataset 'diabetes' uses one hundred classifiers (B = 100), multi-class problems use two hundred classifiers (K x 200 coding matrix), and the face database uses five hundred classifiers (40 x 500 code matrix). The Resilient BackPropagation (RPROP) [22] and Levenberg-Marquardt (LM) [6] algorithms are used. The only observed difference between the two algorithms is in the number of epochs required for optimal performance, indicating that the MLP ensemble is insensitive to the convergence properties of the base algorithm. In the examples presented, RPROP is used for the face database and LM for benchmark data. For further details see reference [12].

5.1 Benchmark Data

Figure 3 shows 'diabetes' 20/80, a difficult dataset that is known to over-fit with increasing classifier complexity. Figure 3 (a) (b) shows base classifier and ensemble test error rates, (c) (d) the base classifier and ensemble OOB estimates and (e) (f) the measures σ', Q defined in equations (5) and (6) for various node-epoch combinations. It may be seen that σ' and base classier OOB are good predictors of base classifier test error rates as base classifier complexity is varied. The correlation between σ' and test error was thoroughly investigated in [12], showing high values of correlation coefficient (0.98) that were significant (95 % confidence when compared with random chance). In [12] it was also shown that bootstrapping did not significantly change the ensemble error rates, actually improving them slightly on average.

The class separability measure σ' shows that the base classifier test error rates are optimised when the number of epochs is chosen to maximize class separability. Furthermore, at the optimal number of epochs Q shows that diversity is minimized. It

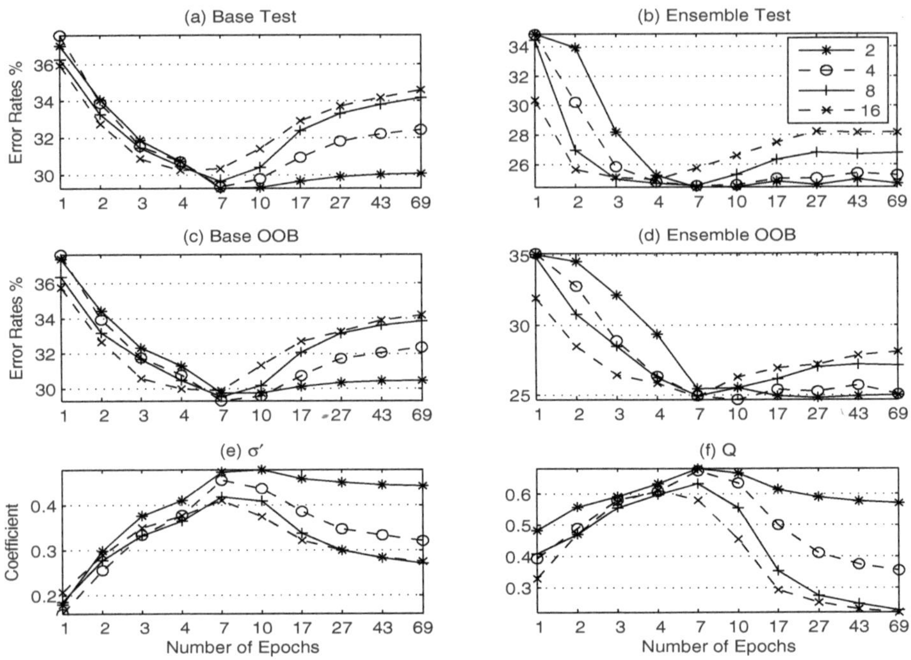

Fig. 3. Mean test error rates, OOB estimates, measures σ', Q for Diabetes 20/80 with [2,4,8,16] nodes

appears that base classifiers starting from random weights increase correlation (reduce diversity) as complexity is increased and correlation peaks as the classifier starts to over-fit the data. A possible explanation of the over-fitting behaviour is that classifiers produce different fits of those patterns in the region where classes are overlapped. From equation (4) σ' is the probability that $\left(\frac{\tilde{N}_n^{11}}{\sum_{m=1}^{\mu} \tilde{N}_m^{11}} - \frac{\tilde{N}_n^{00}}{\sum_{m=1}^{\mu} \tilde{N}_m^{00}} \right) > 0$. It provides an intuitive explanation of why we may expect that σ' correlates well with base classifier test error, and in particular peaks at the same number of training epochs. Consider the example of two overlapping Gaussians representing a two-class problem, with the Bayes boundary assumed to be placed where the probability density curves cross. Let the overlapping region be defined as the tails of the two Gaussians with respect to the Bayes boundary. If base classifiers are capable of approximating the Bayes boundary, by definition an optimal base classifier will incorrectly classify all patterns in the overlapping region and correctly classify all other patterns. Now consider the situation that the complexity of the base classifiers increases beyond optimal, so that some patterns in the overlapping region become correctly classified and some of the remaining patterns become incorrectly classified. The result is that there is greater variability in classification among patterns close to the Bayes boundary, and it is more difficult to separate them. The probability represented by σ' decreases as complexity increases since $\widetilde{N}^{00}$ is more evenly distributed over all patterns, leading to a reduction in positively correlated patterns. However, if the base classifier becomes too powerful, eventually all patterns are correctly classified and $\widetilde{N}^{00} \rightarrow 0$ and $\sigma' \rightarrow 1$, so it is expected that σ' would start to increase.

Note from Figure 3 that the ensemble is more resistant to over-fitting than base classifier for epochs greater than 7, and the ensemble OOB accurately predicts this trend. This experiment was performed for all datasets, and in general the ensemble test error was found to be more resistant to over-fitting. Figure 4 shows similar curves to Figure 3 averaged over all multi-class datasets (for multi-class, the correlation coefficient between σ' and test error is 0.96).

5.2 Face Data

Facial images are a popular source of biometric information since they are relatively easy to acquire. However, automated face recognition systems often perform poorly due to small number of relatively high-dimensional training patterns, which can lead to poor generalisation through over-fitting. Face recognition is an integral part of systems designed for many applications including identity verification, security, surveillance and crime-solving. Improving their performance is known to be a difficult task, but one approach to improving accuracy and efficiency is provided by the method of ECOC ensemble classifiers [23].

A typical face recognition system consists of three functional stages. In the first stage, the image of a face is registered and normalised. Since face images differ in both shape and intensity, *shape alignment* (geometric normalisation) and *intensity correction* (photometric normalisation) can improve performance. The second stage is feature extraction in which discriminant features are extracted from the face region. Finally, there is the matching stage in which a decision-making scheme needs to be designed depending on the task to be performed. In identification, the system classifies a face from a database of known individuals, (while in verification the system should confirm or reject a claimed identity).

There are two main approaches to feature extraction. In the geometric approach, relative positions and other parameters of distinctive features such as eyes, mouth and nose are extracted. The alternative is to consider the global image of the face as with methods such as Principal Component Analysis (PCA). By itself PCA is not adequate for the face recognition task since projection directions only maximise the total scatter across all classes. Therefore PCA may be combined with Linear Discriminant Analysis (LDA), which requires computation of the between-class scatter matrix, S_B and the within-class scatter matrix, S_W. The objective of LDA is to find the transformation matrix, W_{opt}, that maximises the ratio of determinants $\left|w^T S_B w\right| / \left|w^T S_W w\right|$. W_{opt} is known to be the solution of the following eigenvalue problem $S_B - S_W\Lambda = 0$, where Λ is a diagonal matrix whose elements are the eigenvalues of matrix $S_W^{-1} S_B$.

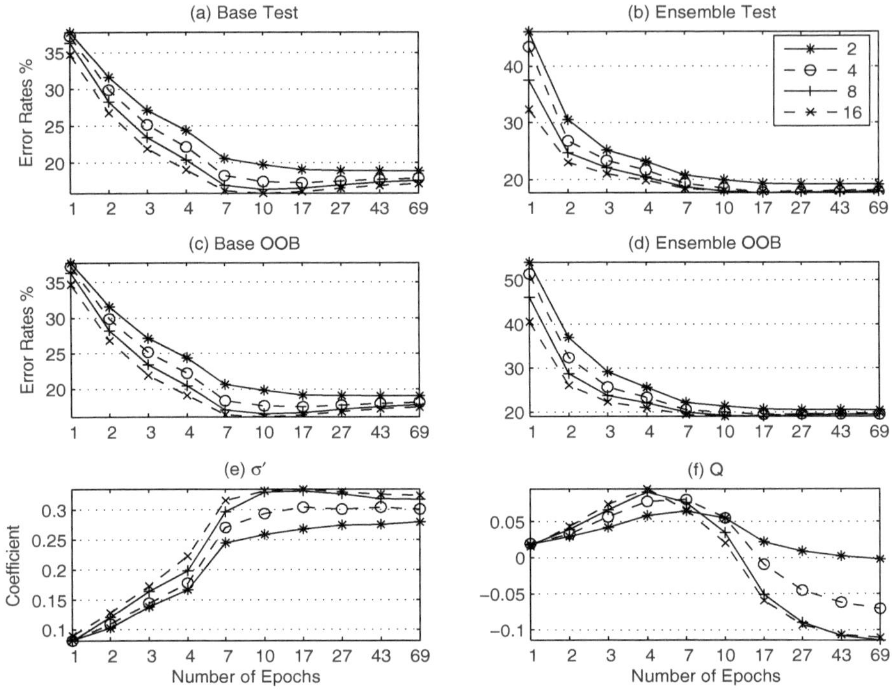

Fig. 4. Mean test error rates, OOB estimates, measures σ', Q over ten multi-class 20/80 datasets with [2,4,8,16] nodes

The face database used is the ORL (Olivetti Research Laboratory http//www.cam-orl.co.uk), consisting of four hundred images of forty individual faces with some variation in lighting, facial expression, facial hair, pose and spectacles. The background is controlled with subjects in an upright frontal position, although small variation in rotation and scale is allowed. The advantage of this database is that it can be used without need for a face detection algorithm or any other pre-processing, so that there is a fairer comparison with the results obtained by other researchers. In our examples, images have been projected to forty-dimensions using PCA and subsequently to a twenty-dimension feature space using LDA. It is treated as a forty-class face identification problem with the four hundred images randomly split into 50/50 training/testing patterns. A comparison of results on this database is given in [24], using 50/50 split, but the number of runs is smaller than used in our examples. As pointed out in [24], some researchers do not specify the split that they have used, and some only base their results on one run, so that care is needed before making any comparison.

Figure 5 shows error rates, σ', Q for 50/50 60/40 70/30 and 80/20 splits with 16 nodes and bootstrapping applied. As with two-class and multi-class benchmark problems, σ' appears to be strongly correlated with test error.

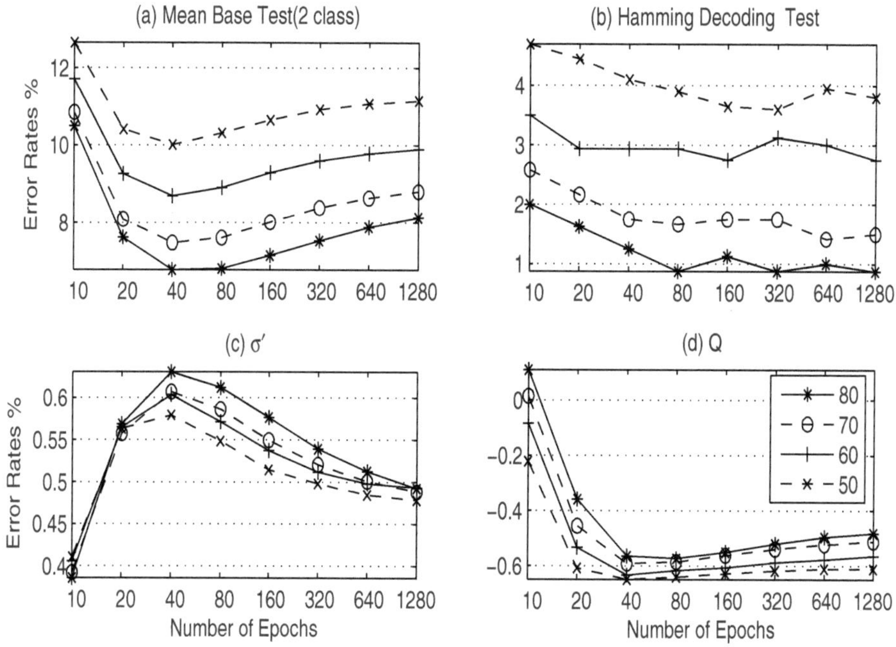

Fig. 5. Test error, σ′, Q for ORL database using 16 hidden-node bootstrapped base classifiers for [50/50,60/40,70/30,80/20] train/test splits

6 Discussion

There is another aspect of ensemble classifier design that needs to be addressed, the problem of feature selection. The aim of feature selection is to find a feature subset from the original set of features such that an induction algorithm that is run on data containing only those features generates a classifier that has the highest possible accuracy [25]. Typically, an exhaustive search is computationally prohibitive, and the problem is known to be NP-hard, so that a greedy search scheme is required. For problems with a large number of features, classical feature selection schemes are not greedy enough, and filter, wrapper and embedded approaches have been developed [26]. One-dimensional feature ranking methods consider each feature in isolation and rank the features according to a scoring function, but are disadvantaged by implicit orthogonality assumptions [26]. They are very efficient but in general have been shown to be inferior to multi-dimensional methods that consider all features simultaneously.

The measures proposed in this chapter have been used to select the optimal number of features of an ensemble of MLP classifiers [27]. In [28], a multi-dimensional feature-ranking criterion based on modulus of MLP weights identifies the least relevant features. It is combined with Recursive Feature Elimination (RFE) to recursively remove the irrelevant features until the measures indicate that test error performance degrades if further features are eliminated.

7 Conclusion

Multi-layer perceptrons (MLP) make powerful classifiers that may provide superior performance compared with other classifiers, but are often criticized for the number of free parameters. In this chapter an ensemble of relatively simple MLP classifiers is proposed, along with some measures that facilitate ensemble design. The proposed measures facilitate the detection of over-fitting as base classifier complexity is varied. Although the measures are only applicable to two-class problems, they may also be applied to multi-class via the two-class decompositions induced by Error-Correcting Output Coding.

References

1. Breiman, L.: Bagging predictors. Machine Learning 24(2), 123–140 (1997)
2. Windeatt, T., Tebbs, R.: Spectral technique for hidden layer neural network training. Pattern Recognition Letters 18(8), 723–731 (1997)
3. Windeatt, T.: Recursive Partitioning for combining multiple classifiers. Neural Processing Letters 13(3), 221–236 (2001)
4. Windeatt, T.: Vote Counting Measures for Ensemble Classifiers. Pattern Recognition 36(12), 2743–2756 (2003)
5. Tikhonov, A.N., Arsenin, V.A.: Solutions of ill-posed problems. Winston & Sons, Washington (1977)
6. Haykin, S.: Neural Networks: A Comprehensive Foundation. Prentice Hall, Englewood Cliffs (1999)
7. Efron, B., Tibshirani, R.J.: An Introduction to the Bootstrap. Chapman & Hall, Boca Raton (1993)
8. Bylander, T.: Estimating generalisation error two-class datasets using out-of-bag estimate. Machine Learning 48, 287–297 (2002)
9. Hansen, L.K., Salamon, P.: Neural Network Ensembles. IEEE Trans. Pattern Analysis and Machine Intelligence 12(10), 993–1001 (1990)
10. Fukunaga, K.: Introduction to statistical pattern recognition. Academic Press, London (1990)
11. Ho, T.K., Basu, M.: Complexity measures of supervised classification problems. IEEE Trans. PAMI 24(3), 289–300 (2002)
12. Windeatt, T.: Accuracy/Diversity and ensemble classifier design. IEEE Trans. Neural Networks 17(5), 1194–1211 (2006)
13. Kuncheva, L.I., Whitaker, C.J.: Measures of diversity in classifier ensembles. Machine Learning 51, 181–207 (2003)
14. Windeatt, T.: Diversity Measures for Multiple Classifier System Analysis and Design. Information Fusion 6(1), 21–36 (2004)
15. Dietterich, T.G., Bakiri, G.: Solving multiclass learning problems via error-correcting output codes. Journal of Artificial Intelligence Research 2, 263–286 (1995)
16. Sejnowski, T.J., Rosenberg, C.R.: Parallel networks that learn to pronounce english text. Journal of Complex Systems 1(1), 145–168 (1987)
17. Windeatt, T., Ghaderi, R.: Coding and Decoding Strategies for Multi-class Learning Problems. Information Fusion 4(1), 11–21 (2003)

18. Allwein, E.L., Schapire, R.E., Singer, Y.: Reducing Multi-class to Binary: A Unifying Approach for Margin Classifiers. Journal of Machine Learning Research 1, 113–141 (2000)
19. Schapire, R.E.: Using Output Codes to Boost Multi-class Learning Problems. In: 14th Int. Conf. of Machine Learning, pp. 313–321. Morgan Kaufmann, San Francisco (1997)
20. Prechelt, L.: Proben1: A Set of Neural Network Benchmark Problems and Benchmarking Rules, Tech Report 21/94, Univ. Karlsruhe, Germany (1994)
21. Merz, C.J., Murphy, P.M.: UCI Repository of Machine Learning Databases (1998), http://www.ics.uci.edu/~mlearn/MLRepository.html
22. Riedmiller, M., Braun, H.: A Direct Adaptive Method for Faster Backpropagation Learning: The {RPROP} Algorithm. In: Proc. Intl. Conf. on Neural Networks, San Francisco, Calif., pp. 586–591 (1993)
23. Kittler, J., Ghaderi, R., Windeatt, T., Matas, J.: Face Verification via Error Correcting Output Codes. Image and Vision Computing 21(13-14), 1163–1169 (2003)
24. Er, M.J., Wu, S., Toh, H.L.: Face Recognition with RBF Neural Networks. IEEE Trans. Neural Networks 13(3), 697–710 (2002)
25. Kohavi, R., John, G.H.: Wrappers for feature subset selection. Artificial Intelligence Journal, special issue on relevance 97(1-2), 273–324 (1997)
26. Guyon, I., Elisseeff, A.: An introduction to variable and feature selection. Journal of Machine Learning Research 3, 1157–1182 (2003)
27. Windeatt, T., Prior, M.: Stopping Criteria for Ensemble-based Feature Selection. In: Haindl, M., Kittler, J., Roli, F. (eds.) MCS 2007. LNCS, vol. 4472. Springer, Heidelberg (2007)
28. Windeatt, T., Prior, M., Effron, N., Intrator, N.: Ensemble-based Feature Selection Criteria. In: Perner, P. (ed.) MLDM 2007. LNCS (LNAI), vol. 4571. Springer, Heidelberg (2007)

Functional Principal Points and Functional Cluster Analysis

Nobuo Shimizu[1] and Masahiro Mizuta[2]

[1] The Institute of Statistical Mathematics,
4-6-7 Minami-Azabu, Minato-ku, Tokyo 106-8569 Japan
[2] Information Initiative Center, Hokkaido University,
Kita 11 Nishi 5, Kita-ku, Sapporo 060-0811, Japan

Abstract. In this chapter, we deal with functional principal points and functional cluster analysis. The k principal points [4] are defined as the set of k points which minimizes the sum of expected squared distances from every points in the distribution to the nearest points of the set, and are mathematically equivalent to centers of gravity by k-means clustering [3]. The concept of principal points can be extended for functional data analysis [16]. We call the extended principal points *functional principal points.*

Random function [6] is defined in a probability space, and functional principal points of random functions have a close relation to functional cluster analysis. We derive functional principal points of polynomial random functions using orthonormal basis transformation. For functional data according to Gaussian random functions, we discuss the relation between the optimum clustering of the functional data and the functional principal points.

We also evaluate the numbers of local solutions in functional k-means clustering of polynomial random functions using orthonormal basis transformation.

Keywords: Functional data analysis, k-means clustering, Gaussian random functions, Orthogonal basis, Optimization.

1 Introduction

Most conventional data analysis methods assume that data are sets of numbers with some structure, *i.e.* a set of vectors or a set of matrices etc. Nowadays, we must often analyze more complex data. One type of the complex data is functional data; data themselves are represented as functions. For functional data, Ramsay and Silverman proposed functional data analysis (FDA) [13][16]. There are many techniques in FDA including functional regression analysis, functional discriminant analysis, functional principal components analysis, functional canonical correlation analysis, functional multidimensional scaling, functional cluster analysis, and so on [1][2][7][9][10][11][14][16][17][19]. Among them, we focus on functional cluster analysis.

We devide methods of cluster analysis into two groups: hierarchical methods and nonhierarchical methods. Hierarchical clustering refers to the formation of a recursive clustering of the objects data: a partition into two clusters, each of

L.C. Jain et al. (Eds.): Comp. Intel. Para.: Innov. Applications, SCI 137, pp. 149–165, 2008.
springerlink.com

which is itself hierarchically clustered. We usually use dataset of dissimilarities; $S = \{s_{ij}; i, j = 1, 2, \cdots, n\}$, where s_{ij} are dissimilarity between objects i and j, and n is a number of the objects. Single linkage is a typical hierarchical clustering method. The results of hierarchical clustering are usually represented by dendrogram. Nonhierarchical clustering partitions the data based on a specific criterion. Most nonhierarchical clustering methods do not deal with a set of dissimilarities directly. They use a set of p-tuples: $X = \{x_i; i = 1, 2, \cdots, n\}$. K-means method is a kind of nonhierarchical clustering and is to choose initial seeds points and assign the cases to them using a minimum the trace of within variances.

From the view points of functional cluster analysis, there are also two groups of methods: hierarchical methods and nonhierarchical methods. Functional k-means method, which is an extension of k-means method for functional data, is a typical one.

Principal points [4] are equivalent to optimal 'cluster means' in a probability distribution by k-means clustering. For example, in the Swiss army, there had been a practical problem of determining optimal sizes and shapes of newly designed protection masks. The proposed solution to the mask fitting problem, which is computationally equivalent to k-means clustering, led to the question of how to find optimal 'cluster means' for a homogeneous population with some probability distribution. Flury gave the optimal 'cluster means' a new name *principal points* [4]. The research to which some typical patterns are taken out of a lot of weather maps by using principal points was done [12].

Tarpey and Kinateder [20] extended the concept of principal points for functional data analysis. We call the extended principal points *functional principal points*. Whereas principal points are defined for a probability distribution, functional principal points are defined for a random function.

In this chapter, we deal with the relation between functional principal points and functional cluster analysis. In Section 2, we introduce the definitions of principal points and functional principal points, functional k-means clustering algorithm and orthonormal basis transformation of random functions. In Section 3, we show some examples of functional principal points for typical random functions. In Section 4, we classify n objects of functional data following the random function ξ into k clusters, and compare the k centers of the optimal clusters with the k functional principal points of ξ. In Section 5, we calculate the numbers of local solutions in functional k-means clustering of n functional data following the random function ξ.

2 Preliminaries

2.1 Definition of Principal Points

We describe the definitions of principal points and the domain of attraction according to Flury [4][5].

Let $W = \{\boldsymbol{y}_1, \cdots, \boldsymbol{y}_k\}, \boldsymbol{y}_j \in R^p$ $(1 \leq j \leq k)$ denote a set of k points in p-dimensional space. The minimal distance between $\boldsymbol{x} \in R^p$ and W is defined as

$$d(\boldsymbol{x}|\boldsymbol{y}_1,\cdots,\boldsymbol{y}_k) = \min_{1\le j\le k}\{(\boldsymbol{x}-\boldsymbol{y}_j)'(\boldsymbol{x}-\boldsymbol{y}_j)\}^{1/2}.$$

Definition 1 (Flury[4]). $\boldsymbol{\xi}_j \in R^p$ $(1 \le j \le k)$ *are called k principal points of p-variate random variable X, which is distributed according to p-variate probability distribution F if*

$$E_F\{d^2(X|\boldsymbol{\xi}_1,\cdots,\boldsymbol{\xi}_k)\} = \min_{\boldsymbol{y}_1,\cdots,\boldsymbol{y}_k} E\{d^2(X|\boldsymbol{y}_1,\cdots,\boldsymbol{y}_k)\}.$$

Let

$$A_j = \{\boldsymbol{x} \in R; (\boldsymbol{x}-\boldsymbol{y}_j)'(\boldsymbol{x}-\boldsymbol{y}_j) < (\boldsymbol{x}-\boldsymbol{y}_i)'(\boldsymbol{x}-\boldsymbol{y}_i), {}^{\forall}i\neq j\} \tag{1}$$

denote *the domain of attraction of* $\boldsymbol{y}_j$ [5].
Flury defined *self-consistent* as follows:

Definition 2 [Flury[5]]. *The set of k points $\{\boldsymbol{y}_1,\cdots,\boldsymbol{y}_k\}$ is called self-consistent (with respect to the p-variate random variable X) if*

$$E[X|X \in A_j] = \boldsymbol{y}_j \quad (1 \le j \le k) \tag{2}$$

where the A_j are defined as in expression (1).

2.2 Definition of Random Functions

Functional principal points are defined for random functions, as well as principal points are defined for probability distributions. In this section, we describe the definition of functional principal points according to Tarpey and Kinateder [20].

Let (Ω, F, P) be a probability space. Let E denote an integration with respect to the probability measure P. Let T be a real interval and we consider $L^2(T)$, the complete separable Hilbert space consisting of all square-integrable measurable functions on T. The associated norm and the inner product on L^2 will be denoted by $||\cdot||$ and $(\cdot,\cdot)$ respectively. Let ν denote the space of measurable maps from Ω to L^2, an element of ν will be called *a random function* [6].

2.3 Definition of Functional Principal Points of Random Functions

Let $y_1,\cdots,y_k$ denote k distinct nonrandom functions in $L^2(T)$ and define $A_j = \{y \in L^2(T); ||y_j - y||^2 < ||y_i - y||^2, {}^{\forall}i\neq j\}$. A_j is called *the domain of attraction of* y_j.

Functional principal points can be defined as the extension of Definition 1 as follows:

Definition 3 (Tarpey and Kinateder [20]). *Let $\xi \in L^2(T)$ denote a continuous random function. For a given set of k distinct nonrandom functions $x_1,\cdots,x_k$, with respective domains of attraction A_j, $j = 1,\cdots,k$, define another random function x as $x = \sum_{j=1}^k x_j I_{\{\xi\in A_j\}}$. The functions $\{x_1,\cdots,x_k\}$ are called the set of k functional principal points of ξ if $E||\xi - x||^2 \le E||\xi - y||^2$ for all y of the form $y = \sum_{j=1}^k y_j I_{\{\xi\in D_j\}}$, with k distinct nonrandom functions $y_1,\cdots,y_k$ and respective domains of attraction $D_1,\cdots,D_k$.*

2.4 *K*-Means Functional Clustering

K-means clustering method [3] is a typical technique for a certain data set when non-hierarchical clustering is analyzed. It is used very frequently because its algorithm is simple. However, solutions of clustering by k-means method depend on the initial values of its algorithm.

For a functional data set, when it is possible to think about the distance between distinct arbitrary functions, we can also apply the non-hierarchical clustering method based on k-means to the functional data set.

We suppose functional data consist of n functions: $f_1(t), \cdots, f_n(t)$, with $t \in [T_L, T_U]$. To cluster the functions using k-means algorithm[3], one begins with k arbitrary and distinct initial cluster means: $m_1(t), \cdots, m_k(t)$. Let C_j denote the cluster formed by centering on $m_j(t)$. Then each function $f_i(t)$ would be assigned to the cluster based on the closest cluster mean according to the squared $L^2[T_L, T_U]$ distance:

$$\min_j [\int_{T_L}^{T_U} (m_j(t) - f_i(t))^2 dt]. \tag{3}$$

After all of the functions $f_i(t)$ have been assigned to clusters, the cluster means are updated as follows:

$$m_j(t) = \sum_{f_i(t) \in C_j} f_i(t)/n_j, \tag{4}$$

where n_j is the number of functions in C_j. This procedure continues until no point (function) is assigned to the other cluster.

When $m_1(t), \cdots, m_k(t)$ minimize

$$q(k) = \sum_{j=1}^{k} \sum_{y_i(t) \in C_j} \int_{T_L}^{T_U} (m_j(t) - f_i(t))^2 dt, \tag{5}$$

$m_1(t), \cdots, m_k(t)$ are called *optimal cluster centers.*

2.5 Orthonormal Basis Transformation of Functions

When $g_1(t), \cdots, g_p(t)$ are defined as p orthonormal bases of $\xi(t)$ in $L^2[T_L, T_U]$, we can write $\xi(t)$ as

$$\xi(t) = \sum_{i=1}^{p} b_i g_i(t) \tag{6}$$

with p coefficients $b_1, \cdots, b_p$. Here, $g_1(t), \cdots, g_p(t)$ satisfy

$$\int_{T_L}^{T_U} \{g_i(t)\}^2 dt = 1 \quad (i = 1, \cdots, p) \tag{7}$$

and

$$\int_{T_L}^{T_U} g_i(t) g_j(t) dt = 0 \quad (i \neq j). \tag{8}$$

By using the orthonormal basis transformation, functional data can be converted as coordinates on the p-dimensional space, and the distance between two distinct functional data corresponds to the distance between two coordinates on the p-dimensional space.

Since the p-dimensional coefficient vector $(b_1, \cdots, b_p)'$ follows p-variate distribution, we can correspond the k principal points in p-dimensional space spanned by $\{g_1(t), \cdots, g_p(t)\}$ to the k functional principal points of $\xi(t)$.

3 Some Examples of Functional Principal Points

Tarpey and Kinateder showed some examples of functional principal points of linear functional data following Gaussian random function [20]. We shall show others.

3.1 The Case That 2-Dimensional Coefficient Vector of Linear Random Function Following Bivariate Normal Distribution

Example 1. Suppose the linear random functions in $L^2[0, 1]$:

$$\xi(t) = a_1 + a_2 t, \tag{9}$$

where $(a_1, a_2)'$ following bivariate normal distribution $N_2((\alpha_1, \alpha_2)', \Phi_2)$ with

$$\Phi_2 = \begin{pmatrix} \sigma_1^2 & \sigma_{12} \\ \sigma_{12} & \sigma_2^2 \end{pmatrix}.$$

Let $g_1(t)$ and $g_2(t)$ denote

$$g_1(t) = 1, \quad g_2(t) = \sqrt{12}(t - 1/2). \tag{10}$$

The linear random function (9) can be rewritten by using (6), with (10) and

$$b_1 = a_1 + a_2/2, \quad b_2 = a_2/\sqrt{12}. \tag{11}$$

The covariance function $C(s, t)$ for the random function ξ can be found as follows:

$$\begin{aligned} C(s,t) &= E[(\xi(s) - (\alpha_1 + \alpha_2 s))(\xi(t) - (\alpha_1 + \alpha_2 t))] \\ &= E[((a_1 - \alpha_1) + (a_2 - \alpha_2)s)((a_1 - \alpha_1) + (a_2 - \alpha_2)t)] \\ &= \sigma_1^2 + (s + t)\sigma_{12} + st\sigma_2^2. \ [20] \end{aligned}$$

To find the eigenvalues and eigenfunctions, we must solve the following:

$$\int_0^1 C(s,t)(c_1 g_1(s) + c_2 g_2(s))ds = \lambda(c_1 g_1(t) + c_2 g_2(t)). \tag{12}$$

We solve (12) and get the matrix form

$$\begin{pmatrix} \sigma_1^2 + \sigma_{12}/2 & \sigma_1^2/2 + \sigma_{12}/3 \\ \sigma_{12} + \sigma_2^2/2 & \sigma_{12}/2 + \sigma_2^2/3 \end{pmatrix} \begin{pmatrix} c_1 - \sqrt{3}c_2 \\ \sqrt{12}c_2 \end{pmatrix} = \lambda \begin{pmatrix} c_1 - \sqrt{3}c_2 \\ \sqrt{12}c_2 \end{pmatrix}. \tag{13}$$

We can rewrite (13) as

$$A_2\Phi_2 B_2^{-1}(c_1, c_2)' = \lambda B_2^{-1}(c_1, c_2)'$$

where

$$A_2 = \begin{pmatrix} 1 & 1/2 \\ 1/2 & 1/3 \end{pmatrix}, \quad B_2 = \begin{pmatrix} 1 & 1/2 \\ 0 & 1/\sqrt{12} \end{pmatrix}.$$

Thus, λ and (c_1, c_2) correspond to an eigenvalue and eigenvector pair for the matrix $B_2\Phi_2 A_2 B_2^{-1}$. That is, the 2-dimensional coefficient vector $(b_1, b_2)'$ follows $N_2((B_2(\alpha_1, \alpha_2)', B_2 A_2 \Phi_2 B_2^{-1})$. B_2 satisfies $(b_1, b_2)' = B_2(a_1, a_2)'$ in (11).

Suppose Φ_2 satisfies

$$B_2 A_2 \Phi_2 B_2^{-1} = \mathrm{diag}(\varsigma_1^2, \varsigma_2^2). \tag{14}$$

If $\varsigma_1^2 = \varsigma_2^2$ is satisfied, k functional principal points in the space spanned by $\{g_1(t), g_2(t)\}$ are not unique, because $(b_1, b_2)'$ follows the bivariate standard normal distribution and the set of k principal points has freedom of rotation that centers on the expectation of the distribution.

So, $k = 2$ functional principal points of ξ can be expressed

$$y = \alpha_1 + \alpha_2 t \pm \sqrt{2/\pi}\varsigma_1^2 \gamma_1(t), \tag{15}$$

where

$$\gamma_1(t) = \cos\theta_0 + \sqrt{12}(t - 1/2)\sin\theta_0. \tag{16}$$

Here, θ_0 is an arbitrary value which satisfies $0 \le \theta_0 < 2\pi$.

And also, $k = 3$ functional principal points of ξ can be expressed

$$y = \alpha_1 + \alpha_2 t + \sqrt{27/8\pi}\varsigma_1^2 \gamma_1(t), \tag{17}$$

where

$$\gamma_1(t) = \cos\theta + \sqrt{12}(t - 1/2)\sin\theta. \tag{18}$$

with $\theta = \theta_0, \theta_0 \pm \frac{2}{3}\pi$.

3.2 The Case That p-Dimensional Coefficient Vector of Polynomial Random Function Following p-Variate Normal Distribution

Example 2. Suppose the polynomial random functions in $L^2[0, 1]$:

$$\xi(t) = a_1 + a_2 t + \cdots + a_p t^{p-1}, \tag{19}$$

where $(a_1, \cdots, a_n)'$ following p-variate normal distribution $N_p((\alpha_1, \cdots, \alpha_p)', \Phi_p)$. Let $g_1(t)$, $\cdots$, $g_p(t)$ denote

$$g_1(t) = 1, \quad g_2(t) = \sqrt{12}(t - 1/2), \quad g_3(t) = \sqrt{180}(t^2 - t + 1/6), \cdots. \tag{20}$$

Then, $b_1, \cdots, b_p$ can be written by using (6) and (20). If B_p satisfies $(b_1, \cdots, b_p)' = B_p(a_1, \cdots, a_p)'$, the p-dimensional coefficient vector $(b_1, \cdots, b_p)'$ follows $N_p(B_p(\alpha_1, \cdots, \alpha_p)', B_p A_p \Phi_p B_p^{-1})$ where

$$A_p = \begin{pmatrix} 1 & 1/2 \cdots & & 1/p \\ 1/2 & 1/3 & & \vdots \\ \vdots & & \ddots & \vdots \\ 1/p & \cdots & \cdots & 1/(2p-1) \end{pmatrix}.$$

Suppose Φ_p satisfy

$$B_p A_p \Phi_p B_p^{-1} = \mathrm{diag}(\varsigma_1^2, \cdots, \varsigma_p^2). \tag{21}$$

If $\varsigma_1^2 > \varsigma_2^2 \geq \cdots \geq \varsigma_p^2$ are satisfied, $k = 2$ functional principal points of ξ can be expressed

$$y = \alpha_1 + \cdots + \alpha_p t^{p-1} \pm \sqrt{2/\pi}\varsigma_1^2. \tag{22}$$

3.3 The Case That Fourier Polynomial Random Function Following Multivariate Normal Distribution

Example 3. Suppose the polynomial random functions in $L^2[0, 2\pi/\omega]$:

$$\xi(t) = \frac{a_0}{\sqrt{2}} + \sum_{j=1}^{p}(a_{2j-1}\cos j\omega t + a_{2j}\sin j\omega t) \tag{23}$$

where $(a_0, \cdots, a_{2p})'$ following $(2p+1)$-variate normal distribution $N_{2p+1}((\alpha_0, \cdots, \alpha_{2p})', \Phi_{2p+1})$. Let $g_0(t)$, $\cdots$, $g_{2p}(t)$ denote

$$g_0(t) = \frac{1}{\sqrt{2\pi/\omega}}, \quad g_{2j-1}(t) = \frac{\cos j\omega t}{\sqrt{\pi/\omega}}, \quad g_{2j}(t) = \frac{\sin j\omega t}{\sqrt{\pi/\omega}} \quad (1 \leq j \leq p). \tag{24}$$

We can write $(b_0, \cdots, b_{2p})' = \sqrt{\pi/\omega}(a_0, \cdots, a_{2p})'$ by using (6) and (24). For $(b_0, \cdots, b_{2p})'$, eigenfunctions of ξ correspond to eigenvalues of $(\pi/\omega)\Phi_{2p+1}$.

If $\lambda_1 > \lambda_2 \geq \cdots \geq \lambda_{2p+1}$ are the eigenvalues of $(\pi/\omega)\Phi_{2p+1}$, $k = 2$ functional principal points can be expressed

$$y = \frac{\alpha_0}{\sqrt{2}} + \sum_{j=1}^{p}(\alpha_{2j-1}\cos j\omega t + \alpha_{2j}\sin j\omega t) \pm \sqrt{2/\pi}\gamma_1(t), \tag{25}$$

where eigenfunction $\gamma_1(t)$ corresponding to λ_1.

4 Optimal Functional Clustering and Functional Principal Points

On functional data analysis, we often have the case that observed functional data are distributed according to an unknown random function. To confirm how the functional principal points of unknown random functions are reliable, we classify nonrandom n functional data following the random function ξ into k clusters.

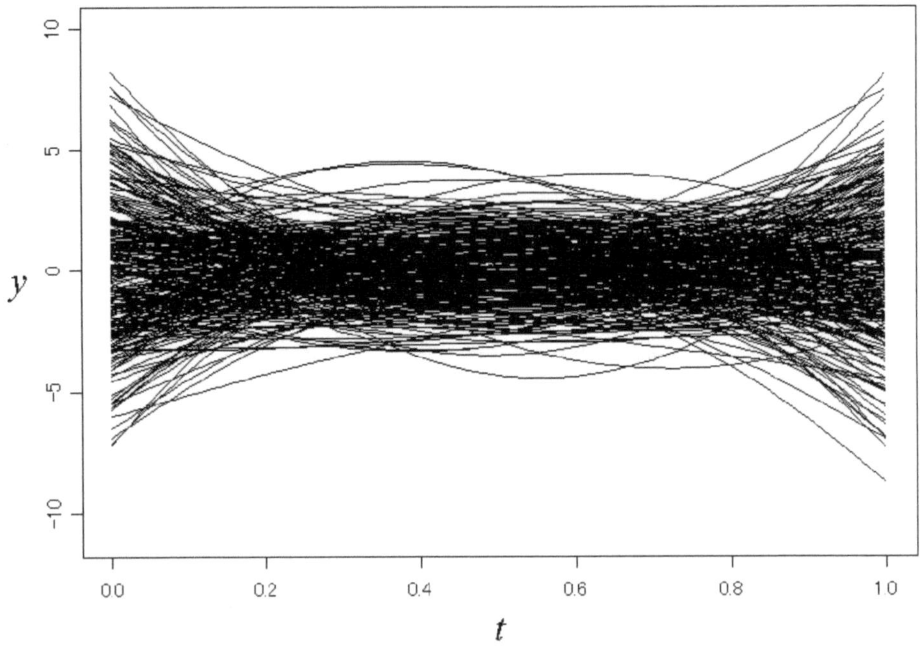

Fig. 1. A simulated set of $n = 250$ functional data following $y = \xi(t) = a_1 + a_2 t + a_3 t^2$

Based on this result, we compare the k center functions of the optimal clusters and the k functional principal points of ξ.

Example 4. Suppose the polynomial random functions in $L^2[0, 1]$:

$$\xi(t) = a_1 + a_2 t + a_3 t^2, \tag{26}$$

where $(a_1, a_2, a_3)'$ following trivariate normal distribution $N_3((0, 0, 0)', \Phi_3)$. Let $g_1(t)$, $g_2(t)$, $g_3(t)$ denote

$$g_1(t) = 1, \quad g_2(t) = \sqrt{12}(t - 1/2), \quad g_3(t) = \sqrt{180}(t^2 - t + 1/6). \tag{27}$$

The polynomial random function (26) can be rewritten by using (6), with (27) and

$$b_1 = a_1 + a_2/2 + a_3/3, \quad b_2 = (a_2 + a_3)/\sqrt{12}, \quad b_3 = a_3/\sqrt{180}. \tag{28}$$

The 3-dimensional coefficient vector $(b_1, b_2, b_3)'$ follows $N_3((0, 0, 0)', B_3 A_3 \Phi_3 B_3^{-1})$ where

$$A_3 = \begin{pmatrix} 1 & 1/2 & 1/3 \\ 1/2 & 1/3 & 1/4 \\ 1/3 & 1/4 & 1/5 \end{pmatrix}, \quad B_3 = \begin{pmatrix} 1 & 1/2 & 1/3 \\ 0 & 1/\sqrt{12} & 1/\sqrt{12} \\ 0 & 0 & 1/\sqrt{180} \end{pmatrix}.$$

Suppose Φ_3 satisfies

$$B_3 A_3 \Phi_3 B_3^{-1} = \mathrm{diag}(\varsigma_1^2, \varsigma_2^2, \varsigma_3^2). \tag{29}$$

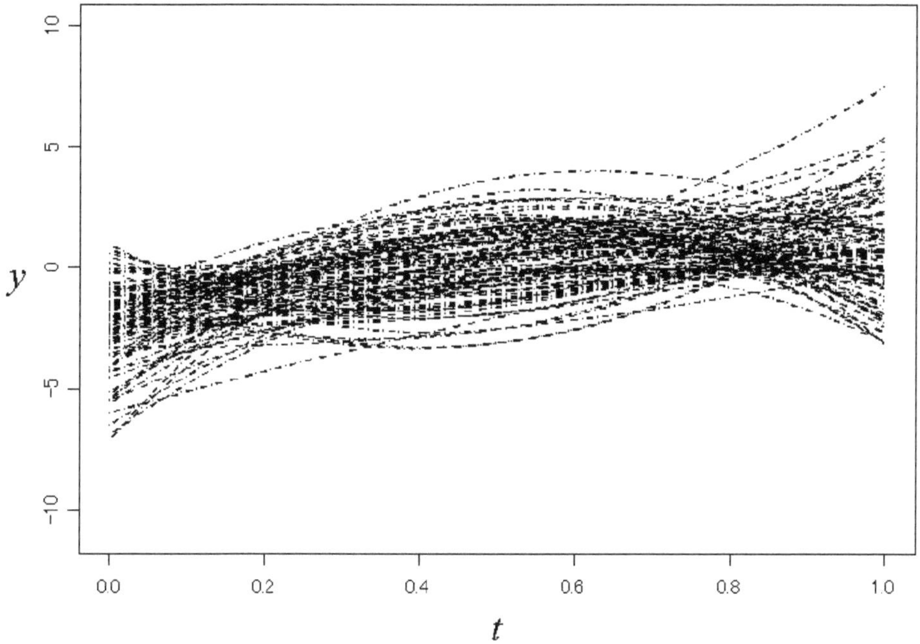

Fig. 2. Functional data in Cluster 1 (at the optimal clusters)

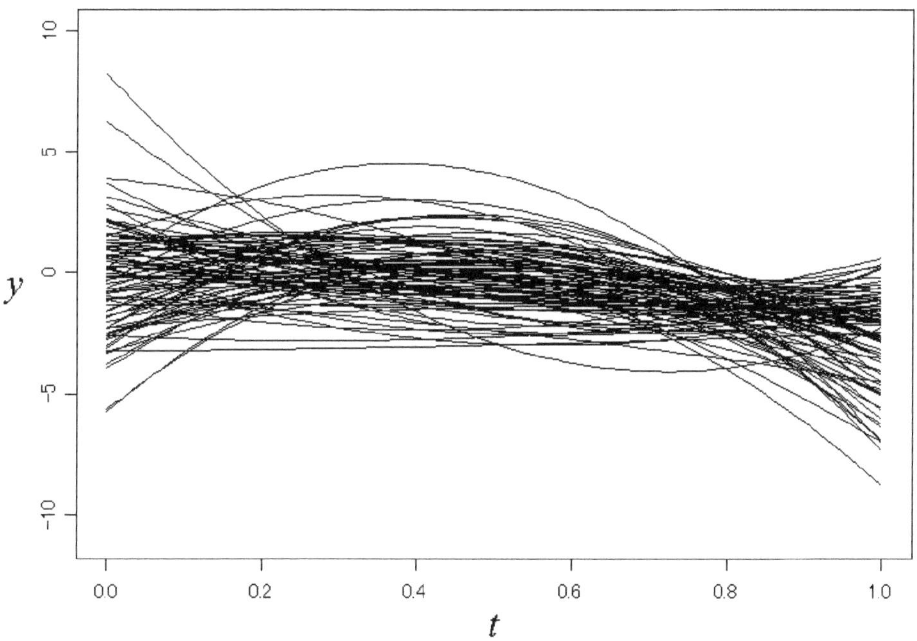

Fig. 3. Functional data in Cluster 2 (at the optimal clusters)

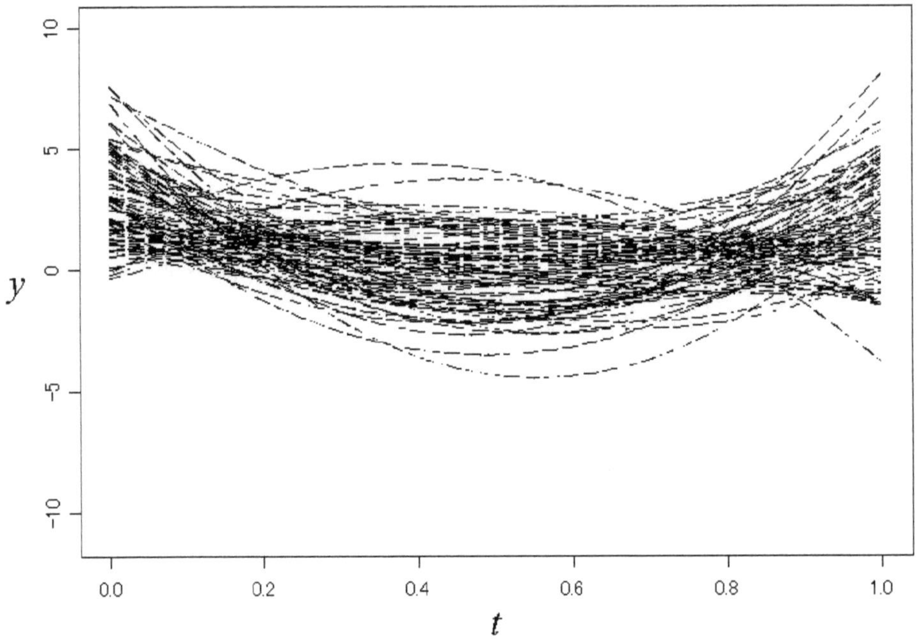

Fig. 4. Functional data in Cluster 3 (at the optimal clusters)

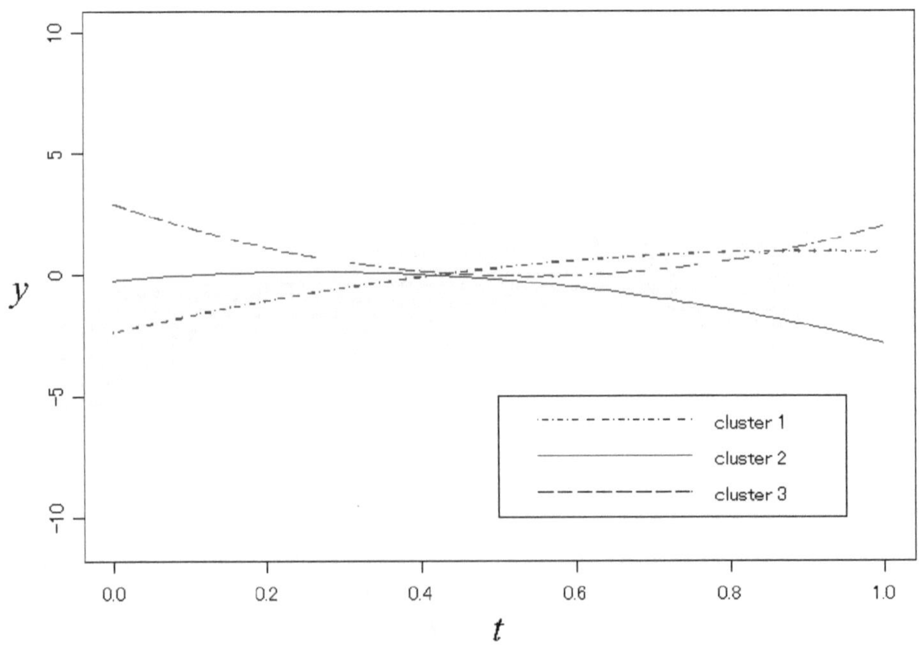

Fig. 5. $k = 3$ center functions of the optimal clusters

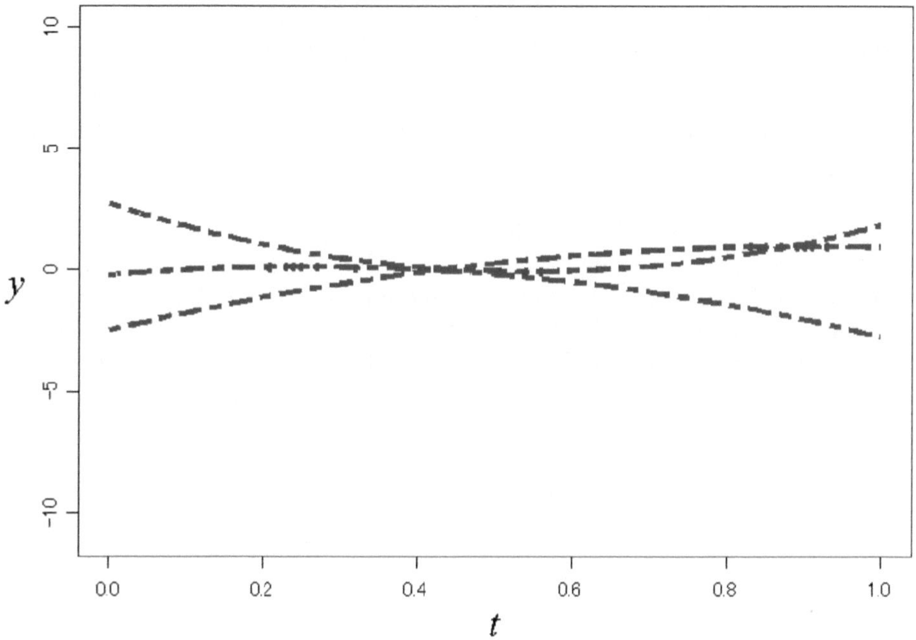

Fig. 6. $k = 3$ functional principal points of $\xi(t)$

We generate n functional data following $\xi(t)$ and classify them into $k = 3$ clusters. In functional k-means clustering, the initial values of $\{m_1(t), \cdots, m_k(t)\}$ have a big influence on the result of the optimal clusters. Then, we prepared $N = 1000$ sets of initial values.

Fig. 1 shows a simulated set of $n = 250$ functional data following $\xi(t)$ with $\varsigma_1^2 = \varsigma_2^2 = \varsigma_3^2 = 1$. We classify the functional data set into $k = 3$ clusters by using functional clustering algorithm (see Section 2.3). We could get many local minima by using different initial values, and chose the clusters which minimized loss function $q(k)$ (see (5)). Fig. 2, 3, 4 are the optimal clusters of the functional data by k-means functional clustering. We show $k = 3$ center functions of the optimal clusters in Fig. 5.

Let (26) transform into (6) with (27) and (28). Then, k functional principal points are the k functions corresponding to k principal points in the space spanned by $\{g_1(t), g_2(t), g_3(t)\}$. If $\varsigma_1^2 = \varsigma_2^2 = \varsigma_3^2 = 1$ are satisfied, $k = 3$ principal points in the space spanned by $\{g_1(t), g_2(t), g_3(t)\}$ are not unique, because $(b_1, b_2, b_3)'$ follows the trivariate standard normal distribution and the set of $k = 3$ principal points has freedom of rotation that centers on the expectation of the distribution. Fig. 6 shows an example set of $k = 3$ functional principal points of ξ as the nearest $k = 3$ center functions of the optimal clusters in Fig. 5. The 3 functions in Fig. 5 and the 3 functional principal points in Fig. 6 are very close. Therefore, when nonrandom functional data following ξ are classified into

k clusters, the k center functions of the optimal clusters can be considered a good approximation of the k functional principal points of ξ.

5 The Numbers of Local Solutions in Functional k-Means Clustering

On functional data analysis, we often have the case that observed functional data are distributed according to an unknown random function. To evaluate the number of local solutions, we classify nonrandom n functional data following the Gaussian random function ξ into k clusters with initial values. Based on this result, we count the numbers of local solutions of functional k-means clustering.

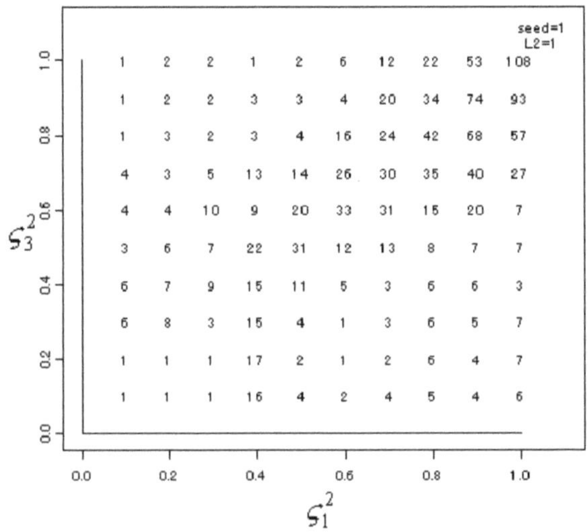

Fig. 7. The numbers of local solutions in functional k-means clustering ($\varsigma_2^2 = 1.0$)

Example 5. As well as the case in Example 4, suppose the polynomial random functions in $L^2[0,1]$:

$$\xi(t) = a_1 + a_2 t + a_3 t^2, \tag{30}$$

where $(a_1, a_2, a_3)'$ following trivariate normal distribution $N_3((0,0,0)', \Phi_3)$. Let $g_1(t)$, $g_2(t)$, $g_3(t)$ denote

$$g_1(t) = 1, \quad g_2(t) = \sqrt{12}(t - 1/2), \quad g_3(t) = \sqrt{180}(t^2 - t + 1/6). \tag{31}$$

The polynomial random function (26) can be rewritten by using (6), with (27) and

$$b_1 = a_1 + a_2/2 + a_3/3, \quad b_2 = (a_2 + a_3)/\sqrt{12}, \quad b_3 = a_3/\sqrt{180}. \tag{32}$$

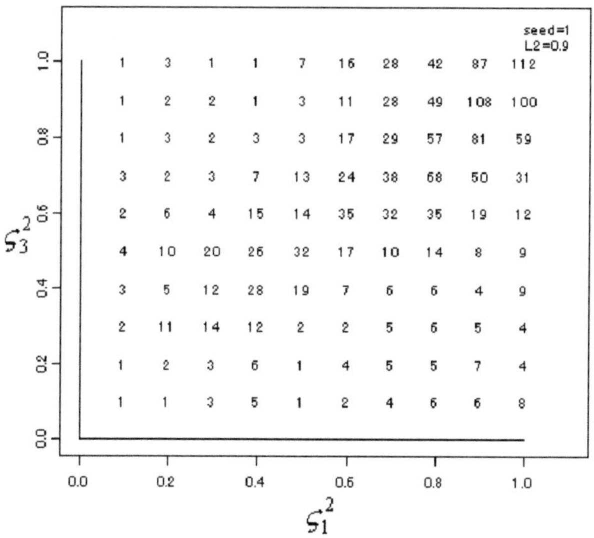

Fig. 8. The numbers of local solutions in functional k-means clustering ($\varsigma_2^2 = 0.9$)

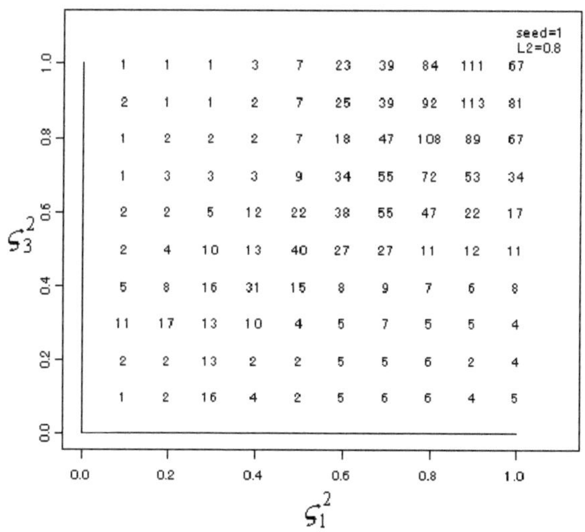

Fig. 9. The numbers of local solutions in functional k-means clustering ($\varsigma_2^2 = 0.8$)

The 3-dimensional coefficient vector $(b_1, b_2, b_3)'$ follows $N_3((0,0,0)', B_3 A_3 \Phi_3 B_3^{-1})$ where

$$A_3 = \begin{pmatrix} 1 & 1/2 & 1/3 \\ 1/2 & 1/3 & 1/4 \\ 1/3 & 1/4 & 1/5 \end{pmatrix}, \qquad B_3 = \begin{pmatrix} 1 & 1/2 & 1/3 \\ 0 & 1/\sqrt{12} & 1/\sqrt{12} \\ 0 & 0 & 1/\sqrt{180} \end{pmatrix}.$$

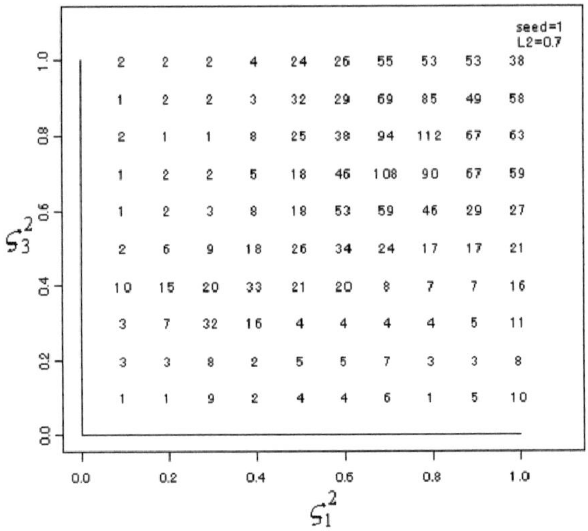

Fig. 10. The numbers of local solutions in functional k-means clustering ($\varsigma_2^2 = 0.7$)

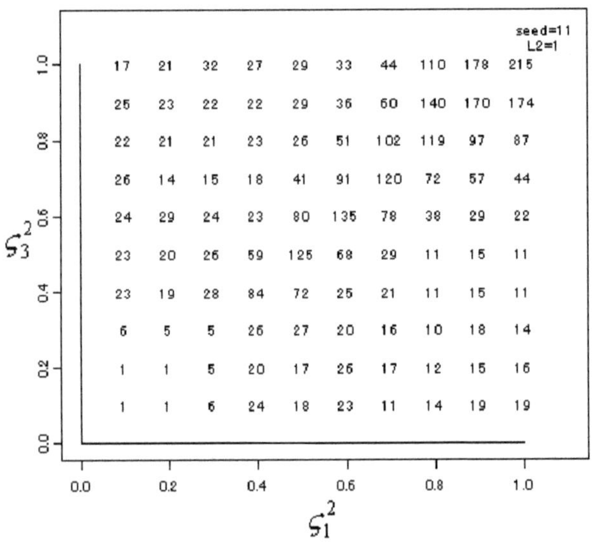

Fig. 11. The numbers of local solutions in functional k-means clustering ($\varsigma_2^2 = 1.0$, seed=11)

Suppose Φ_3 satisfies

$$B_3 A_3 \Phi_3 B_3^{-1} = \text{diag}(\varsigma_1^2, \varsigma_2^2, \varsigma_3^2). \tag{33}$$

We generate n functional data following $\xi(t)$ and classify them into $k = 3$ clusters with the initial values of $\{m_1(t), \cdots, m_k(t)\}$. In functional k-means clustering, we prepared $N = 1000$ sets of initial values.

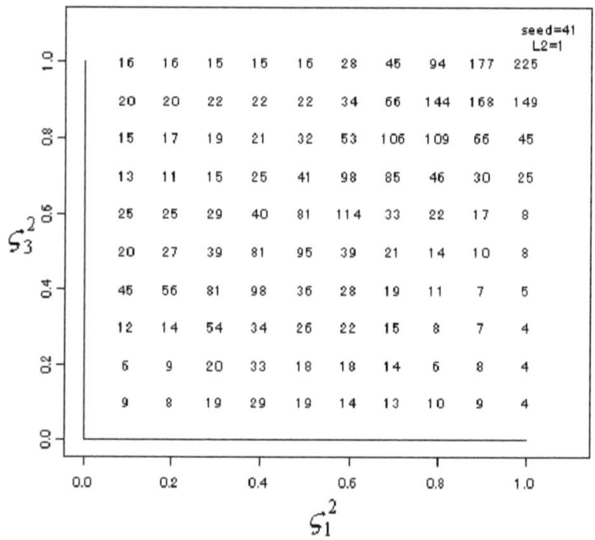

Fig. 12. The numbers of local solutions in functional k-means clustering ($\varsigma_2^2 = 1.0$, seed=41)

Fig. 7 shows a set of the numbers of local solutions in functional k-means clustering of a simulated set of $n = 250$ functional data following $\xi(t)$ with $\varsigma_2^2 = 1.0$, and ς_1^2 and ς_3^2 independently take the value from 0.1 to 1.0 respectively. For example, we could get 108 local solutions in functional k-means clustering when $\varsigma_1^2 = \varsigma_2^2 = \varsigma_3^2 = 1.0$.

As well as Fig. 7, Fig. 8,9 and 10 show each set of the numbers of local solutions in functional k-means clustering of a simulated set of $n = 250$ functional data following $\xi(t)$ with $\varsigma_2^2 = 0.9, 0.8, 0.7$.

We also generate two simulated sets of $n = 250$ functional data following $\xi(t)$ with $\varsigma_2^2 = 1.0$ besides the set used in Fig. 7. Fig. 11 and 12 show each set of the numbers of local solutions in functional k-means clustering.

By these results, when both ς_1^2 and ς_3^2 are smaller enough than ς_2^2, it becomes easy for the optimal solution to generate without depending on initial values. On the other hand, when both ς_1^2 and ς_3^2 are close to ς_2^2, more local solutions are generated. Therefore, many sets of initial values should be given to obtain the optimal clusters when all the eigenvalues of the orthonormalized covariance matrix $B_3 \Phi_3 A_3 B_3^{-1}$ are almost same.

6 Summary

K-means functional clustering is 'equivalent' to clustering the coefficients of the orthonormal basis transformation of the functions [20]. However, as for theoretical researches of functional clustering and functional principal points, more detailed researches are requested.

Tarpey and Kinateder [20] applied precipitation data to functional k-means clustering and estimated functional principal points under the supposition that the data are following Gaussian random function. We showed a close relationship between k functional principal points of a Gaussian random function and k center functions of the optimal clusters of nonrandom functional data following random functions. This means that for observed functional data according to an unknown random function, k-means functional clustering to the functional data can estimate functional principal points of the unknown random function.

Moreover, we evaluated the numbers of local solutions in functional k-means clustering of functional data sets following the Gaussian random function ξ, and showed the relationship between the numbers of local solutions and the eigenvalues of the orthonormalized covariance matrix of ξ. The number of local solutions in functional k-means clustering of the functional data can estimate 'closeness' of all the eigenvalues of the orthonormalized covariance matrix of the unknown Gaussian random function.

References

1. Araki, Y., Konishi, S., Imoto, S.: Functional discriminant analysis for microarray gene expression data via radial basis function networks. In: Proceedings of COMPSTAT, pp. 613–620 (2004)
2. Araki, Y., Konishi, S.: Functional regression modeling via reguralized basis expansions and model selection, MHF Preprint Series, Kyushu University (2005)
3. Forgy, E.: Cluster analysis of multivariate data: Efficiency versus interpretability of classification. Biometrics 21, 768–780 (1965)
4. Flury, B.: Principal points. Biometrika 77(1), 33–41 (1990)
5. Flury, B.: Estimation of principal points. Journal of the Royal Statistical Society, Series C, Applied Statistics 42(1), 139–151 (1993)
6. Ibragimov, I.A., Rozanov, Y.A.: Gaussian Random Processes. Springer, Heidelberg (1978)
7. James, G.M., Hastie, T.J., Sugar, C.A.: Principal component models for sparse functional data. Biometrika 87, 587–602 (2000)
8. Linde, Y., Buzo, A., Gray, R.M.: An algorithm for vector quantizer design. IEEE Transactions on Communications 28, 84–95 (1980)
9. Mizuta, M.: Functional multidimensional scaling. In: Proceedings of the 10th Japan and Korea Joint Conference of Statistics, pp. 77–82 (2000)
10. Mizuta, M.: Cluster analysis for functional data. In: Proceedings of the 4th Conference of the Asian Regional Section of the International Association for Statistical Computing, pp. 219–221 (2002)
11. Mizuta, M.: K-means method for functional data. Bulletin of International Statistical Institute, 54th Session, Book 2, pp. 69–71 (2003)
12. Muraki, C., Ohtaki, M., Mizuta, M.: Principal points analysis of daily weather maps at the Far East region in the summer seasons of 1993-1995 (in Japanese). Japanese Journal of Applied Statistics 27(1), 17–31 (1998)
13. Ramsay, J.O.: When the data are functions. Psychometrika 47, 379–396 (1982)
14. Ramsay, J.O., Silverman, B.W.: Functional Data Analysis. Springer, Heidelberg (1997)

15. Ramsay, J.O., Silverman, B.W.: Applied Functional Data Analysis: Methods and Case Studies. Springer, Heidelberg (2002)
16. Ramsay, J.O., Silverman, B.W.: Functional Data Analysis, 2nd edn. Springer, Heidelberg (2005)
17. Rossi, F., Conan-Guez, B., El Golli, A.: Clustering functional data with the SOM algorithm. In: Proceedings of European Symposium on Artificial Neural Networks 2004, pp. 305–312 (2004)
18. Shimizu, N., Mizuta, M., Sato, Y.: Some properties of principal points (in Japanese). Japanese Journal of Applied Statistics 27(1), 1–16 (1998)
19. Shimokawa, M., Mizuta, M., Sato, Y.: An expansion of functional regression analysis (in Japanese). Japanese Journal of Applied Statistics 29(1), 27–39 (2000)
20. Tarpey, T., Kinateder, K.: Clustering functional data. Journal of Classification 20, 93–114 (2003)

Clustering with Size Constraints

Frank Höppner[1] and Frank Klawonn[2]

[1] Department of Economics
University of Applied Sciences Braunschweig /Wolfenbüttel
Robert Koch Platz 10-14
38440 Wolfsburg, Germany
[2] Department of Computer Science
University of Applied Sciences Braunschweig /Wolfenbüttel
Salzdahlumer Str. 46/48
38302 Wolfenbüttel, Germany
{f.hoeppner,f.klawonn}@fh-wolfenbuettel.de

Abstract. We consider the problem of partitioning a data set of n data objects into c homogeneous subsets or clusters (that is, data objects in the same subset should be similar to each other) with constraints on the number of data per cluster. The proposed techniques can be used for various purposes. If a set of items, jobs or customers has to be distributed among a limited number of resources and the workload for each resource shall be balanced, clusters of approximately the same size would be needed. If the resources have different capacities, then clusters of the corresponding sizes need to be found. We also extend our approach to avoid extremely small or large clusters in standard cluster analysis. Another extension offers a measure for comparing different prototype-based clustring results.

1 Introduction

Cluster analysis is a widely used technique that seeks for groups in data. The result of such an analysis is a set of groups or clusters where data in the same group are similar (homogeneous) and data from distinct groups are different (heterogeneous) [1]. In this paper, we consider variations of the clustering problem, namely the problem of subdividing a set X of n objects into c homogeneous groups with constraints on some or all clusters concerning the number of data they can contain. In contrast to the standard clustering problem, we abandon the heterogeneity between groups partly and introduce constraints on the number of data per cluster.

Applications of *uniform clustering* where all clusters should contain approximately the same number of data include for instance: (a) The distribution of n students into c groups of equal strength to obtain fair class sizes and with homogeneous abilities and skills to allow for teaching methods tailored to the specific

L.C. Jain et al. (Eds.): Comp. Intel. Para.: Innov. Applications, SCI 137, pp. 167–180, 2008.
springerlink.com

needs of each group. (b) The distribution of n jobs to c machines or workers such that every machine has an identical workload and as similar jobs as possible to reduce the configuration time. (c) The placement of c sites such that goods from n locations can be transported to the c sites, while the total covered distance is minimized and queuing at the sites is avoided, that is, approximately the same number of goods should arrive at each site.

More general applciations include problems where the resources, for instance the lecturing halls in the above example (a), the machines in example (b) and the size of the sites in example (c), differ in capacity, so that the clusters still should have fixed size corresponding to the capcities, but not necessarily the same size for all clusters.

Due to the similarity of our problem with traditional clustering problems, we are going to modify an existing clustering algorithm, which will be reviewed in section 2. This objective function-based clustering algorithm – a variant of k-means – transforms the discrete, combinatorial problem into a continuous one, such that numerical problem solving methods can be applied. As proposed in [5], we modify the objective function such that the equi-sized clusters are considered in section 3 and discuss the results in section 4. Section 5 extends the approach to clusters with different size constraints. In section 6 our approach is exploited to define a similarity measure to compare different clustering results.

2 The FCM Algorithm

The fuzzy c-means (FCM) clustering algorithm partitions a data set $X := \{x_1, ..., x_n\} \subset \mathbf{R}^d$ into c clusters. A cluster is represented by a prototype $p_i \in \mathbf{R}^d$, $1 \leq i \leq c$. The data-prototype relation is not binary, but a membership degree $u_{ij} \in [0, 1]$ indicates the degree of belongingness of data object x_j to prototype p_i or cluster number i. All membership degrees form a membership matrix $U \in \mathbf{R}^{c \times n}$. We can interpret the membership degrees as "probabilistic memberships", since we require

$$\forall 1 \leq j \leq n : \qquad \sum_{i=1}^{c} u_{ij} = 1 . \tag{1}$$

The clustering process is carried out by minimizing the objective function

$$J_m = \sum_{j=1}^{n} \sum_{i=1}^{c} u_{ij}^m d_{ij} \quad \text{with} \quad d_{ij} = \|x_j - p_i\|^2 . \tag{2}$$

under constraint (1). If the Euclidean distance between datum x_j and prototype p_i is high, J_m is minimized by choosing a low membership degree near 0. If the distance is small, the membership degree approaches 1. J_m is effectively minimized by alternating optimisation, that is, we alternatingly minimize (2) with respect to the prototypes (assuming memberships to be constant) and then with respect to the membership degrees (assuming prototypes to be constant). In both minimization steps, we obtain closed form solutions, for the prototypes:

$$\forall 1 \leq i \leq c: \qquad p_i = \frac{\sum_{j=1}^{n} u_{ij}^m x_j}{\sum_{j=1}^{n} u_{ij}^m} \tag{3}$$

and for the membership degrees:

$$u_{ij} = \begin{cases} \frac{1}{\sum_{l=1}^{c} \left(\frac{\|x_j - p_i\|^2}{\|x_j - p_l\|^2} \right)^{\frac{1}{m-1}}} & \text{in case } I_j = \emptyset \\ \frac{1}{|I_j|} & \text{in case } I_j \neq \emptyset, i \in I_j \\ 0 & \text{in case } I_j \neq \emptyset, i \notin I_j \end{cases} \tag{4}$$

where $I_j = \{k \in \mathbf{N}_{\leq c} \,|\, x_j = p_k\}$. The FCM algorithm is depicted in Fig. 1. For a more detailed discussion of FCM and examples we refer to the literature, e.g. [2, 3].

---

choose $m > 1$ (typically $m = 2$)
choose termination threshold $\varepsilon > 0$
initialize prototypes p_i (randomly)
repeat
  update memberships using (4)
  update prototypes using (3)
until change in memberships drops below ε

Fig. 1. The FCM algorithm

3 Equi-sized Clusters

It is often said that the k-means (as well as the FCM) algorithm seeks for clusters of approximately the same size, but this is only true if the data density is uniform. As soon as the data density varies, a single prototype may very well cover a high-density cluster and thereby gains many more data objects than the other clusters. This leads to large differences in the size of the clusters. Examples for this phenomenon are shown in Fig. 2 for two data sets: On the left image, there is a very high density cluster in the top left corner, on the right image, the density decreases from left to right, so the rightmost cluster has only some data.

The idea of our modification is to include an additional constraint in the objective function (2) that forces the clusters to cover the same number of data objects. The size of cluster i (number of data objects) corresponds to the sum of the membership values $\sum_{j=1}^{n} u_{ij}$. In fact, since we have continuous membership degrees we may require

$$\sum_{j=1}^{n} u_{ij} = \frac{n}{c} \tag{5}$$

for all $i \in \{1, \ldots, c\}$ even if n is not a multitude of c. This additional constraint (5) is – together with the constraint (1) – integrated into the objective function (2) via Lagrange multipliers. We then solve for the cluster prototypes and Lagrange multipliers by setting the partial derivatives to zero. This turns out to be a difficult problem for the general case of an arbitrary value of m, therefore we restrict ourselves to the case of $m = 2$, which is the most frequently used value of m in FCM. Given our Lagrange function

$$L = \sum_{i=1}^{c}\sum_{j=1}^{n} u_{ij}^2 d_{ij} + \sum_{j=1}^{n} \alpha_j \left(1 - \sum_{i=1}^{c} u_{ij}\right) + \sum_{i=1}^{c} \beta_i \left(\frac{n}{c} - \sum_{j=1}^{n} u_{ij}\right) \tag{6}$$

we obtain as partial derivatives

$$\frac{\partial L}{\partial u_{ij}} = 2u_{ij}d_{ij} - \alpha_j - \beta_i = 0 \tag{7}$$

These equations, together with the constraints (1) and (5), lead to the following system of $(c \cdot n + c + n)$ linear equations for the variable u_{ij}, α_i and β_j ($i \in \{1, \ldots, c\}$, $j \in \{1, \ldots, n\}$). Empty entries indicate the value zero, RHS stands for the right hand side of the equation.

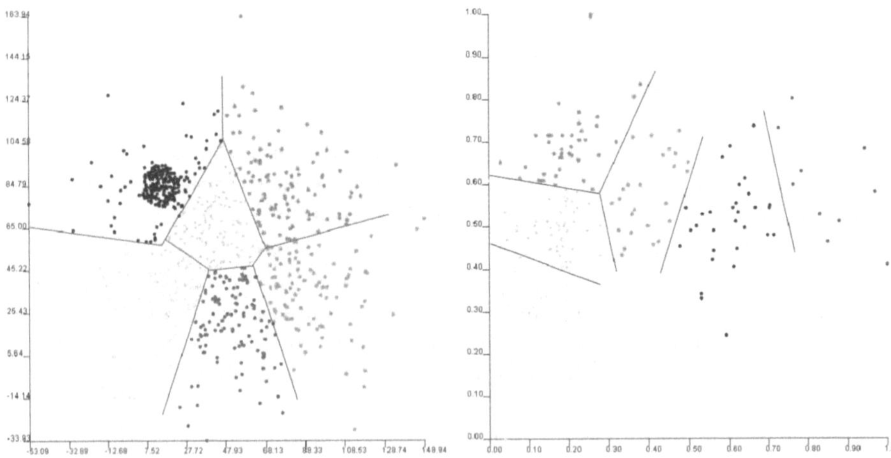

Fig. 2. Results of the FCM algorithm on two data sets

	$u_{1,1}$	$\dots$	$u_{1,n}$	$\dots$	$u_{c,1}$	$\dots$	$u_{c,n}$	α_1	$\dots$	α_n	β_1	$\dots$	β_c	RHS
$\frac{\partial L}{\partial u_{1,1}}$	$2d_{1,1}$							-1			-1			
$\vdots$		$\ddots$							$\ddots$		$\vdots$			
$\frac{\partial L}{\partial u_{1,n}}$			$2d_{1,n}$							-1	-1			
$\vdots$				$\ddots$					$\ddots$			$\ddots$		
$\frac{\partial L}{\partial u_{c,1}}$					$2d_{c,1}$			-1					-1	
$\vdots$						$\ddots$			$\ddots$				$\vdots$	
$\frac{\partial L}{\partial u_{c,n}}$							$2d_{c,n}$			-1			-1	
$\sum u_{i,1}$	1			$\dots$	1									1
$\vdots$		$\ddots$				$\ddots$								$\vdots$
$\sum u_{i,n}$			1				1							1
$\sum u_{1,j}$	1	$\dots$	1											n/c
$\vdots$				$\ddots$										$\vdots$
$\sum u_{c,j}$					1	$\dots$	1							n/c

In principle, this system of linear equations could be solved by a suitable numerical algorithm. Even for small data sets with 200 data objects and 5 clusters, this would mean that we have to solve a system of 1205 equations in each iteration step of the clustering algorithm, which is not acceptable in terms of computational costs. However, it is possible to solve this system of equations in a more efficient way.

When multiplying the equations for $u_{k1}, \dots, u_{kn}$ by $\frac{1}{2d_{k1}}, \dots, \frac{1}{2d_{kn}}$, respectively, and then subtracting the resulting equations from the equation for $\sum_j u_{kj}$, we obtain

$$\sum_{j=1}^{n} \frac{\alpha_j}{2d_{kj}} + \beta_k \sum_{j=1}^{n} \frac{1}{2d_{kj}} = \frac{n}{c}. \tag{8}$$

From equation (7), we obtain

$$u_{ij} = \frac{\alpha_j + \beta_i}{2d_{ij}}. \tag{9}$$

Taking constraint (1) into account, yields

$$1 = \sum_{i=1}^{c} u_{ij} = \frac{\alpha_j}{2} \sum_{i=1}^{c} \frac{1}{d_{ij}} + \frac{1}{2} \sum_{i=1}^{c} \frac{\beta_i}{d_{ij}},$$

leading to

$$\alpha_j = \frac{2 - \sum_{i=1}^{c} \frac{\beta_i}{d_{ij}}}{\sum_{i=1}^{c} \frac{1}{d_{ij}}}. \tag{10}$$

Inserting (10) into (8), we obtain

$$\sum_{j=1}^{n} \frac{2 - \sum_{i=1}^{c} \frac{\beta_i}{d_{ij}}}{2 \sum_{i=1}^{c} \frac{d_{kj}}{d_{ij}}} + \beta_k \sum_{j=1}^{n} \frac{1}{2d_{kj}} = \frac{n}{c}$$

and thus

$$-\sum_{j=1}^{n} \frac{\sum_{i=1}^{c} \frac{\beta_i}{d_{ij}}}{2 \sum_{i=1}^{c} \frac{d_{kj}}{d_{ij}}} + \beta_k \sum_{j=1}^{n} \frac{1}{2d_{kj}} = \frac{n}{c} - \sum_{j=1}^{n} \frac{1}{\sum_{i=1}^{c} \frac{d_{kj}}{d_{ij}}}. \tag{11}$$

This induces a system of c linear equations for the β_k with coefficients

$$a_{k\ell} = \begin{cases} -\sum_{j=1}^{n} \frac{\frac{1}{d_{\ell j}}}{2\sum_{i=1}^{c} \frac{d_{kj}}{d_{ij}}} & \text{if } k \neq \ell \\ -\sum_{j=1}^{n} \frac{\frac{1}{d_{\ell j}}}{2\sum_{i=1}^{c} \frac{d_{kj}}{d_{ij}}} + \sum_{j=1}^{n} \frac{1}{2d_{kj}} & \text{if } k = \ell. \end{cases} \tag{12}$$

This system of linear equations can be solved by a suitable numerical algorithm. The computation time is acceptable, since the number of equations is equal to the number of clusters and therefore independent of the number of data. Once the β_i have been determined, we can compute the α_j using equation (10) and finally obtain the membership degrees based on equation (9). After all, we arrive at the clustering algorithm depicted in Fig. 3.

```
choose termination threshold ε > 0
initialise prototypes p_i
repeat
  solve linear equation system (11) for β
  using β, calculate α using (10), update memberships using (9)
  update prototypes using (3)
until change in memberships drops below ε
```

Fig. 3. The proposed algorithm

Note that the boundedness of the membership degrees $u_{ij} \in [0,1]$ represents an additional constraint on the objective function of FCM as well as the objective function of our new algorithm. In the original FCM, however, it was not necessary to consider it explicitly, because one can easily see from the resulting membership degrees (4) that this condition is satisfied. It is not possible to conclude this boundedness for the new membership degrees (9). It is clear, however, that the influence of negative memberships will be rather small: Since the objective function (2) and (6) uses only positive weights u_{ij}^2, large negative values cannot help in the minimization. We will comment on this in the following section.

4 Examples for Uniform Clustering

To illustrate the impact of our modified objective function, we show the results of the new algorithm for the data sets shown in Fig. 2, where the standard FCM algorithm yielded a result with high variation in the cluster size. The results are shown in the left images of Figs. 4 and 6. By comparison to Fig. 2 we see, that the high-density cluster has been split into two clusters (Fig. 4) and that the data on the left of Fig. 6 is now distributed among four rather than three clusters, such that the rightmost cluster gains more data. As expected, the sum of membership degrees for each individual cluster equals $\frac{n}{c}$.

Regarding the boundedness of the membership degrees u_{ij} it turned out that they actually take negative values. This is, of course, an undesired effect, because then the interpretation of $\sum_{j=1}^{n} u_{ij}$ as the *size* or number of data objects is not quite correct. As conjectured in the previous section, it turned out on closer examination that the total sum of negative weights is rather small. In both data sets, the sum of all negative membership degrees was below 0.5% of the total data set size n.

We want to illustrate the kind of situation in which negative membership degrees occur with the help of the data set shown in Fig. 5. Consider the data set is partitioned into three clusters. Since the leftmost cluster has an additional data object x in the middle, it is not obvious how to distribute the data among all clusters in equal shares.

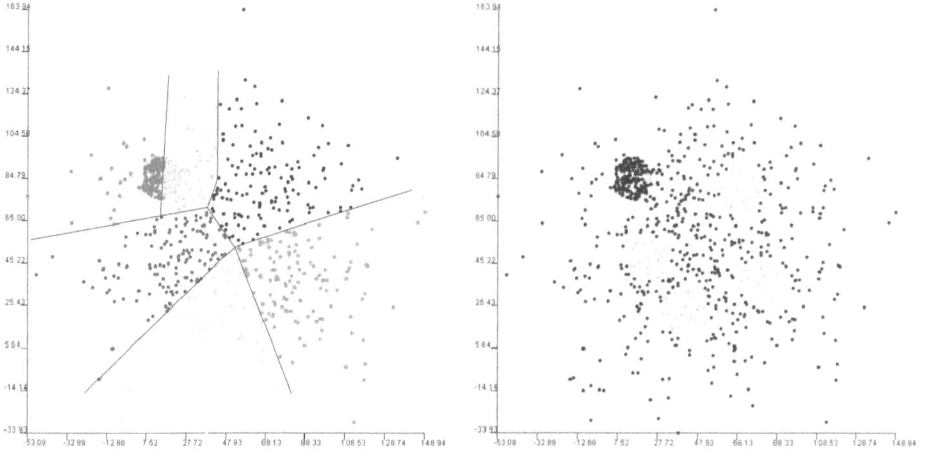

Fig. 4. Results of the new algorithm

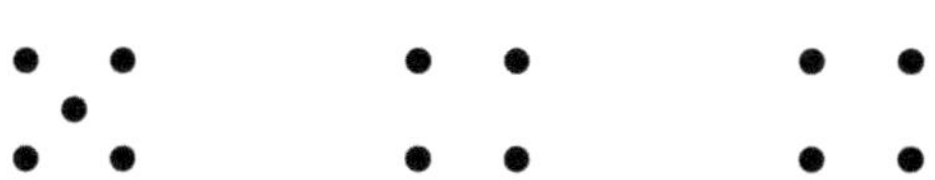

Fig. 5. A 'difficult' example data set

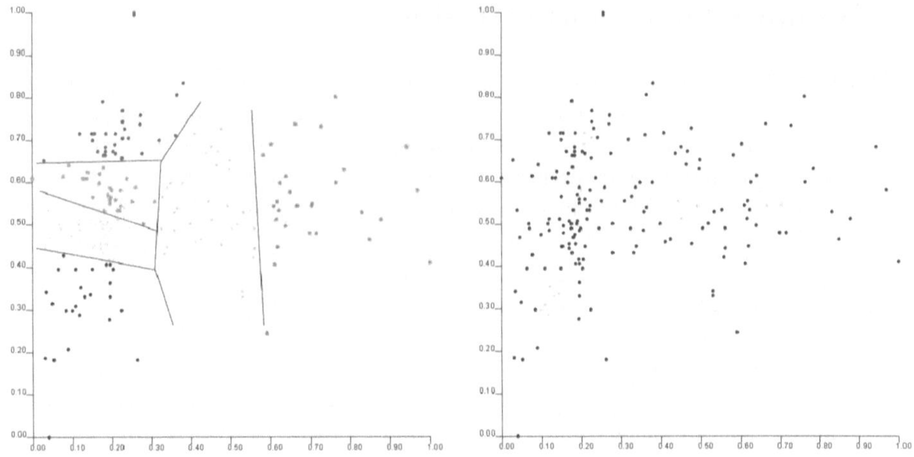

Fig. 6. Results of the new algorithm

Regarding the minimization of the sum of weighted distances, it would be optimal to assign high membership degrees to all five data objects. This would, however, violate the constraint that all clusters must share the same size. To get membership degrees as high as possible, the membership of all other data objects (middle and right cluster) to this cluster are chosen slightly negative. Since the sum of all membership values is constrained to be one, negative values for the middle and right data allow us to have slightly higher degrees for the leftmost data. On the other hand, having a negative membership degrees for some data object x on the right forces us to increase other membership degrees of x (to guarantee a sum of 1). This is possible almost at no cost, if x is close to the centre of another cluster, because then we have a small distance value and increasing the membership degree to this cluster does no harm in the minimization of (2). (For a detailed discussion of the general influence of the membership weight u_{ij}^m see [4].)

To summarise: In a situation where an equi-sized partition is difficult to obtain while minimizing at the same time the sum of weighted distances (2), the cluster with too many data objects 'borrows' some membership from data near the centres of the other clusters. Figures 4 and 6 show this effect for the two example data sets. The data for which negative membership values occur are shown in a lighter shading. These data objects are all close to the respective cluster prototype. And there is always one cluster without any negative membership degrees, which corresponds to the rightmost cluster in our example data set in Fig. 5.

In all our experiments, the side effects of this trade off between minimizing (2) and satisfying (5) were quite small, so we do not consider this as a major drawback of our approach. We can even make use of this information: By analysing which cluster has no negative membership degrees at all, we can find out which cluster tends to be 'too big'. When breaking ties in the final assignment of data to clusters, it should be this cluster that gets more data objects than the other.

5 Limiting the Size of Single Clusters Only

Instead of forcing all clusters to have the same size of $\frac{n}{c}$, it is also possible to have different requirements for each cluster individually. We introduce size parameters s_i and replace (5) by

$$\sum_{j=1}^{n} u_{ij} = s_i \tag{13}$$

The derivation of the linear equation systems still holds, we just have to replace the fraction $\frac{n}{c}$ in (11) by s_i.

For some of the clusters, we may have no constraint on their size. In such a case the condition (13) needs not to be guaranteed by means of a Lagrange multiplier as in equation (6). So we consider a linear equation system with a reduced number of Lagrange multipliers (or force the respective β_i to be zero, such that it has no influence on the remaining equations).

Two examples are given to illustrate the performance of the algorithm on two low-dimensional data sets. Figure 7 shows a set consisting of basically two clusters (50 data points each) and three outliers in the lower left corner. When clustered with $c = 3$ clusters, FCM assigns the three outliers to a single cluster (cf. figure 7, left). Then, we force two clusters to have a size of $s_i = 34$. This forces the outlier cluster to move closer to one of the larger clusters (cf. figure 7, right).

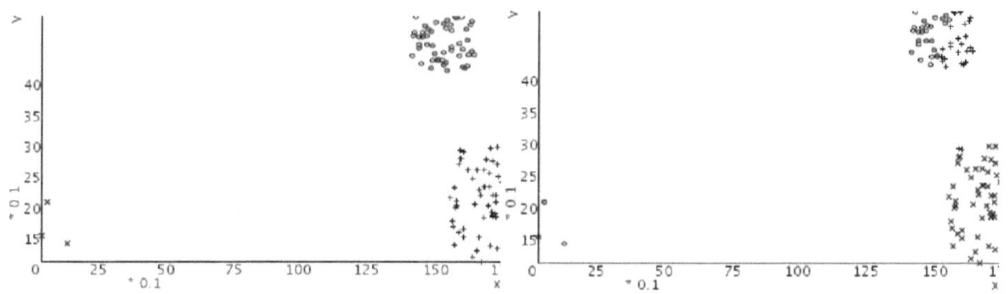

Fig. 7. Example 1. Left: Result of FCM. Right: Result of Size-Constrained FCM.

The second example in figure 8 shows the case of two long-stretched clusters of 100 data points each. Using 4 clusters FCM covers each of the two long-stretched clusters by 2 prototypes (figure 8, left). Now we force one of the bottom clusters to have exactly a size of 100 data points. To achieve this, the other prototype (supported by 50 data points) has to be repelled completely from the clusters and the data from the second cluster must now be shared among three prototypes.

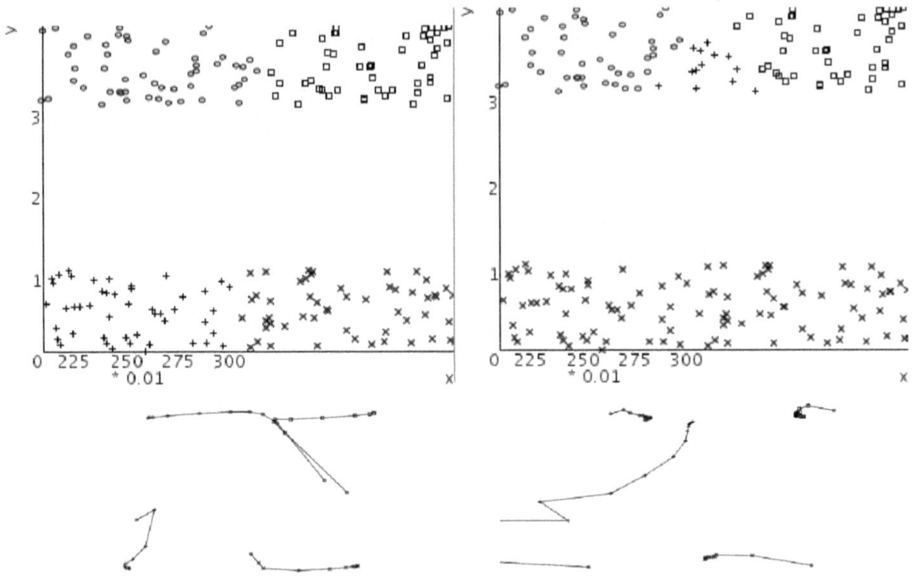

Fig. 8. Example 2: Left: Result of FCM. Right: Result of Size-Constrained FCM. (top: resulting partition, bottom: trace of prototypes to their final location).

---

choose minimal and maximal cluster size $s_{\min}$ and $s_{\max}$.
initialize size constraint vector $s_i = (0, .., 0)$ (no size constraint for any cluster)
run traditional Fuzzy c-Means (cf. figure 1)
while (at least one cluster sizes a_i does not satisfy $s_{\min} \le a_i \le s_{\max}$.
    for all clusters i with $s_i = 0$
        if $a_i < s_{\min}$, set $s_i = s_{\min}$
        if $a_i > s_{\max}$, set $s_i = s_{\max}$
    end
    run algorithm in figure 3 using current constraints s_i (with $s_i \neq 0$)
end

Fig. 9. Clustering with constraints on cluster size

This is exactly the case in figure 8 (right). The figure also shows how the position of the prototypes evolves over the iterations.

It is also possible to apply our technique to avoid extremely small or large clusters in a cluster analysis. We simply formulate a fuzzy clustering algorithm with upper and lower bounds on the size of a cluster as shown in figure 9. We first run the original FCM and then impose constraints on cluster sizes in case a cluster's size is below the lower or above the upper bound on cluster sizes.

It should be mentioned, that the *size* s_i *of a cluster* has been interpreted in a fuzzy sense: It *may* happen that a prototype achieves quite a large size $\sum_{j=1}^{n} u_{i,j}$

although only in very few cases this prototype receives the highest membership degree. From our experiences such situations are more likely to occur, if the partition is very stable and the number of clusters is chosen wrongly.

6 A Distance Measure for Sets of Prototypes

In this section we consider the problem of defining a distance measure between two sets of prototypes. Figure 10 illustrates the problem by an example: Prototype set A consists of three prototypes $\{a, b, c\}$, set B consists of four prototypes $\{d, e, f, g\}$. We need a measure of divergence between the two sets. Intuitively we have to match each of the prototypes in set A to one (or more) prototype(s) in set B and then account for the distances between the associated pairs of prototypes. An intuitive matching is illustrated by dotted lines in figure 10.

This problem is quite similar to clustering a small data set with $c \approx n$. If we interpret the membership degree $u_{i,j}$ as the strength of the relationship between prototype i (set A) and prototype j (set B), cf. dotted lines in figure 10, the sum of weighted distances (2) becomes a good indicator for the distance between the sets.

The membership constraints (1) and (5) assure that every data object / prototype has the same overall weight in the assignment, that is, every data object (or prototype, resp.) must be assigned to a prototype (or data objects, resp.) with equal shares. This is illustrated on the left side of figure 11. Since clusters may have different sizes, forcing each cluster to share the same number of data objects is a quite unrealistic requirement, therefore we have introduced cluster size s_i in the previous section. Since we want to measure the distance between two sets of clusters, we also replace the data objects in our membership matrix by cluster prototypes. Again, while it was reasonable to assign a constant weight of 1 to each of the data points, now that we have clusters of different sizes it makes more sense to consider their size as a constraint for the sum of membership degrees.

We consider the membership matrix U as a data distribution matrix, where $u_{i,j}$ denotes how many data objects of cluster i of partition A shall be assigned

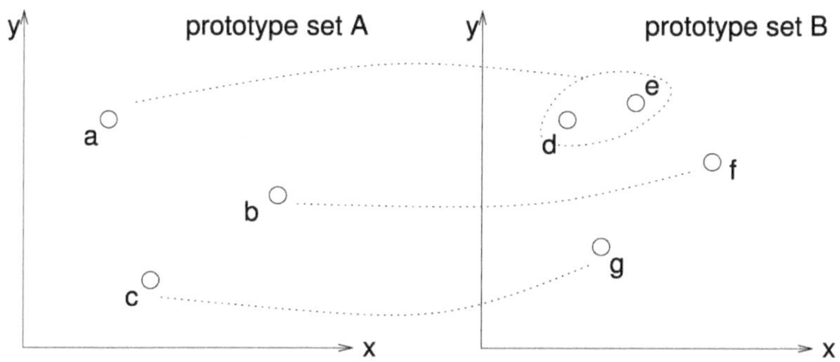

Fig. 10. Constraints on the membership matrix

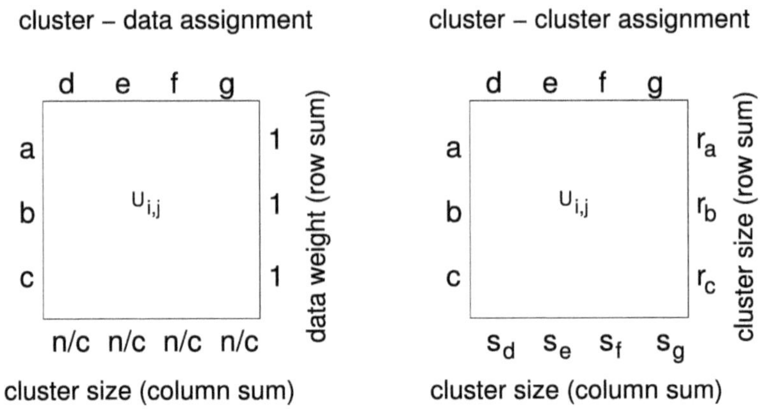

Fig. 11. Constraints on the membership matrix. Left: case of uniform clusters. Right: case of individual weights for data and prototypes.

to cluster j in partition B. Denoting the cluster size of set A by r_j, the cluster sizes r_j and s_i provide us with marginal distributions of the matrix. We are interested in deriving the joint distribution from the marginal distribution given that the total assignment costs are minimized. (Note that due to the possibility of negative entries for $u_{i,j}$ as discussed in section 4, the interpretation as a frequency distribution is not fully justified.)

Since we want to interpret the cluster sizes s_i and r_j as marginal distributions of the same matrix, we apparently require $S := \sum_i s_i = \sum_j r_j =: R$. If the prototypes are derived from data sets of different size such that $R \neq S$, we proportionally adjust the sizes of one data set, e.g. scale s_i by a factor of $\frac{R}{S}$.

To find the (unique) solution, we only have to replace constraint (1) by new constraints

$$\sum_{i=1}^{c} u_{i,j} = r_j \tag{14}$$

and we obtain a slightly modified linear system $A\beta = b'$ with identical $a_{k\ell}$ coefficients but new left hand side

$$b'_k = s_k - \sum_{j=1}^{n} \frac{r_j}{\sum_{i=1}^{c} \frac{d_{kj}}{d_{ij}}} \tag{15}$$

Determining the (minimal) distance between two sets of prototypes, consisting of their position and size, therefore involves solving this linear equation system and calculating the weighted sum of distances via the membership degrees (9) using

$$\alpha_j = \frac{2r_j - \sum_{i=1}^{c} \frac{\beta_i}{d_{ij}}}{\sum_{i=1}^{c} \frac{1}{d_{ij}}}. \tag{16}$$

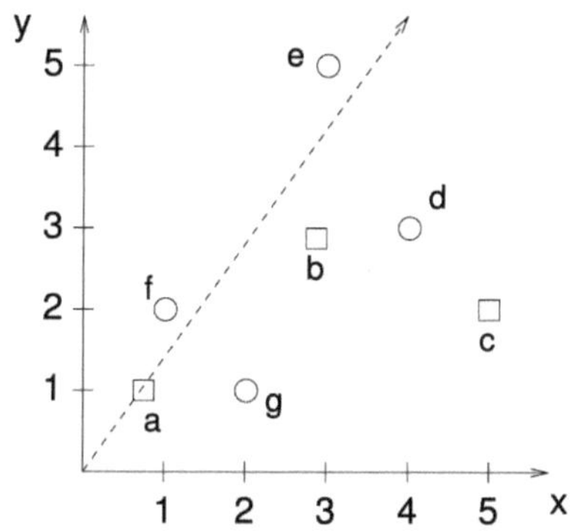

Fig. 12. Example data set. The prototype on the lower left will be moving along the dotted line towards the upper right corner.

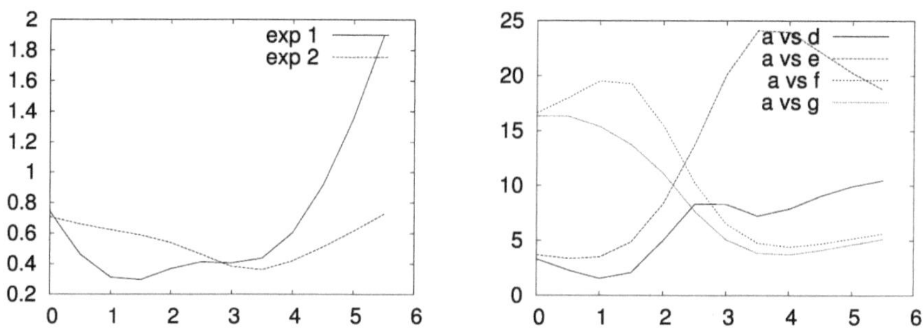

Fig. 13. Left: Distance values for both experiments. Right: distribution of the large cluster a (experiment 1) to clusters in partition $\{d, e, f, g\}$ $(u_{a,*})$.

We illustrate the performance of this distance measure using an example given in figure 12. One set of clusters consists of 4, the other of 3 prototypes. We examine the influence of the position of cluster a, which will move along the indicated line. All cluster sizes were 20, except the three clusters in the lower left corner. In experiment 1 the moving cluster has size 40 (and the two adjacent clusters of the other partition have size 20), in experiment 2 the moving cluster has size 10 (and the other two have size 5).

Figure 13 shows the distance values depending on the x-coordinate of the moving cluster a. For experiment 1 (size 40) the best match is obtained if the prototype is close to the two smaller clusters f and g (cf. figure 13, right). If the size of clusters f and g is rather small (experiment 2), a good match for f and g is less valuable than a good match of cluster e: at $x \approx 3.5$ prototype a is very close to prototype e and the distance becomes minimal. This example shows that the consideration of the cluster sizes adds valuable information in the comparison.

7 Conclusions

In this paper, we have considered the problem of subdividing a data set into homogeneous groups with constraints on the number of data per group. Finding homogeneous groups is a typical task for clustering algorithms, however, if the data density is not uniform, such algorithms usually tend to deliver clusters of varying size, which is inappropriate for some applications. We have proposed an algorithm that outperforms a popular variant of k-means in that respect. We have discussed the case of equi-sized clusters as well as clusters of different sizes. Another interesting aspect of the proposed approach is the extension to the comparison of clustering results in general.

References

1. Jain, A.K., Dubes, R.C.: Algorithms for Clustering Data. Prentice-Hall, Englewood Cliffs (1988)
2. Bezdek, J.C.: Pattern Recognition with Fuzzy Objective Function Algorithms. Plenum Press, New York (1981)
3. Höppner, F., Klawonn, F., Kruse, R., Runkler, T.A.: Fuzzy Cluster Analysis. John Wiley & Sons, Chichester (1999)
4. Klawonn, F., Höppner, F.: What is fuzzy about fuzzy clustering? – Understanding and improving the concept of the fuzzifier. In: R. Berthold, M., Lenz, H.-J., Bradley, E., Kruse, R., Borgelt, C. (eds.) IDA 2003. LNCS, vol. 2810, pp. 254–264. Springer, Heidelberg (2003)
5. Klawonn, F., Höppner, F.: Equi-sized, homogeneous partitioning. In: Gabrys, B., Howlett, R.J., Jain, L.C. (eds.) KES 2006. LNCS (LNAI), vol. 4252, pp. 70–77. Springer, Heidelberg (2006)

Cluster Validating Techniques in the Presence of Duplicates

Ravi Jain[1] and Andy Koronios[2]

[1] School of Computer and Information Sciences,
University of South Australia, Australia
Ravi.Jain@unisa.edu.au

[2] School of Computer and Information Sciences,
University of South Australia, Australia
andy.koronios@unisa.edu.au

Abstract. To detect database records containing approximate and exact duplicates because of data entry error or differences in the detailed schemas of records from multiple databases or for some other reasons is an important line of research. Yet no comprehensive comparative study has been performed to evaluate the effectiveness of Silhouette width, Calinski & Harbasz index (pseudo F-statistics) and Baker & Hubert index (γ index) algorithms for exact and approximate duplicates. In this chapter, a comparative study and effectiveness of these three cluster validation techniques which involve measuring the stability of a partition in a data set in the presence of noise, in particular, approximate and exact duplicates are presented. Silhouette width, Calinski & Harbasz index and Baker & Hubert index are calculated before and after inserting the exact and approximate duplicates (deliberately) in the data set. Comprehensive experiments on glass, wine, iris and ruspini database confirms that the Baker & Hubert index is not stable in the presence of approximate duplicates. Moreover, Silhouette width, Calinski and Harbasz index and Baker & Hubert indice do not exceed the original data indice in the presence of approximate duplicates.

Keywords: Clustering algorithms, Silhouette width, Calinski & Harbasz index, Baker & Hubert indices.

1 Introduction

Detection of approximate and exact duplicates in a database is a difficult task. This type of problem can arise when multiple databases are combined and several records may refer to the same real world entity. A number of character, token, phonetic and numeric similarity metrics based duplicate detection tools (such as The Febrl system, TAILOR, WHIRL,BigMatch) have been developed to find duplicate records. Determining the correct number of clusters within the data set and cluster validation is still a major challenging problem in cluster analysis. Different strategies are used to alleviate this problem, but generally

L.C. Jain et al. (Eds.): Comp. Intel. Para.: Innov. Applications, SCI 137, pp. 181–193, 2008.
springerlink.com

different runs of the program with different values of k need to be evaluated with a quality cluster index to decide about the optimal number of clusters. Nevertheless, this strategy is computationally expensive. The best partition into clusters is the one which optimises the objective function. Selection of an optimal clustering scheme and compactness and cohesion are two main criteria proposed for clustering evaluation. The quality of clusters depends on the type of data set, clustering algorithms [10], and similarity measure used. Silhouette width and Calinski and Harbasz index [13] are considered to be two of the most successful methods for validation in the literature. However, the main disadvantages of these validity indices is that (i) they cannot measure the arbitrary shaped clusters as they usually choose a representative point from each cluster and they calculate distance of the representative points and calculate some other parameter based on these points and (ii) computational time.

The main contribution of this chapter is to assess and compare Silhouette width, Calinski and Harbasz, and Baker & Hubert validation techniques in the presence of exact and approximate duplicate records. The indices were calculated before and after inserting duplicates in the data set. This chapter is organized as follows. The clustering algorithms used in this chapter are described in Section 2. Section 3, presents a brief description Silhouette, Calinski & Harabasz and Baker & Hubert validation techniques. Section 4, presents experimental results followed by conclusion and discussion.

1.1 Outline of Our Approach

Our approach consists of the following two steps. In the first step, we apply clustering algorithms to find the best clusters in the data set and in the second step clustering validation techniques are applied on (i) data without duplicates (ii) data with approximate duplicates and (iii) data with exact duplicates.

2 Clustering Algorithms

A number of clustering algorithms have been developed [6]. Clustering algorithms do not define the proper number of clusters by themselves. These algorithms need the pre-specification of the number of clusters. Clustering algorithms determines the number of clusters in the data set and cluster validation is a measure to evaluate the results of clustering algorithms and provides best fit partition in a data set. We have compared K-means, Partitioning Around Medoids (PAM), Clustering LARge Applications (CLARA), and Fuzzy algorithms [1]. The advantages and disadvantages of these algorithms are also presented in Table 1. In our experiments, the best cluster results were recorded using PAM compare to Kmeans, Clara, and fuzzy because its more robust and insenstive to outliers. Kmeans, Clara, and fuzzy yielded low stability scores. The following section presents a brief description of PAM.

Table 1. Advantages and Disadvantages of various Clustering Algorithms

	KMEANS	PAM	CLARA	FUZZY	SOM
Advantages	KMEANS is a simple clustering method that, when appropriate, shows optimal results. Complexity in time O(nkl), and in space is O(k) where n number of patterns, k number of clusters and l is the number of iterations. Applicable to fairly large data sets.	PAM is more robust than k-means in the presence of noise and outliers because a medoid is less influenced by outliers or other extreme values than a mean. It produces representative prototypes. Complexity $O(k(n-k)^2$ for each iteration where n is # of data, k is # of clusters.	CLARA is fast because it searches in randomly sampled subspaces. Complexity $O(k(40+k)^2+k(n-k))$	FUZZY converges rapidly. Can be used on non quantitative data. Can be used for hard clustering of data. Complexity O(n) Can learn membership functions. Non unique solution.	SOM data is easily interpreted and understood. The reduction of dimensionality and grid clustering makes it easy to observe similarities in the data. SOMs are capable of handling several types of classification problems while providing a useful, interactive, and intelligable summary of the data. Complexity O (2n)
Disadvantages	Sensitive to initial centers. Applicable only when mean is defined. Need to specify k, the number of clusters, in advance. Not suitable to discover clusters with non-convex shapes. The implementation of k-means assumes all attributes to be independent and normally distributed. A lack of explanation requires additional analysis by a supervised learning model. Converges to a local optimum.	PAM works efficiently for small data sets but does not scale well for large data sets.	CLARA is only performing searches in subspaces. The true global minimum is a particular set of k objects that serve as minimum energy medoid. With random sampling, it is unlikely that Clara will find these particular points.	Sensitivity to the presence of noise and outliers in the data. Sensitivity to the initial guess (speed, local minima).	The major disadvantage of a SOM, is that it requires necessary and sufficient data in order to develop meaningful clusters. The weight vectors must be based on data that can successfully group and distinguish inputs. Lack of data or extraneous data in the weight vectors will add randomness to the groupings. Finding the correct data involves determining which factors are relevant and can be a difficult or even impossible task in several problems. The ability to determine a good data set is a deciding factor in determining whether to use a SOM or not. Another problem with SOMs is that it is often difficult to obtain a perfect mapping where groupings are unique within the map. Instead, anomalies in the map often generate where two similar groupings appear in different areas on the same map. Clusters will often get divided into smaller clusters, creating several areas of similar neurons.

2.1 Partitioning Around Medoids (PAM)

PAM [1, 2] is a generalization of the K-Means clustering Algorithm. PAM accepts any dissimilarity matrix and more robust than k-means in the presence of noise and outliers because it minimises a sum of dissimilarities instead of a sum of squared Euclidian distance and a medoid is less influenced by outliers or other extreme values than a mean. The objective of PAM is to determine a representative object or medoid for each cluster that is most centrally located objects within clusters. The algorithms is divided into two phases. In the first phase, called BUILT, the k representative objects are chosen. The second phase, called SWAP, is attempted to improve the set of representative objects that was chosen in the first phase. For a pre-specified number of clusters k, it begins by selecting an object as medoid for each cluster. After finding a set of k medoids, k clusters are constructed by assigning each observation to the nearest medoids. The medoids minimise the sum of the distances of the observations to their closed medoids.

$$E = \sum_{i=1}^{i=k} \sum_{p \in C_i} d(p - o_i)^2 \tag{1}$$

3 Validation Techniques

There are several types of cluster validation techniques [4, 5, 6, 7, 8, 9, 11, 12, 13, 14, 15, 16]. Cluster validation are often based on internal, external and relative criteria [10]. Internal criteria assess the degree to which a clustering structure, produced by a clustering algorithm, and matches the proximity matrix of the corresponding data set and measures how compact and well-separated the clusters are. To obtain clusters with these characteristics, the dispersion measure for each cluster needs to be as small as possible, while the dissimilarity measure between clusters need to be large. This section presents a brief summary of well-known Silhouette, Calinski & Harabasz and Baker & Hubert validation techniques.

3.1 Silhouette Index

The silhouette validation technique [1, 2] combine ideas of both cohesion and separation for individual points. It calculates the *silhouette width* for each sample, *average silhouette width* for each cluster and overall *average silhouette width* for a total data set and requires at least 2 clusters [1, 2]. Using this approach each cluster is represented by so called silhouette, which is based on the comparison of its tightness and separation. The *average silhouette width* is applied for evaluation of clustering validity and can be used to decide how good the number of selected clusters are. For a given cluster $X_k, (k = 1...c)$, this method assigns to each sample of X_k a quality measure, $s(i)(i = 1...m)$ known as the Silhouette width. The Silhouette width is a confidence indicator on the membership of the ith sample in cluster X_k. The Silhouette width for the ith sample in cluster X_k is defined as follows.

$$s(i) = \frac{b(i) - a(i)}{max(a(i), b(i))} \tag{2}$$

where $a(i)$ is the average distance between the ith sample and all of the samples of X_k."max" is the maximum operator and $b(i)$ is the minimum average distance between the ith sample and all of the samples clustered in $X_k, k \neq c$. Clearly it shows $-1 \leq s(i) \leq 1$. When $a(i)$ is close to 1, the within dissimilarity $a(i)$ is much smaller than the smallest between dissimilarity $b(i)$ that is, the ith sample has been "well clustered" [1]. When $a(i)$ is -1, implies that $a(i)$ is much larger than $b(i)$ that is it is misclassified and it can be assigned again. When $a(i)$ is close to zero, $a(i)$ and $b(i)$ have similar values and the ith sample can be assigned to the nearest neighboring cluster that is lies equally far away from both the clusters. Thus, for a given cluster $X_k(k = 1...c)$, Silhouette width S_j is calculated which characterises the heterogeneity and isolation properties of such a cluster:

$$S_j = \frac{1}{n} \sum_{i=1}^{n} s(i) \tag{3}$$

where n is number of samples in S_j. For any partition $U \leftrightarrow X : X_1 \cup X_2 \cup \ldots X_c$ a *Global Silhouette width,* GS_u, can be used as an effective validity index for U.

$$GS_u = \frac{1}{c} \sum_{i=1}^{c} S_j \tag{4}$$

The most appropriate number of clusters are represented by equation (4) for U and the partition with the maximum S_u is considered as the optimal partition. The average silhouette over all elements has been used to evaluate and compare clustering results, including selecting the number of clusters k by maximizing average silhouette over a range of possible values for k [1].

3.2 Calinski and Harabasz Index

Calinski and Harabas [12, 16] proposed the following index to assesses the quality of well isolated and coherent clusters.

$$CH(k) = \frac{B(k)(k-1)}{W(k)(n-k)} \tag{5}$$

where k denotes the number of clusters and $B(k)$ and $W(k)$ denote the between (separation) and within (cohesion) cluster sums of squares of the partition, respectively. Separation is measured by

$$BSS = \sum_{i} \sum_{x \epsilon C_i} (x - m_i)^2 \tag{6}$$

Cohesion is measured by

$$WSS = \sum_{i} |C_i|(m - m_i)^2 \tag{7}$$

where $|C_i|$ is the size of cluster i. An optimal number of clusters is then defined as a value of k that maximizes $CH(k)$. In essence, the higher the value of this index, the greater the separation between clusters.

3.3 Baker and Hubert Index

The Baker and Huberts index [14, 18] measures the correlation between the matrices, X and Y, of dimension N×N, drawn independently of each other. Here comparisons are made between all cluster pairwise dissimilarities and all between-clusters pairwise dissimilarities. A comparison is defined as consistent if a within cluster dissimilarity is strictly less than a between-clusters dissimilarity. The Γ index is computed as

$$\Gamma = \frac{\Gamma_+ - \Gamma_-}{\Gamma_+ + \Gamma_-} \tag{8}$$

where Γ_+ and Γ_- represents the number of consistent and inconsistent comparisons respectively. The maximum value of the Γ index indicates the correct number of clusters. The absolute maximum of that index is 1.

4 Experimental Evaluation

To demonstrate the effectiveness of Silhouette, Calinski & Harabasz and Baker & Hubert validation indices we have performed comprehensive tests on iris, wine, glass, and ruspini data set from UCI Machine Learning Repository [19] after applying PAM clustering algorithms presented in Section 2. For simplicity, we present results for iris and ruspini data sets.

Data Description
The iris database is the most often used data set in pattern recognition, statistics, data analysis, and machine learning. The iris data contain four quantitative measurements sepal length and width, and petal length and width for 150 flowers, 50 irises from each of three species setosa, versicolor, and virginica. Petal-length and petal-width are very good discriminators of the three classes.

The Ruspini data were first used by Ruspini (1970) to illustrate fuzzy clustering techniques. The data are two dimensional and consist of 75 points containing four clusters. The following three cases were considered. Figure 2 and 3 shows box plot for iris and ruspini data.

Case 1: when no attributes were duplicated.
Case 2: when some attributes were duplicated (Approximate Duplicates).
Case 3: when all the attributes were exactly duplicated.

4.1 Coefficient Relibility wrt Record Based Duplicates

Table 2, 3 and 4 provides a summary of comparison results for silhouette width (Sil), Calinski and Harbasz index ($G1$) and Baker & Hubert indices ($G2$) for iris

Table 2. Silhouette, Calinski and Harabasz & Baker and Hubert indices for clean iris data

Data	# of Clusters	Silhouette	G1	G2
Iris	3	**0.51**	**357.55**	**0.90**
	4	0.44	311.07	0.89
	5	0.41	262.00	0.87
	6	0.32	262.56	0.86
	7	0.30	237.61	0.85
	8	0.30	223.37	0.85
	9	0.31	235.80	0.89
	10	0.29	217.33	0.90

Table 3. Silhouette, Calinski and Harabasz & Baker and Hubert indices after inserting 25% duplicate iris data

Data	# of Clusters	Sil	G1	G2
Iris	3	**0.40**	**365.20**	0.82
	4	0.38	333.13	0.84
	5	0.36	310.80	0.85
	6	0.36	310.10	0.87
	7	0.34	286.70	0.87
	8	0.34	273.47	0.88
	9	0.30	276.97	**0.89**
	10	0.29	262.17	0.89

Table 4. Silhouette, Calinski and Harabasz & Baker and Hubert indices for after inserting 100% exact iris data

Data	# of Clusters	Sil	G1	G2
Iris	3	**0.51**	**722.41**	**0.90**
	4	0.45	630.66	0.89
	5	0.43	533.04	0.87
	6	0.34	536.07	0.87
	7	0.33	486.86	0.86
	8	0.32	459.33	0.86
	9	0.33	486.66	**0.90**
	10	0.31	450.19	**0.90**

data with no duplicates, approximate and exact duplicates respectively. Table 5 presents silhouette width Sil, Calinski and Harbasz and Baker & Hubert indices for several subsets of 25% value based duplicates of iris data. Table 6, 7 and 8 provides a summary of comparison results for ruspini data set. Approximate duplicates were synthetically generated values were padded in the real data. All synthetic data were generated using R [3] function *runif*. We simulated with 25%,

Table 5. Silhouette, Calinski and Harabasz & Baker and Hubert indices after inserting 25% duplicate iris data for various subsets

Data	Duplicate Attributes	Sil	G1	G2
Iris	A1	0.19	37.83	0.50
	A2	0.23	51.05	0.59
	A3	0.23	44.45	0.58
	A4	0.16	32.51	0.45
	A1A2	0.27	155.06	0.62
	A1A3	0.29	182.77	0.68
	A1A4	0.30	188.40	0.68
	A2A3	0.30	180.58	0.70
	A2A4	0.28	172.51	0.68
	A3A4	0.33	241.18	0.75
	A1A2A3	0.37	266.17	0.77
	A1A3A4	**0.38**	**349.90**	**0.81**
	A2A3A4	0.38	349.90	0.81
	A1A2A4	0.32	229.94	0.73

Table 6. Silhouette, Calinski and Harabasz & Baker and Hubert indices for clean Ruspini data

Data	# of Clusters	Sil	G1	G2
Ruspini	2	0.53	89.38	0.83
	3	0.60	98.67	0.83
	4	**0.74**	357.12	**1.00**
	5	0.64	307.78	0.97
	6	0.61	322.10	0.98
	7	0.52	341.87	0.96
	8	0.51	353.23	0.96
	9	0.51	347.03	0.96
	10	0.51	**354.37**	0.96

50% and 75% duplicated data. The results presented in the following tables for 10 runs for clusters ranging from 3 to 10.

Table 2 presents the results of Sil, G1, G2 indices for clusters ranging from 3 to 10 for cleaned iris data. From table 2, we can see that the highest value for Sil, G1 and G2 value are 0.51, 357.55 and 0.90 respectively for 3 clusters resulting in correct classification.

Table 3 presents the results of Sil, G1, G2 indices after injecting 25% duplicate in iris data set. From table 3, we can see that the highest value for Sil, and G1 value are 0.40, 365.20 respectively for 3 clusters but G2 is 0.89 for 9 clusters. From table 3, we can clearly see the difference of 0.11 in Silhoutte index and G1 is dramatically increased by 7.65 compared to table 2 and G2 is unreliably predicting 9 clusters.

Table 7. Silhouette, Calinski and Harabasz & Baker and Hubert indices after inserting 25% duplicates in Ruspini data

Data	# of Clusters	Sil	G1	G2
Ruspini	2	0.53	89.38	0.83
	2	0.46	132.78	0.73
	3	0.49	151.47	0.81
	4	0.56	255.48	0.95
	5	0.55	245.96	0.95
	6	**0.49**	**246.22**	0.94
	7	0.48	229.74	0.94
	8	0.49	239.13	0.96
	9	0.49	246.23	**0.97**
	10	0.45	232.06	0.96

Table 8. Silhouette, Calinski and Harabasz & Baker and Hubert indices for after inserting 100% exact duplicates in Ruspini data

Data	# of Clusters	Sil	G1	G2
Ruspini	2	0.53	89.38	0.83
	2	0.54	181.21	0.83
	3	0.61	201.45	0.83
	4	**0.75**	734.36	**1.00**
	5	0.66	637.54	0.97
	6	0.63	672.20	0.98
	7	0.54	718.93	0.96
	8	0.54	748.63	0.96
	9	0.54	741.39	0.96
	10	0.55	**763.26**	0.96

Table 4 presents the results of Sil, G1 and G2 indices for exact duplicates of iris data. From table 4, we can see that the highest value for Sil, and G1 value are 0.51, 722.20 respectively for 3 clusters but G2 is 0.90 for 3, 9 and 10 clusters. Notice here a remarkable difference in G1 which is jumped to twice from previous tables 2 and 3 leading to unstable and unreliable results.

4.2 Coefficient Reliability wrt Value Based Duplicates

Table 5 presents the results of Sil, G1, G2 indices after inserting 25% duplicates in iris data. In previous tables, record based duplicates were injected. In table 5 value-based duplicates were injected instead of record based duplicates. Here every attribute value in a tuple were slightly altered while other values remains the same for several subsets. From Table 5, we can see that the highest Sil, G1 and G2 value were 0.38, 349.90.20 and 0.81 for 3 clusters respectively. Notice that higher the number of duplicates the higher the values of the indices. The

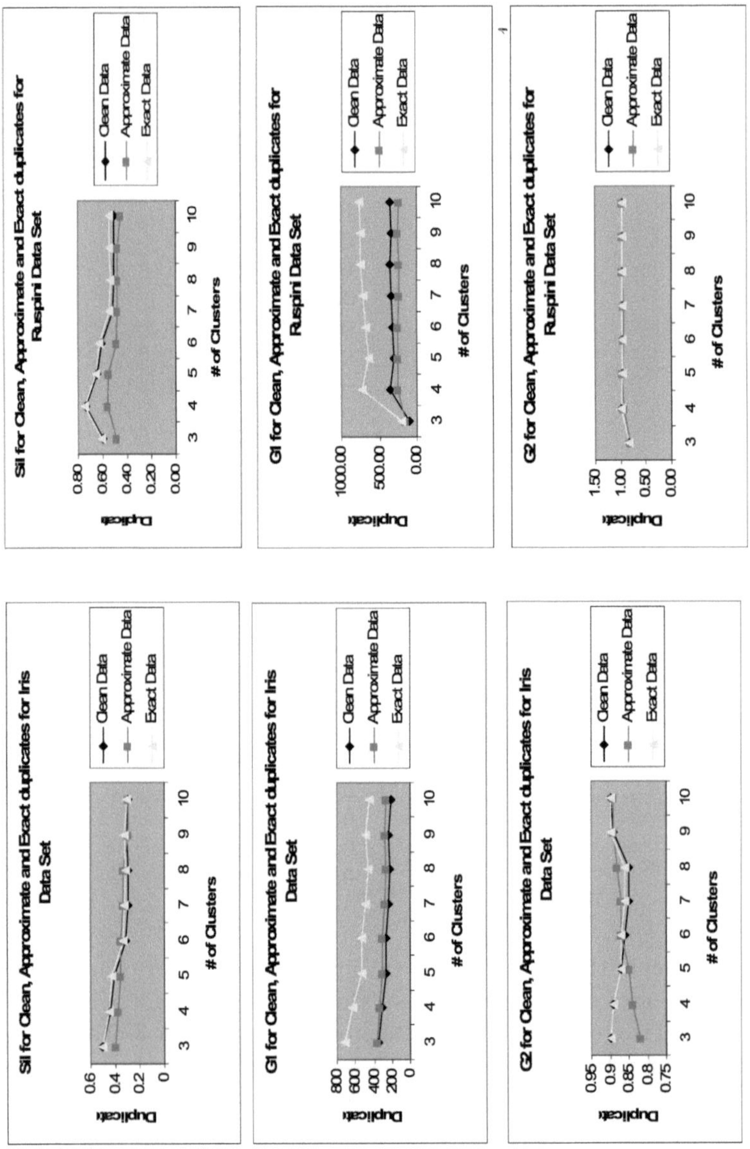

Fig. 1. Sil, G1 and G2 for Iris & Ruspini Data set

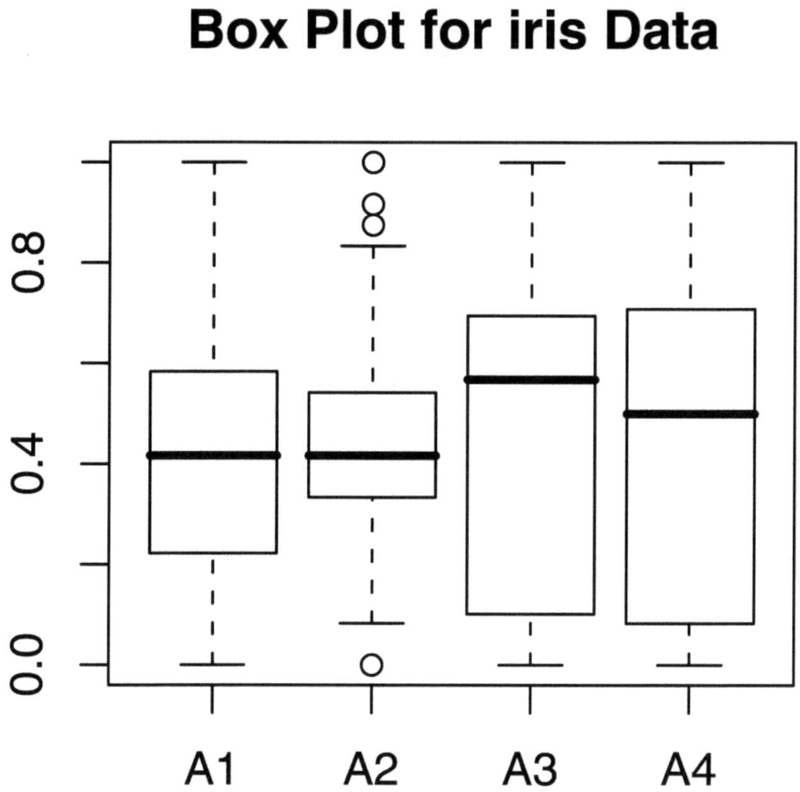

Fig. 2. Box plot for iris data

highest Sil, G1 and G2 were 0.38, 349.90 and 0.81 respectively for two duplicate subsets $\{A1, A3, A4\}$ and $\{A2, A3, A4\}$. From these experiments we can conclude that techniques failed to correctly classify approximate duplicates in the data.

Table 6 presents the results of Sil, G1, G2 indices without duplicates in Ruspini data. From table 6, we can see that the highest value for Sil, and G2 value are 0.74, and 1 respectively for 4 clusters. But G2 is 354.37 for 10 clusters leading to incorrect classification.

Table 7 presents the results of Sil, G1, G2 indices after inserting 25% duplicates in Ruspini data. From table 7, we can see that the highest value for Sil and G1 value are 0.49, 246.22 for 6 clusters but G2 is 0.97 respectively for 9 clusters. Once again G2 leading to incorrect classification.

Table 8 presents the results of Sil, G1, G2 indices inserting 100% for exact ruspini data. From table 8, we can see that the highest Sil and G2 value are 0.75, respectively for 4 clusters and G2 is 763.26 for 10 clusters. Notice here a remarkable difference in G1 which is jumped to twice from previous tables 2 and 3 leading to unstable and unreliable results.

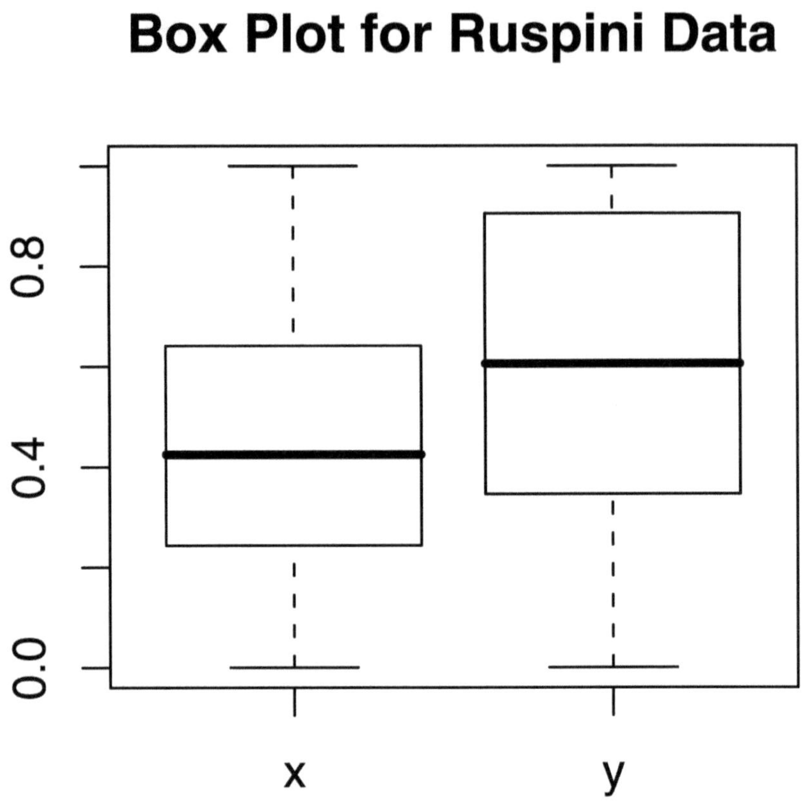

Fig. 3. Box Plot for ruspini data

5 Conclusion

In this chapter, we have presented a brief description and comparative performances of three cluster validation techniques, Silhouette width, Calinski and Harbasz index and Baker and Hubert index on iris and ruspini data set. We have conducted several comprehensive experiments on iris, glass, wine, and ruspini data set with record and tuple based duplicates.

From our experiments, we conclude that Baker and Hubert index is not stable leading to unstable and incorrect classification in the presence of approximate duplicates. We also presented advantages and disadvantages of various clustering algorithms. Also, we found that in the presence of approximate data these indices always remain low compare to the original data set. But failed when 100% data is duplicated which may not occur in reality. Though, a series of extensive experiments on large data set are needed. Our future goals to develop a new algorithm that can handle duplicates data.

Acknowledgements

This work was performed during my employment with Institut de recherche en informatique et systmes alatoires (IRISA), FRANCE in 2007. I am grateful to Professor Laure Berti for the opportunity, innovative ideas, and constructive feedback during my stay in her research laboratory in FRANCE.

References

[1] Rousseeuw, P.J.: Finding groups in data: an introduction to cluster analysis. Wiley, New York (1990)
[2] Rousseeuw, P.J.: Silhouettes: A graphical aid to the interpretation and validation of cluster analysis. Journal of Comp App. Math. 20, 53–65 (1987)
[3] R Development Core Team R: A language and environment for statistical computing. R Foundation for Statistical Computing, Vienna, Austria, http://www.r-project.org/
[4] Hirano, S., et al.: Comparison of clustering methods for clinical databases. Journal of Information Sciences, 155–165 (2004)
[5] Halkidi, M., et al.: On Clustering validation techniques. Journal of Intelligent Information Systems 17(2/3), 107–145 (2001)
[6] Jain, A., et al.: Data clustering: a review. ACM Computing Surveys 31(3), 264–323 (1999)
[7] Halkidi, M., et al.: Cluster validity methods: Part 1. Sigmod Record 31(2), 40–45 (2002)
[8] Halkidi, M., et al.: Cluster validity methods: Part 2. Sigmod Record 31(3), 19–27 (2002)
[9] Halkidi, M., et al.: On clustering validation techniques. Journal of Intelligent Information Systems 17(2/3), 107–145 (2001)
[10] MacQueen, J.B.: Some Methods for classification and analysis of multivariate observations. In: Proceedings of 5-th Berkeley Symposium on Mathematical Statistics and Probability, vol. 1, pp. 281–297. University of California Press, Berkeley (1967)
[11] Bolshakova, N., Azuaje, F.: Cluster validation techniques for genome expression data 83(4), 825–833 (2003)
[12] Tibshirani, et al.: Estimating the number of clusters in a data set via the gap statistic. Journal R. Stat. Soc. Ser. B 63, 411–423 (2001)
[13] Milligan, G.W., Cooper, M.C.: An examination of procedures for determining the number of clusters in a data set. Psychometrika 50(2), 159–179 (1985)
[14] Baker, F.B., Hubert, L.J.: Measuring the power of hierarchical cluster analysis. Journal of the American Statistical Association, 31–38 (1975)
[15] Stein, B., et al.: On cluster validity and the information need of users. In: 3rd IASTED Int. Conference on Artificial Intelligence and Applications (AIA 2003), pp. 216–221 (2003)
[16] Ahmed, K., et al.: Duplicate record detection: A survey. IEEE Transactions on Data and Knowledge and Engineering 19(1), 1–16 (2007)
[17] Calinski, R.B., Harabasz, J.: A dendrite method for cluster analysis. Communications in Statistics 3, 1–27 (1974)
[18] Hubert, L., Arabie, P.: Comparing partitions. Journal of Classification 2, 193–218 (1985)
[19] Blake, C.L., et al.: UCI repository of machine learning databases (1998)

Fuzzy Blocking Regression Models

Mika Sato-Ilic

Faculty of Systems and Information Engineering, University of Tsukuba
Tsukuba, Ibaraki 305-8573, Japan
mika@risk.tsukuba.ac.jp

Abstract. Regression analysis is a well known and a widely used technique in multivariate data analysis. The efficiency of it is extensively recognized. Recently, several proposed regression models have exploited the spatial classification structure of data. The purpose of this inclusion of the spatial classification structure is to set a heterogeneous data structure to homogeneous structure in order to adjust the heterogeneous data structure to a single regression model. One such method is a blocking regression model. However, the ordinal blocking regression model can not reflect the complex classification structure satisfactorily. Therefore, the fuzzy blocking regression models are offered to represent the classification structure by using fuzzy clustering methods. This chapter's focus is on the methods of the fuzzy clustering based blocking regression models. They are extensions of the conventional blocking regression model.

1 Introduction

Fuzzy clustering techniques are widely used in a large number of areas, since this clustering can obtain a result describing real-world phenomenon that is substantiated by robust solutions and low calculation cost when compared with hard clustering. Hard clustering means classifying the given observation into exclusive clusters. This way, we can discriminate clearly whether an object belongs to a cluster or not. However, such a partition is not sufficient to represent many real situations. So, a fuzzy clustering method is offered to allow clusters to have uncertainty boundaries, this method allows one object to belong to some overlapping clusters with some grades [2], [9], [13].

By exploiting such advantages of fuzzy clustering, several hybrid methods have been proposed in the area of multivariate data analysis [14]. In particular, the regression methods use the result of fuzzy clustering as spatial weights of data structure and represent the nonlinearity of data by using the weights.

Fuzzy c-regression model [8], geographically weighted regression model [3], and fuzzy cluster loading models [14] are typical examples. These models use weights which show the degree of contribution of objects to clusters and represent the distribution of the weights by using fuzzy membership function or probabilistic density function. The estimates of the regression coefficients are obtained by the weighted least squares method. That is, these models are implemented by

L.C. Jain et al. (Eds.): Comp. Intel. Para.: Innov. Applications, SCI 137, pp. 195–217, 2008.
springerlink.com

multiplying the weights to the data and attempts to set a heterogeneous data structure to homogeneous structure.

On the other hand, a regression model which attempts to solve this problem by adding different intercepts according to different clusters (or blocks) [4]. In this model, identification of clusters (or blocks) are represented by dummy variables and the values of the dummy variables are given in advance by using external information to identify the blocks. Considering the substantial reliance on data itself, we have proposed a fuzzy blocking regression model [15] by using fuzzy clustering for identifying the clusters (or blocks). In this model, we exploit the feature of fuzzy clustering so that we can obtain the result as continuous values. Usually, data is observed as numerical continuous values, if we replace the result of conventional hard clustering with the dummy values, then we need to analyze the mixed data including continuous values and discrete values. Therefore, the fitness is decreased. However, the fuzzy blocking regression model can avoid this problem, since this regression model is based on a result of fuzzy clustering.

Using the idea of fuzzy intercept, we also have proposed a different fuzzy blocking regression model involving fuzzy block intercepts by recognizing that the conventional blocking regression model is reduced to a model in which the difference of the blocks are shown by the difference of intercepts [17].

Moreover, we have proposed an extended fuzzy blocking regression model involving weights which can show the variable based fuzzy clustering structure for the data [18]. That is, in this model, we consider not only the weight of data over all variables, but also the weights for each variable. The weight of data over all variables is estimated by using conventional fuzzy clustering. Weighting for each variable can be done in which the classification structure is clearer, which makes the variable more significant. That is, this weight works to represent how each variable contributes to the classification by using the dissimilarity structure at each variable.

In this method, we use the variable based fuzzy clustering method [16] in order to estimate the weights for each variable. This method is a fuzzy clustering method considering weights of variable-based dissimilarities over objects in the subspace of the object's space. In order to estimate the weights, we propose two methods. One is a method in which a conventional fuzzy clustering method is directly used for the variable-based dissimilarity data. The other is to use a new objective function. Exploiting the weights, we define a dissimilarity assigned with the significance of each variable for the classification and reduce the number of variables. We can implement a fuzzy clustering method under the intrinsically weighted classification structure on the subspace of data.

The key merits of the fuzzy regression blocking models when compared to an ordinal blocking regression model are as follows:

1. Regression coefficients of independent variables and regression coefficients of all fuzzy clusters for each object can be obtained simultaneously, that is we can capture the relationship between the fuzzy clusters and dependent variable and the relationship between independent variables and dependent variable at the same time.

2. We do not need any external information in order to identify the clusters (or blocks).
3. We can obtain a better performance of fitness for the proposed model.

This chapter consists of nine sections. The next section describes fuzzy clustering methods. In section 3, we describe the variable-based fuzzy clustering methods. In section 4, we briefly mention an ordinal blocking regression analysis. In sections 5 and 6, we state a fuzzy blocking regression analysis and a regression model using fuzzy block intercepts. Section 7 describes the variable based fuzzy blocking regression model. Section 8 shows several numerical examples and section 9 states conclusions.

2 Fuzzy Clustering

Suppose a set of n objects with respect to p variables is as follows:

$$S = \{\boldsymbol{x}_1, \cdots, \boldsymbol{x}_n\},\ \boldsymbol{x}_i = (x_{i1}, \cdots, x_{ip}),\ i = 1, \cdots n. \tag{1}$$

The purpose of clustering is to classify n objects into K clusters.

The state of fuzzy clustering is represented by a partition matrix $U = (u_{ik})$ whose elements show the grade of belongingness of the objects to the clusters, $u_{ik},\ i = 1, \cdots, n,\ k = 1, \cdots, K$. In general, u_{ik} satisfies the following conditions:

$$u_{ik} \in [0, 1],\ \sum_{k=1}^{K} u_{ik} = 1. \tag{2}$$

Fuzzy c-means (FCM) [1] is one of the fuzzy clustering methods. FCM is the method which minimizes the weighted within-class sum of squares:

$$J(U, v) = \sum_{i=1}^{n} \sum_{k=1}^{K} u_{ik}^{m} \sum_{a=1}^{p} (x_{ia} - v_{ka})^2, \tag{3}$$

where $v = (v_{ka}),\ k = 1, \cdots, K,\ a = 1, \cdots, p$ shows a matrix of centers whose element v_{ka} denotes the value of the center of a cluster k with respect to a variable a. x_{ia} is i-th object with respect to a variable a. The exponent m which determines the degree of fuzziness of the clustering is chosen from $[1, \infty)$ in advance. The purpose is to obtain the solutions U and v which minimize equation (3). From conditions shown in equation (2), the local extrema of equation (3) can be obtained as follows:

$$u_{ik} = 1 / \sum_{t=1}^{K} \left[\frac{\sum_{a=1}^{p} (x_{ia} - v_{ka})^2}{\sum_{a=1}^{p} (x_{ia} - v_{ta})^2} \right]^{\frac{1}{m-1}}, \tag{4}$$

$$v_{ka} = \sum_{i=1}^{n} u_{ik}^m x_{ia} / \sum_{i=1}^{n} u_{ik}^m, \tag{5}$$

$$i = 1, \cdots, n, \quad k = 1, \cdots, K, \quad a = 1, \cdots, p.$$

If we assume equation (5), then the minimizer of equation (3) is shown as:

$$J(U) = \sum_{k=1}^{K} \left(\sum_{i=1}^{n} \sum_{j=1}^{n} u_{ik}^m u_{jk}^m d_{ij} / (2 \sum_{l=1}^{n} u_{lk}^m) \right), \tag{6}$$

where $d_{ij} = \sum_{a=1}^{p} (x_{ia} - x_{ja})^2$. When $m = 2$, equation (6) is the objective function of the FANNY algorithm [10].

3 Variable Based Fuzzy Clustering

Do to the increasing amount of data growth, recently, we are faced with the challenge of analyzing data which has a larger number of variables than the number of objects. If we treat such data as the target of clustering, there are problems of instability of the solution and calculation cost. In order to solve these problems, variable selection has been used as the pre-processing data considering correlation between variables or variance for each variable. The use of correlation is based on the idea that similar explainable variables are redundant and consideration of the variance is based on an idea that we should delete variables which have low reliability. That is, the problems have been dealt with by using a subspace of data. Recently, several clustering methods which include the idea of subspaces of data have been discussed [6], [12]. Also, associated with the clusters and weights of variables, several data analyses are discussed [7], [14].

In this section, we state a clustering method on an obtained subspace based on the weights estimated as a result of fuzzy clustering. In this method, we define a new dissimilarity between objects. This dissimilarity is based on a weight which can show how each subspace spanned by each variable contributes to the classification of objects over all variables.

3.1 Variable Based Dissimilarity

Suppose

$$X = \begin{pmatrix} \boldsymbol{x}_1 \\ \vdots \\ \boldsymbol{x}_n \end{pmatrix}, \quad \boldsymbol{x}_i = (x_{i1}, \cdots, x_{ip}), \quad i = 1, \cdots, n \tag{7}$$

is a $n \times p$ data matrix.

We define the dissimilarity $\tilde{d}_{ij}$ between objects i and j as follows:

$$\tilde{d}_{ij} = \sum_{a=1}^{p} w_a d_{ija}, \quad i, j = 1. \cdots, n. \tag{8}$$

Where d_{ija} is (i,j)-th element of a dissimilarity matrix D_a and shows dissimilarity between objects i and j with respect to a variable a. We define this as follows:

$$D_a = (d_{ija}), \quad d_{ija} = (x_{ia} - x_{ja})^2, \quad i,j = 1, \cdots, n, \quad a = 1, \cdots, p. \tag{9}$$

w_a shows a weight of variable a. In particular, conventional squared Euclidean distance

$$d_{ij} = \sum_{a=1}^{p} (x_{ia} - x_{ja})^2, \; i,j = 1, \cdots, n$$

is the case when

$$w_a = 1, \; \forall a$$

in equation (8). This means a special case of squared Euclidean distance when we assume the same weight for all of the variables.

In equation (8), we define the weight as follows:

$$w_a = \frac{1}{n} \sum_{i=1}^{n} \sum_{k=1}^{K} u_{ika}^2, \; a = 1, \cdots, p. \tag{10}$$

Where u_{ika} shows degree of belongingness of an object i to a cluster k with respect to a variable a and is assumed to satisfy the following conditions:

$$u_{ika} \in [0,1], \quad \sum_{k=1}^{K} u_{ika} = 1. \tag{11}$$

K satisfies $1 < K < n$. Equation (10) shows a partition coefficient [2] of a classification using a dissimilarity matrix with respect to a fixed variable. That is, a value of w_a becomes larger when the classification structure is clearer and becomes smaller when the classification structure is less clear. w_a satisfies $\frac{1}{K} \leq w_a \leq 1, \; \forall a$. In order to normalize the weight w_a into $[0,1]$, we put w_a as follows:

$$w_a \equiv \frac{w_a - \frac{1}{K}}{1 - \frac{1}{K}}. \tag{12}$$

Substituting equations (9) and (10) for equation (8), dissimilarity considering weights of variables based on fuzzy clustering is shown as follows:

$$\tilde{d}_{ij} = \frac{1}{n} \sum_{a=1}^{p} \sum_{l=1}^{n} \sum_{k=1}^{K} u_{lka}^2 (x_{ia} - x_{ja})^2, \quad i,j = 1. \cdots, n. \tag{13}$$

Also, substituting equations (9) and (12) for the right hand side of equation (8) and using equation (13), dissimilarity considering normalized weights of variables based on fuzzy clustering, $\tilde{\tilde{d}}_{ij}$, is shown as follows:

$$\tilde{\tilde{d}}_{ij} = \frac{K}{n(K-1)} \tilde{d}_{ij} - \frac{1}{K-1} d_{ij}, \quad i,j = 1. \cdots, n. \tag{14}$$

Once data is given, then d_{ij} is a constant, the properties of $\tilde{d}_{ij}$ related with the weights is essentially the same as $\tilde{\tilde{d}}_{ij}$ shown in equation (14). The purpose of the variable based fuzzy clustering is to classify objects by using the dissimilarities $\tilde{d}_{ij}$ or $\tilde{\tilde{d}}_{ij}$ shown in equations (13) or (14), respectively.

3.2 Variable Based Fuzzy Clustering Methods

We describe two methods to estimate the degree of belongingness shown in equation (11). One of them is a method to obtain the degree of belongingness of objects to clusters by applying the dissimilarity data at each variable shown in equation (9) directly into a fuzzy clustering method in which the target data is dissimilarity data. The other is a method to estimate using a new criterion.

When considering a fuzzy clustering method in which the target data is dissimilarity data, we use the FANNY algorithm [10]. The objective function of this algorithm is defined in equation (6). In equation (6), we redefine the objective function with respect to a variable a by using equations (9) and (11) as follows:

$$J(U_a) = \sum_{k=1}^{K} \left(\sum_{i=1}^{n} \sum_{j=1}^{n} u_{ika}^m u_{jka}^m d_{ija} / (2 \sum_{s=1}^{n} u_{ska}^m) \right), \quad a = 1, \cdots, p. \tag{15}$$

Therefore, u_{ika} can be estimated by minimizing equation (15).

For the second method, we define the following objective function.

$$F(U_a, v) = \sum_{i=1}^{n} \sum_{k=1}^{K} \sum_{a=1}^{p} u_{ika}^m (x_{ia} - v_{ka})^2. \tag{16}$$

Where, v_{ka} shows the center of a cluster k with respect to a variable a. If $\exists i, k$, u_{ika} is a constant for all a, then this is the same objective function as fuzzy c-means shown in equation (3). The purpose is to estimate $U_a = (u_{ika})$ and $v = (v_{ka})$ which minimize equation (16). From conditions shown in equation (11), the local extrema of equation (16) can be obtained as follows:

$$u_{ika} = 1 / \sum_{t=1}^{K} \left[\frac{x_{ia} - v_{ka}}{x_{ia} - v_{ta}} \right]^{\frac{2}{m-1}}, \tag{17}$$

$$v_{ka} = \sum_{i=1}^{n} u_{ika}^m x_{ia} / \sum_{i=1}^{n} u_{ika}^m, \tag{18}$$

$$i = 1, \cdots, n, \quad k = 1, \cdots, K, \quad a = 1, \cdots, p.$$

Therefore, we can estimate the degree of belongingness by alternating optimization using equations (17) and (18).

The algorithm of the proposed fuzzy clustering which considers the weights of variables is as follows:

[Step 1] Calculate $D_a,\ a = 1, \cdots, p$ by using $X = (x_{ia})$ shown in equation (7) and equation (9).

[Step 2] Find $U_a = (u_{ika})$ by using D_a which is obtained at Step 1 and equation (15). Otherwise, find $U_a = (u_{ika})$ by using $X = (x_{ia})$ shown in equation (7) and equation (17).

[Step 3] Find w_a by using $U_a = (u_{ika})$ which are obtained at Step 2 and equations (10) and (12).

[Step 4] If $w_a < \varepsilon$, then $w_a = 0$.

[Step 5] Find $\tilde{d}_{ij}$ by using w_a which are obtained at Steps 3 and 4, D_a which is obtained at Step 1, and equation (8).

[Step 6] Apply $\tilde{d}_{ij}$ which is obtained at Step 5 into equation (6) and find $\hat{U} = (\hat{u}_{ik})$.

4 Blocking Regression Model

Suppose X shown in equation (7) is a $n \times p$ data matrix with respect to p independent variables and

$$\boldsymbol{y} = (y_1, \cdots, y_n)^t$$

shows values of a dependent variable. If an observed data has K distinct blocks, then we must consider the blocks in the regression analysis. To take account of the fact that there are K distinct blocks in the data, a blocking regression model [4] is defined as follows:

$$\boldsymbol{y} = \hat{X}\hat{\boldsymbol{\beta}} + \hat{\boldsymbol{e}}, \tag{19}$$

where the $n \times (p + K + 1)$ matrix $\hat{X}$ is defined as:

$$\hat{X} = (\mathbf{1} \mid \bar{X} \mid X).$$

$\mathbf{1}$ is a $n \times 1$ vector whose elements are all of 1. The $n \times K$ matrix $\bar{X}$ shows K dummy variables which represent K blocks of data. That is,

$$\bar{X} = (\bar{\boldsymbol{x}}_1, \cdots, \bar{\boldsymbol{x}}_K), \quad \bar{\boldsymbol{x}}_k = (\bar{x}_{1k}, \cdots, \bar{x}_{nk})^t,$$

$$\bar{x}_{ik} \in \{0, 1\}, \quad \sum_{k=1}^{K} \bar{x}_{ik} = 1, \quad \forall i, k. \tag{20}$$

In equation (20), for example, if there are three distinct blocks in the data, each row of $(\bar{\boldsymbol{x}}_1, \bar{\boldsymbol{x}}_2, \bar{\boldsymbol{x}}_3)$ has values of the following three types: $(1, 0, 0)$ indicates data which belongs to the first block, $(0, 1, 0)$ is the data for the second block, and $(0, 0, 1)$ shows the data of the third block. The blocks of data are determined by the external information of the data.

$$\hat{\boldsymbol{\beta}} = (\hat{\beta}_0, \hat{\beta}_1, \cdots, \hat{\beta}_K, \hat{\beta}_{K+1}, \cdots, \hat{\beta}_{p+K})^t, \tag{21}$$

shows the regression coefficients. In equation (21), $\hat{\beta}_1, \cdots, \hat{\beta}_K$ are regression coefficients for K blocks and $\hat{\beta}_{K+1}, \cdots, \hat{\beta}_{p+K}$ are regression coefficients for p

independent variables. $\hat{\mathbf{e}} = (\hat{e}_1, \cdots, \hat{e}_n)^t$ shows residuals. The estimate of the regression coefficient shown in equation (21) can be obtained as follows:

$$\hat{\boldsymbol{\beta}} = (\hat{X}^t \hat{X})^{-1} \hat{X}^t \boldsymbol{y}.$$

For each object i, a blocking regression model shown in equation (19) can be rewritten as follows under the conditions shown in equation (20):

$$y_i = \hat{\beta}_0 + \hat{\beta}_l + \hat{\beta}_{K+1} x_{i1} + \cdots + \hat{\beta}_{p+K} x_{ip} + \hat{e}_i, \tag{22}$$

$$\text{if } \boldsymbol{x}_i \in C_l = \{\boldsymbol{x}_i \mid \bar{x}_{il} = 1\}, \quad l = 1, \cdots, K, \quad i = 1, \cdots, n.$$

Comparing the conventional linear regression model shown as

$$y_i = \hat{\beta}_0 + \hat{\beta}_{K+1} x_{i1} + \cdots + \hat{\beta}_{p+K} x_p + \hat{e}_i, \quad i = 1, \cdots, n, \tag{23}$$

it can be seen that equation (22) adds $\hat{\beta}_l$ which shows the property of cluster l. So, in the blocking regression model, the difference of blocks is represented by the difference of intercepts $\hat{\beta}_l$ in equation (22).

If we assume independency for each cluster (or block), we can redefine equation (22) as follows:

$$\boldsymbol{y} = \hat{\tilde{X}}^{(l)} \hat{\boldsymbol{\beta}}^{(l)} + \hat{\mathbf{e}}^{(l)}, \quad l = 1, \cdots, K, \tag{24}$$

where the $n \times (p+2)$ matrix $\hat{\tilde{X}}^{(l)}$ is defined as:

$$\hat{\tilde{X}}^{(l)} = (\mathbf{1} \mid \bar{\boldsymbol{x}}_l \mid X), \tag{25}$$

$\hat{\boldsymbol{\beta}}^{(l)}$ and $\hat{\mathbf{e}}^{(l)}$ are as follows:

$$\hat{\boldsymbol{\beta}}^{(l)} = (\hat{\beta}_0^{(l)}, \hat{\beta}_l, \hat{\beta}_{K+1}^{(l)}, \cdots, \hat{\beta}_{p+K}^{(l)})^t, \tag{26}$$

$$\hat{\mathbf{e}}^{(l)} = (\hat{e}_1^{(l)}, \cdots, \hat{e}_n^{(l)})^t, \quad l = 1, \cdots, K. \tag{27}$$

In equation (24), the estimate of the regression coefficient shown in equation (26) can be obtained as follows:

$$\hat{\boldsymbol{\beta}}^{(l)} = (\hat{\tilde{X}}^{(l)^t} \hat{\tilde{X}}^{(l)})^{-1} \hat{\tilde{X}}^{(l)^t} \boldsymbol{y}, \quad l = 1, \cdots, K. \tag{28}$$

5 Fuzzy Blocking Regression Model

A fuzzy blocking regression model [15] is defined as:

$$\boldsymbol{y} = \tilde{X} \tilde{\boldsymbol{\beta}} + \tilde{\mathbf{e}}, \tag{29}$$

where the $n \times (p + K + 1)$ matrix $\tilde{X}$ is defined as:

$$\tilde{X} = (\mathbf{1} \mid U \mid X).$$

U is obtained as a result of fuzzy clustering of X shown in equation (2).

$$\tilde{\boldsymbol{\beta}} = (\tilde{\beta}_0, \tilde{\beta}_1, \cdots, \tilde{\beta}_K, \tilde{\beta}_{K+1}, \cdots, \tilde{\beta}_{p+K})^t, \tag{30}$$

is the regression coefficients and K shows the number of clusters. In equation (30),

$$\tilde{\beta}_1, \cdots, \tilde{\beta}_K$$

are regression coefficients for K fuzzy clusters and

$$\tilde{\beta}_{K+1}, \cdots, \tilde{\beta}_{p+K}$$

are regression coefficients for p independent variables.

$$\tilde{\boldsymbol{e}} = (\tilde{e}_1, \cdots, \tilde{e}_n)^t \tag{31}$$

shows residuals. The estimate of the regression coefficient shown in equation (29) can be obtained as follows:

$$\tilde{\boldsymbol{\beta}} = (\tilde{X}^t \tilde{X})^{-1} \tilde{X}^t \boldsymbol{y}. \tag{32}$$

For each object i, the fuzzy blocking regression model shown in equation (29) can be rewritten as follows:

$$y_i = \tilde{\beta}_0 + \tilde{\beta}_1 u_{i1} + \cdots + \tilde{\beta}_K u_{iK} + \tilde{\beta}_{K+1} x_{i1} + \cdots + \tilde{\beta}_{p+K} x_{ip} + \tilde{e}_i, \quad i = 1, \cdots, n. \tag{33}$$

If an object i completely belongs to a cluster l, that is $u_{il} = 1$, under the condition (2), then equation (33) is identical to the ordinal blocking regression model shown in equation (22). Therefore, the fuzzy blocking regression model shown in equation (29) is an extended model of the ordinal blocking regression model shown in equation (19).

The fuzzy blocking regression model shown in equation (29) implements a more complex classification situation to reflect the real data, using a result of fuzzy clustering. In the result of a fuzzy clustering, since $u_{ik} \in [0, 1]$ shown in equation (2), all of the K terms, $\tilde{\beta}_1 u_{i1}, \cdots, \tilde{\beta}_K u_{iK}$, in equation (33) usually do not vanish without a special case when $u_{ik} \in \{0, 1\}$. Using the proposed model, we affect a more reliable classification structure of real data into a regression model.

6 Fuzzy Blocking Regression Model Using Fuzzy Intercepts

As a natural extension of the model (22), we propose the following model using fuzzy intercepts:

$$y_i = \beta_0 + \beta_l u_{il} + \beta_{K+1} x_{i1} + \cdots + \beta_{p+K} x_{ip} + e_i, \quad l = 1, \cdots, K, \tag{34}$$

where

$$u_{il} \equiv \max_{1 \le k \le K} u_{ik}, \quad i = 1, \cdots, n, \quad \exists l. \tag{35}$$

Then if we assume independency for each cluster (or block), we can redefine as follows:

$$\boldsymbol{y} = \tilde{\tilde{X}}^{(l)}\boldsymbol{\beta}^{(l)} + \boldsymbol{e}^{(l)}, \ \ l = 1, \cdots, K, \tag{36}$$

where the $n \times (p+2)$ matrix $\tilde{\tilde{X}}^{(l)}$ is defined as:

$$\tilde{\tilde{X}}^{(l)} = (\boldsymbol{1} \mid \tilde{\boldsymbol{u}}_l \mid X), \ \ \tilde{\boldsymbol{u}}_l = (\tilde{u}_{1l}, \cdots, \tilde{u}_{nl})^t. \tag{37}$$

$\tilde{\boldsymbol{u}}_l$ shows l-th column vector of a $n \times K$ matrix $\tilde{U}$ which is defined as

$$\tilde{U} = (\tilde{u}_{ik}), \ \ i = 1, \cdots n, \ \ k = 1, \cdots, K. \tag{38}$$

In equation (38), $\tilde{u}_{ik}$ is assumed to be satisfied as:

$$\begin{cases} \tilde{u}_{il} = u_{il} \text{ if } u_{il} = \max\limits_{1 \leq k \leq K} u_{ik} \\ \tilde{u}_{ik} = 0 \ \ \text{otherwise} \end{cases}. \tag{39}$$

$\boldsymbol{\beta}^{(l)}$ and $\boldsymbol{e}^{(l)}$ are as follows:

$$\boldsymbol{\beta}^{(l)} = (\beta_0^{(l)}, \beta_l, \beta_{K+1}^{(l)}, \cdots, \beta_{p+K}^{(l)})^t, \tag{40}$$

$$\boldsymbol{e}^{(l)} = (e_1^{(l)}, \cdots, e_n^{(l)})^t, \ \ l = 1, \cdots, K. \tag{41}$$

In equation (36), the estimate of the regression coefficient shown in equation (40) can be obtained as follows:

$$\boldsymbol{\beta}^{(l)} = (\tilde{\tilde{X}}^{(l)^t}\tilde{\tilde{X}}^{(l)})^{-1}\tilde{\tilde{X}}^{(l)^t}\boldsymbol{y}, \ \ l = 1, \cdots, K. \tag{42}$$

The difference between equations (22) and (34) are intercepts, $\hat{\beta}_l$ or $\beta_l u_{il}$. In equation (34), we use the intercepts multiplied by the fuzzy grade which are obtained as a result of fuzzy clustering.

Moreover, in equation (34), β_l shows the regression coefficient between the obtained fuzzy cluster and dependent variable. So, in this model, we can capture the relationship between the fuzzy clusters and a dependent variable and the relationship between independent variables and a dependent variable simultaneously. Also, the clusters (or blocks) in equation (34) are obtained objectively using a numerical fuzzy clustering. However, in equation (22), we need external information in order to identify the clusters (or blocks). Moreover, we apply a regression model to the mixed data $\tilde{\tilde{X}}^{(l)}$ shown in equation (25) and this causes a decrease of the fitness. However, in equation (37), the fuzzy clustering result, $\tilde{u}_{il}$, can be obtained as continuous values so we can obtain a better performance of fitness for the proposed model.

For the identification of solutions shown in equation (42), the following are assumed to be satisfied:

$$p \leq n-2, \ \ n \geq 3.$$

7 Variable Based Fuzzy Blocking Regression Model

An extended model of the fuzzy blocking regression model shown in equation (34) is defined as follows:

$$y_i = \beta_0 + \beta_l u_{il} + \beta_{K+1} w_1 x_{i1} + \cdots + \beta_{p+K} w_p x_{ip} + e_i, \;\; l = 1, \cdots, K, \quad (43)$$

where

$$u_{il} \equiv \max_{1 \leq k \leq K} u_{ik}, \;\; i = 1, \cdots, n, \;\; \exists l.$$

In equation (43), w_a, $a = 1, \cdots, p$ are the weights shown in equation (10). Then if we assume independence for each cluster (or block), we can redefine as follows:

$$\boldsymbol{y} = \tilde{X}^{(l)} \tilde{\boldsymbol{\beta}}^{(l)} + \tilde{\boldsymbol{e}}^{(l)}, \;\; l = 1, \cdots, K, \quad (44)$$

where the $n \times (p+2)$ matrix $\tilde{X}^{(l)}$ is defined as:

$$\tilde{X}^{(l)} = (\boldsymbol{1} \mid \tilde{\boldsymbol{u}}_l \mid XW), \;\; \tilde{\boldsymbol{u}}_l = (\tilde{u}_{1l}, \cdots, \tilde{u}_{nl})^t, \;\; W = \mathrm{diag}(w_1, \cdots, w_p), \quad (45)$$

where $\tilde{u}_{ik}$ satisfy the condition shown in equation (39). $\tilde{\boldsymbol{\beta}}^{(l)}$ and $\tilde{\boldsymbol{e}}^{(l)}$ are as follows:

$$\tilde{\boldsymbol{\beta}}^{(l)} = (\tilde{\beta}_0^{(l)}, \tilde{\beta}_l, \tilde{\beta}_{K+1}^{(l)}, \cdots, \tilde{\beta}_{p+K}^{(l)})^t, \quad (46)$$

$$\tilde{\boldsymbol{e}}^{(l)} = (\tilde{e}_1^{(l)}, \cdots, \tilde{e}_n^{(l)})^t, \;\; l = 1, \cdots, K. \quad (47)$$

In equation (44), the estimate of the regression coefficient shown in equation (46) can be obtained as follows:

$$\tilde{\boldsymbol{\beta}}^{(l)} = (\tilde{X}^{(l)^t} \tilde{X}^{(l)})^{-1} \tilde{X}^{(l)^t} \boldsymbol{y}, \;\; l = 1, \cdots, K. \quad (48)$$

If $w_a = 1, \;\; \forall a$, then the variable based fuzzy blocking regression model shown in equation (43) is the same as the fuzzy blocking regression model shown in equation (34). That is, if we assume the same weight for each variable, then the variable based fuzzy blocking regression model is the same as the conventional fuzzy blocking regression model. Therefore, the variable based fuzzy blocking regression model is an extension of the fuzzy blocking regression model. However, the variable based fuzzy blocking regression model can consider not only the classification structure with respect to all of the variables, but also consider the degree of contribution for the classification at each variable.

8 Numerical Examples

We use diabetes patient records obtained from two sources: an automatic electronic recording device and paper records. [11] Patients with insulin dependent diabetes mellitus are insulin deficient. The data consists of measurements of blood glucose and some kinds of insulin dose at four times as follows: breakfast (08:00), lunch (12:00), dinner (18:00), and bedtime (22:00). The data is observed for 70 patients over several periods. We use only electronic recording device patients data out of the 70 patients data and exclude the patients data which have missing values for the measurements of blood glucose for four times. 10 patients data remain after the screening. Total of the days of records over the 10 patients is 447. For the independent variables, we use eight variables of regular insulin dose at 8:00, NPH insulin dose at 8:00, regular insulin dose at 12:00, NPH insulin dose at 12:00, regular insulin dose at 18:00, NPH insulin dose at 18:00, regular insulin dose at 22:00, and NPH insulin dose at 22:00. The difference between regular insulin and NPH insulin is on its own characteristic time of onset of effect (O), time of peak action (P) and effective duration (D). For regular insulin, O is 15-45 minutes, P is 1-3 hours, and D is 4-6 hours. For NPH insulin, O is 1-3 hours, P is 4-6 hours, and D is 10-14 hours. For a dependent variable, we use value of blood glucose as follows: pre-breakfast blood glucose measurement at 8:00, pre-lunch blood glucose measurement at 12:00, pre-supper blood glucose measurement at 18:00, and unspecified blood glucose measurement at 22:00, respectively. The purpose of this analysis is to find the effectiveness of the treatment of several kinds of insulin treatment at different times for each blood glucose measurement.

Figures 1 and 2 show a result of a fuzzy clustering named FANNY shown in equation (6) when $m = 2$ for a data of independent variables which shows the insulin treatment. Since we use 10 patients, the number of clusters is assumed to be 10. Figure 1 shows the results of clusters 1 - 5 and figure 2 shows results of clusters 6 - 10. In these figures, the abscissa shows the number of 447 records over 10 patients and ordinate shows the degree of belongingness to the obtained clusters. Each different line shows each different cluster. Table 1 shows correspondence between the record numbers and the patient numbers.

From these figures, we can see that patient 1 belongs to clusters 1 and 2, patients 2 and 3 have higher degree of belongingness to a cluster 3 when compared to other clusters. Patient 4 has a higher degree of belongingness to cluster 4 and patient 10 almost belongs to the cluster 4. Patient 5 belongs to cluster 5, patient 7 belongs to cluster 7, and patient 9 belongs to cluster 10. Also, patient 6 belongs to clusters 3 and 6, and patient 8 belongs to clusters 8 and 9. Patients 6 and 8 are patients who change the treatment in their record periods. According to the difference of dose and time of insulin injection over the 10 patients, the result seems to be adaptable.

Using the result of the fuzzy clustering shown in figures 1 and 2, we apply the fuzzy blocking regression model shown in equation (29). In equation (29), $\boldsymbol{y}$ is pre-breakfast blood glucose measurement at 8:00, pre-lunch blood glucose measurement at 12:00, pre-supper blood glucose measurement at 18:00, and unspecified blood glucose measurement at 22:00, respectively. $\tilde{X}$ is created by using the fuzzy clustering result shown in figures 1 and 2 and the insulin dose data. The result of regression coefficients are shown in figures 3 - 7. As a comparison, we show the results of regression coefficients using conventional linear regression analysis shown in equation (23). In figures 3 - 6, the abscissa shows eight independent variables: v_1 shows regular insulin dose at 8:00, v_2 is NPH insulin dose at 8:00, v_3 is regular insulin dose at 12:00, v_4 is NPH insulin dose at 12:00, v_5 is regular insulin dose at 18:00, v_6 is NPH insulin dose at 18:00, v_7 is regular insulin dose at 22:00, and v_8 is NPH insulin dose at 22:00. The ordinate shows values of regression coefficients. The solid line shows the values of regression coefficients of the fuzzy blocking regression analysis and the dotted line shows values of regression coefficients of conventional linear regression analysis. Figure 3 shows the result of regression coefficients when the dependent variable is pre-breakfast blood glucose measurement. Figure 4 shows the result of regression coefficients when the dependent variable is pre-lunch blood glucose measurement. Figure 5 shows the result of regression coefficients when the dependent variable is pre-supper blood glucose measurement and figure 6 shows the result of regression coefficients when the dependent variable is unspecified blood glucose measurement.

From these figures, we can see the similar tendency of regression coefficients between the proposed fuzzy blocking regression analysis and the conventional linear regression analysis without the results shown in figure 6. From figure 3, it can be seen that the pre-breakfast blood glucose measurement has high correlation with regular insulin dose at 8:00. Figure 4 shows that pre-lunch blood glucose measurement has positive correlation with regular insulin dose at 12:00 and negative correlation with NPH insulin dose at 18:00. Figure 5 shows that pre-supper blood glucose measurement has negative correlation with regular insulin dose at 12:00 and positive correlation with regular insulin dose at 18:00. From figure 6, we can not see any significant correlation in the result of linear regression analysis and can see the clear negative correlation with regular insulin dose at 12:00 and positive correlation with regular insulin dose at 22:00.

Figure 7 shows the result of regression coefficients for clusters, that is results of $\tilde{\beta}_1, \cdots, \tilde{\beta}_K$ in equation (30). In this figure, the abscissa shows ten clusters and the ordinate shows the values of regression coefficients. Each different line shows each fuzzy blocking regression analysis when dependent variable is pre-breakfast blood glucose measurement at 8:00, pre-lunch blood glucose measurement at 12:00, pre-supper blood glucose measurement at 18:00, and unspecified blood glucose measurement at 22:00, respectively. From this result, we can see the weights of each cluster for each regression analysis. For example, in the case of regression for unspecified blood glucose measurement, the weights of clusters 8 and 9 are large. From figure 2, the degree of belongingness for clusters 8 and 9 are found in patients 8 and 9 and both of the patients have records of injection of NPH insulin at 22:00. Other patients do not inject NPH insulin at 22:00. So, the weights reflect the adequate difference of the treatment.

Figure 8 shows the comparison of values of multiple correlation coefficients between the fuzzy blocking regression analysis and linear regression analysis over the four cases. The abscissa shows each case that is PB shows the case of regression analysis in which the dependent variable is pre-breakfast blood glucose measurement, PL is the case of pre-lunch blood glucose measurement, PS is pre-supper blood glucose measurement, and US shows the case of unspecified blood glucose measurement. The solid line shows the result of multiple correlation

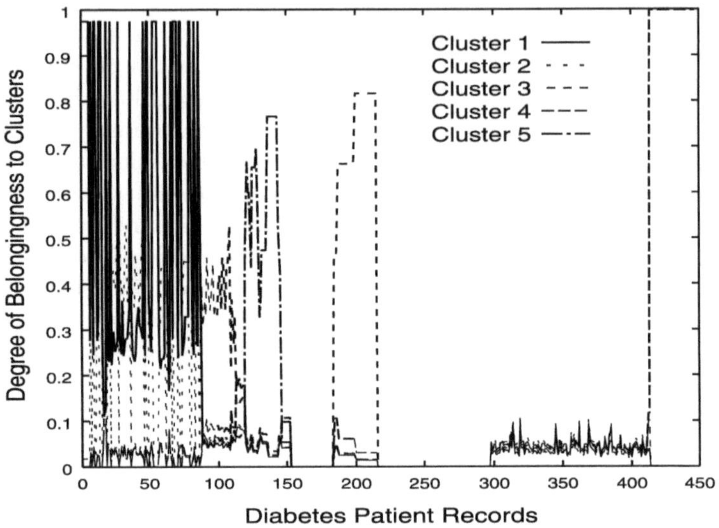

Fig. 1. Fuzzy Clustering of Insulin Dose Data for Clusters 1 - 5

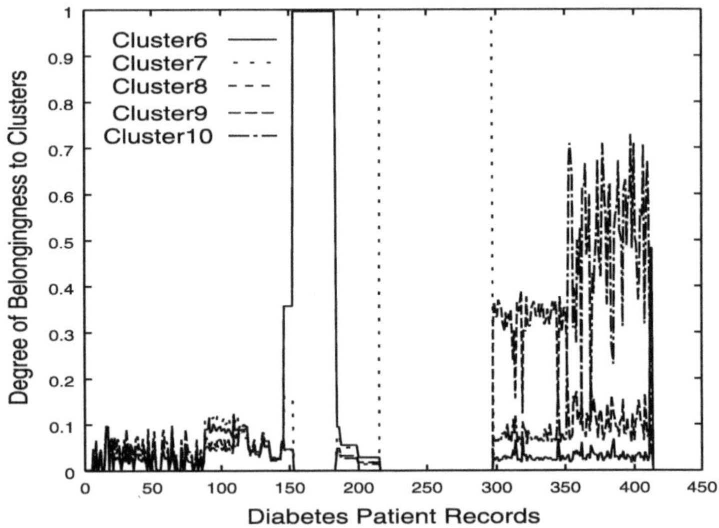

Fig. 2. Fuzzy Clustering of Insulin Dose Data for Clusters 6 - 10

coefficients of fuzzy blocking regression analysis and the dotted line shows the result of multiple correlation coefficients of linear regression analysis. For all of the cases, the proposed fuzzy blocking regression analysis obtains better results of multiple correlation coefficients.

Table 1. Correspondence between Records number and Patients Number

Records Number	Patients Number
1 - 87	1
88 - 95	2
96 - 112	3
113 - 119	4
120 - 145	5
146 - 215	6
216 - 297	7
298 - 351	8
352 - 411	9
412 - 447	10

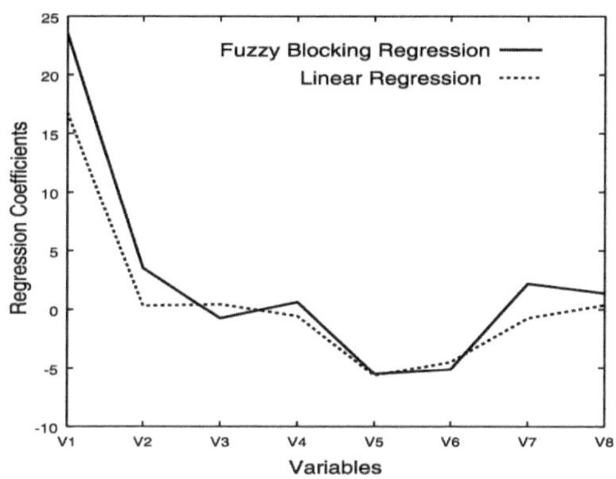

Fig. 3. Fuzzy Blocking Regression Analysis and Linear Regression Analysis for Pre-Breakfast Blood Glucose Measurement

Figure 9 shows a comparison of values of residuals between the fuzzy blocking regression analysis and linear regression analysis over the four cases. That is, we compare the values of $\tilde{\boldsymbol{e}}^t\tilde{\boldsymbol{e}}$ and $\hat{\boldsymbol{e}}^t\hat{\boldsymbol{e}}$, where $\tilde{\boldsymbol{e}}$ is shown in equation (31) and $\hat{\boldsymbol{e}} = (\hat{e}_1, \cdots, \hat{e}_n)^t$ is shown in equation (23). The solid line shows the result of residuals of fuzzy blocking regression analysis and the dotted line shows the result of residuals of linear regression analysis. For all of the cases, the proposed fuzzy blocking regression analysis obtained better results of residuals.

The next data is Fisher iris data [5] which consists of 150 samples of iris flowers with respect to four variables, sepal length, sepal width, petal length, and petal width. The samples are observed from three kinds of iris flowers, iris setosa, iris versicolor, and iris virginica. We use the data with respect to three variables, sepal width, petal length, and petal width for the data of independent variables and the data of sepal length is used for a dependent variable. The number of clusters is assumed to be 3.

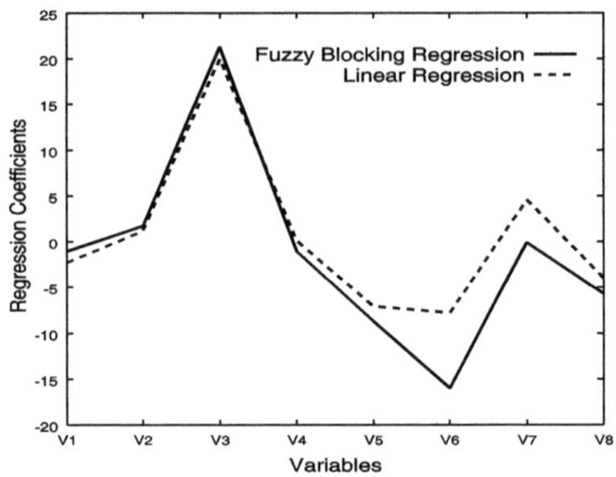

Fig. 4. Fuzzy Blocking Regression Analysis and Linear Regression Analysis for Pre-Lunch Blood Glucose Measurement

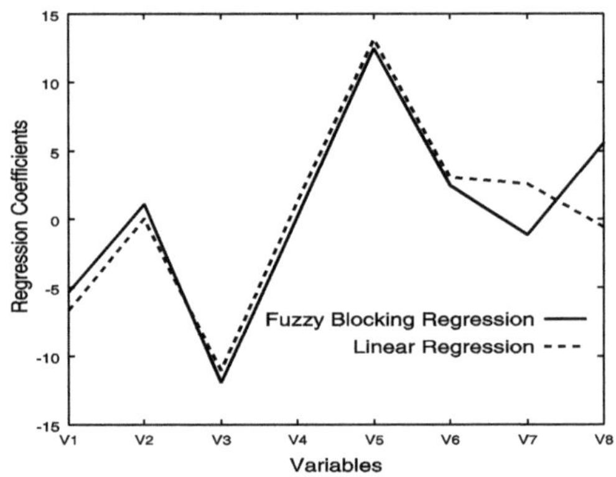

Fig. 5. Fuzzy Blocking Regression Analysis and Linear Regression Analysis for Pre-Supper Blood Glucose Measurement

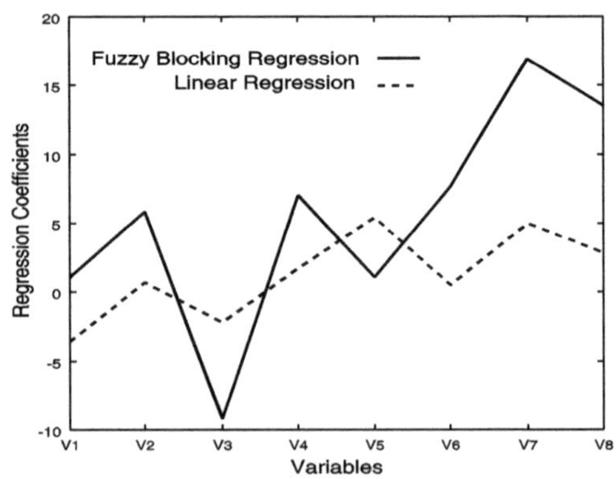

Fig. 6. Fuzzy Blocking Regression Analysis and Linear Regression Analysis for Unspecified Blood Glucose Measurement

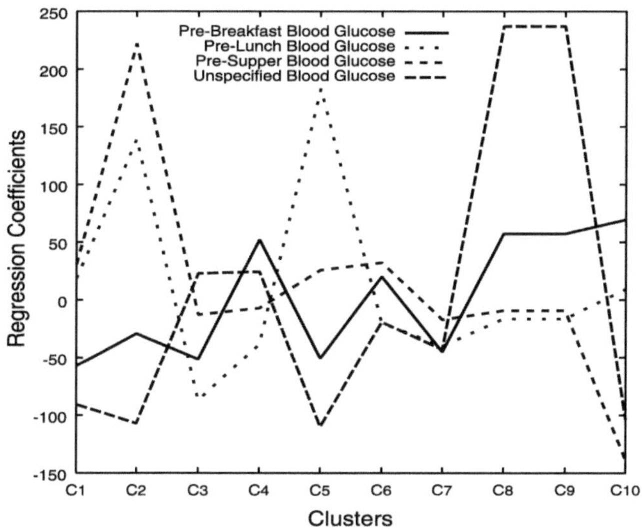

Fig. 7. Regression Coefficients of Clusters in Fuzzy Blocking Regression Analysis for Diabetes Patient Records Data

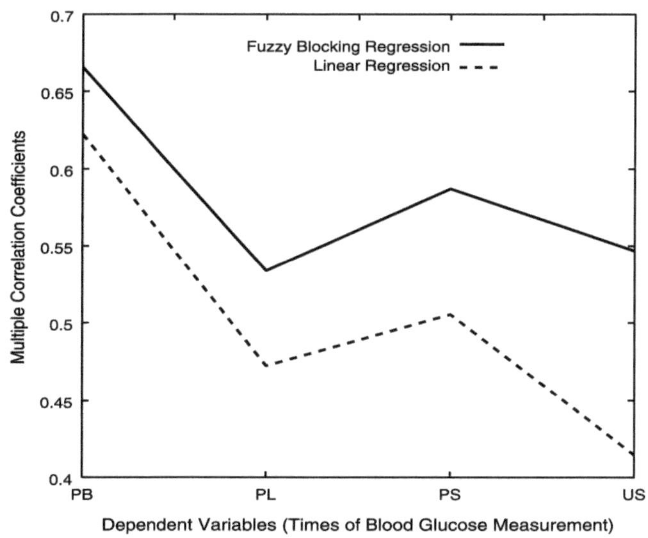

Fig. 8. Comparison of Multiple Correlation Coefficients

We obtain the result of the fuzzy c-means method shown in equation (4) for the data of independent variables and use the maximum values of degree of belongingness for a fixed object over the three fuzzy clusters. That is, we use the following degree of belongingness:

$$\max_{1 \le k \le 3} u_{ik}, \quad \exists i, \quad i = 1, \cdots, 150. \tag{49}$$

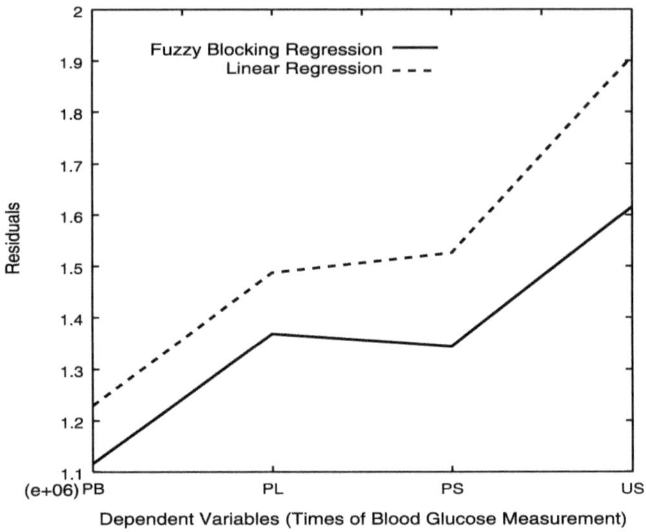

Fig. 9. Comparison of Residuals

Using the result of fuzzy c-means shown in equation (49), we show in figures 10 and 11 the results of estimates of the regression coefficients of the blocking regression model and the fuzzy blocking regression model with the fuzzy intercept shown in equations (28) and (42), respectively.

In equation (24), if the number of clusters is 3, then the regression model (24) can be rewritten as follows:

$$y_i = \hat{\beta}_0^{(1)} + \hat{\beta}_1 + \hat{\beta}_4^{(1)} x_{i1} + \cdots + \hat{\beta}_{p+3}^{(1)} x_{ip} + \hat{e}_i^{(1)},$$
$$\text{if } \boldsymbol{x}_i \in C_1 = \{\boldsymbol{x}_i \mid (\bar{x}_{i1}, \bar{x}_{i2}, \bar{x}_{i3}) = (1,0,0)\},$$

$$y_i = \hat{\beta}_0^{(2)} + \hat{\beta}_2 + \hat{\beta}_4^{(2)} x_{i1} + \cdots + \hat{\beta}_{p+3}^{(2)} x_{ip} + \hat{e}_i^{(2)},$$
$$\text{if } \boldsymbol{x}_i \in C_2 = \{\boldsymbol{x}_i \mid (\bar{x}_{i1}, \bar{x}_{i2}, \bar{x}_{i3}) = (0,1,0)\},$$

$$y_i = \hat{\beta}_0^{(3)} + \hat{\beta}_3 + \hat{\beta}_4^{(3)} x_{i1} + \cdots + \hat{\beta}_{p+3}^{(3)} x_{ip} + \hat{e}_i^{(3)},$$
$$\text{if } \boldsymbol{x}_i \in C_3 = \{\boldsymbol{x}_i \mid (\bar{x}_{i1}, \bar{x}_{i2}, \bar{x}_{i3}) = (0,0,1)\}, \quad i = 1, \cdots, n. \tag{50}$$

In figure 10, the abscissa shows each regression coefficient in equation (50): $b0$ shows an ordinal intercept $\hat{\beta}_0^{(l)}$. $b1$ shows intercepts for each cluster which are shown by using $\hat{\beta}_1$, $\hat{\beta}_2$, and $\hat{\beta}_3$. $b2 - b4$ are three independent variables shown by $\hat{\beta}_4^{(l)}$, $\hat{\beta}_5^{(l)}$, and $\hat{\beta}_6^{(l)}$. That is, $b2$ shows regression coefficient for sepal width, $b3$ is regression coefficient for petal length, and $b4$ is for petal width. The ordinate shows values of the regression coefficients and each cluster is shown by a different line.

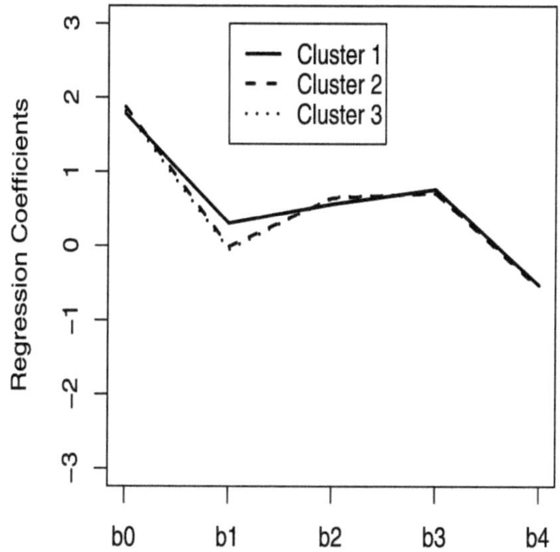

Fig. 10. Result of Regression Coefficients for Blocking Regression Model

In equation (36), if the number of clusters is 3, then the regression model (36) can be rewritten as follows:

$$y_i = \beta_0^{(l)} + \beta_l u_{il} + \beta_4^{(l)} x_{i1} + \cdots + \beta_{p+3}^{(l)} x_{ip} + e_i^{(l)}, \quad l = 1, \cdots, 3, \tag{51}$$

where

$$\max_{1 \le k \le 3} u_{ik} \equiv u_{il}, \quad i = 1, \cdots, n, \quad \exists l.$$

In figure 11, the abscissa shows each regression coefficient in equation (51): $b0$ shows an ordinal intercept $\beta_0^{(l)}$. $b1$ shows intercepts for each fuzzy cluster which are shown by using β_1, β_2, and β_3. $b2 - b4$ are three independent variables shown by using $\beta_4^{(l)}$, $\beta_5^{(l)}$, and $\beta_6^{(l)}$. That is, $b2$ shows regression coefficient for sepal width, $b3$ is the regression coefficient for petal length, and $b4$ is for petal width. The ordinate shows values of the regression coefficients and each cluster is represented by a different line.

From both of the figures, we obtain almost the same results over the three clusters. However, we can see the difference of the regression coefficients on variable $b1$ which show $\hat{\beta}_1, \hat{\beta}_2, \hat{\beta}_3$ in equation (50) and $\beta_1, \beta_2, \beta_3$ in equation (51). That is, these coefficients show the difference of relationship between the dependent variable and each cluster. In particular, cluster 1 is more associated with the dependent variable when compared to the other clusters.

Figure 12 shows the comparison of residuals between the blocking regression model shown in equation (27) and the fuzzy blocking regression model with fuzzy intercept shown in equation (41) over the three clusters. That is we compare the residuals shown in equations (24) and (36). In this figure, the abscissa shows the

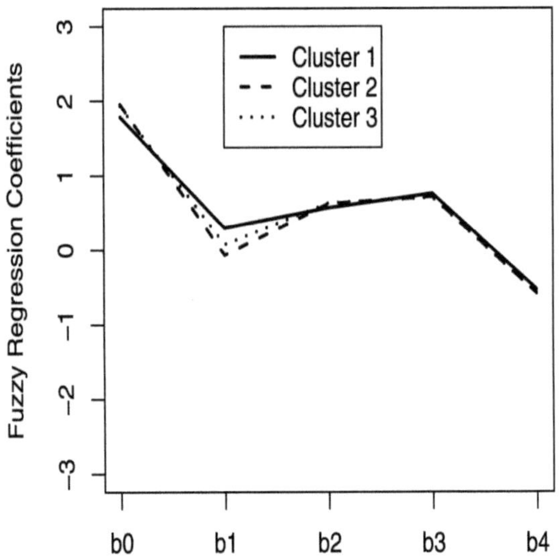

Fig. 11. Result of Regression Coefficients for Fuzzy Blocking Regression Model using Fuzzy Intercepts

three clusters and the ordinate shows values of sum of squared residuals, that is we calculate the following by using equations (24) and (36):

$$\hat{s}^{(l)} \equiv \hat{\boldsymbol{e}}^{(l)^t} \hat{\boldsymbol{e}}^{(l)} = \hat{e}_1^{(l)^2} + \cdots + \hat{e}_n^{(l)^2}, \ \ l = 1, \cdots, K. \tag{52}$$

$$s^{(l)} \equiv \boldsymbol{e}^{(l)^t} \boldsymbol{e}^{(l)} = e_1^{(l)^2} + \cdots + e_n^{(l)^2}, \ \ l = 1, \cdots, K. \tag{53}$$

Solid line shows a regression model using fuzzy block intercepts shown in equation (53) and the dotted line shows the model using hard block intercepts shown in equation (52). From this figure, in the case of clusters 1 and 3, a regression model using fuzzy block intercepts obtain better results. However, in the case of cluster 2, an ordinal model using hard block intercepts obtained a better result.

Table 2 shows the summation of the three residuals for both of the models. That is we compare the values of

$$\hat{s} \equiv \hat{s}^{(1)} + \hat{s}^{(2)} + \hat{s}^{(3)} \tag{54}$$

and

$$s \equiv s^{(1)} + s^{(2)} + s^{(3)} \tag{55}$$

by using equations (52) and (53). From this table, it can be seen that our proposed fuzzy model can obtain a better result.

Next, using the result of variable based fuzzy c-means shown in equation (17) when $m = 2$ for the iris data with respect to three independent variables, sepal width, petal length, and petal width, we obtain the weights shown in equation

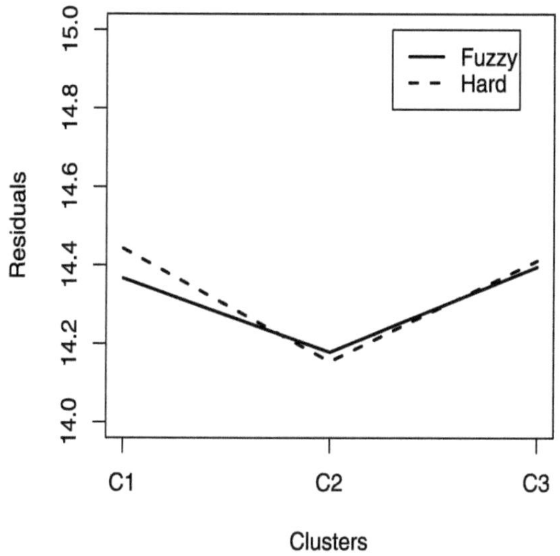

Fig. 12. Comparison of Residuals over Three Clusters

Table 2. Comparison of Residuals

Fuzzy Model	Hard Model
42	43

Table 3. Comparison of Residuals

Variable based Fuzzy Model	Hard Model
42	43

(10) as follows: $w_1 = 0.85$, $w_2 = 0.90$, $w_3 = 0.95$. Using these weights and the result of fuzzy c-means shown in equation (4), we apply this data to the variable based fuzzy blocking regression model shown in equation (44).

Table 3 shows the comparison of the residuals of the blocking regression model and the variable based fuzzy blocking regression model shown in equations (27) and (47). That is we compare the values of $\hat{s}$ shown in equation (54) and

$$\tilde{s} = \tilde{s}^{(1)} + \tilde{s}^{(2)} + \tilde{s}^{(3)},$$

$$\tilde{s}^{(l)} \equiv \tilde{\boldsymbol{e}}^{(l)^t} \tilde{\boldsymbol{e}}^{(l)} = \tilde{e}_1^{(l)^2} + \cdots + \tilde{e}_n^{(l)^2}, \ l = 1, 2, 3,$$

by using equation (47). From this table, it can be seen that the variable based fuzzy blocking regression model can obtain a better result.

9 Conclusion

This chapter presented several fuzzy blocking regression models which are extensions of the ordinal blocking regression model. In the case of the ordinal blocking regression model, the classification structure of the data is represented by using exclusive clusters given using the external information. Since the fuzzy blocking regression models are based on the results of fuzzy clustering, spatially distributed classification structure can be involved into the blocking regression model. Therefore, we can reflect a more complicated data structure.

Moreover, fuzzy blocking regression models can avoid the decrease of fitness caused by fitting to mixed data consisting of continuous and discrete values.

Also, we can obtain two kinds of regression coefficients in the case of fuzzy blocking regression models. One is the regression coefficients between a dependent variable and fuzzy clusters. The other is the regression coefficients between a dependent variable and independent variables. These two kinds of regression coefficients are mathematically comparable. Therefore, we can obtain two kinds of comparable relationships of fuzzy clusters and independent variables via the dependent variable. It is this feature that provides a new spectrum with which to see the latent structure of the data.

References

1. Bezdek, J.C.: Pattern Recognition with Fuzzy Objective Function Algorithms. Plenum Press (1981)
2. Bezdek, J.C., Keller, J., Krisnapuram, R., Pal, N.R.: Fuzzy Models and Algorithms for Pattern Recognition and Image Processing. Kluwer Academic Publishers, Dordrecht (1999)
3. Brunsdon, C., Fotheringham, S., Charlton, M.: Geographically Weighted Regression Modeling Spatial Non-Stationarity. Journal of the Royal Statistical Society 47, 431–443 (1998)
4. Draper, N.R., Smith, H.: Applied Regression Analysis. John Wiley & Sons, Chichester (1966)
5. Fisher, R.A.: The Use of Multiple Measurements in Taxonomic Problems. Annals of Eugenics 7, 179–188 (1936)
6. Friedman, J.H., Meulman, J.J.: Clustering Objects on Subsets of Attributes. Journal of the Royal Statistical Society, Series B 66, 815–849 (2004)
7. Frigui, H., Caudill, J.: Unsupervised Image Segmentation and Annotation for Content-Based Image Retrieval. In: IEEE International Conference on Fuzzy Systems, pp. 274–279 (2006)
8. Hathaway, R.J., Bezdek, J.C.: Switching Regression Models and Fuzzy Clustering. IEEE Trans. Fuzzy Syst. 1(3), 195–204 (1993)
9. Hoppner, F., Klawonn, F., Kruse, R., Runkler, T.: Fuzzy Cluster Analysis. John Wiley & Sons, Chichester (1999)
10. Kaufman, L., Rousseeuw, P.J.: Finding Groups in Data. John Wiley & Sons, Chichester (1990)

11. Newman, D.J., Hettich, S., Blake, C.L., Merz, C.J.: UCI Repository of Machine Learning Databases, Department of Information and Computer Science. University of California, Irvine, CA (1998), http://www.ics.uci.edu/~mlearn/MLRepository.html
12. O'Callaghan, L., Mishra, N., Meyerson, A.: Streaming-Data Algorithms for High-Quality Clustering. In: The 18th International Conference on Data Engineering (2002)
13. Sato, M., Sato, Y., Jain, L.C.: Fuzzy Clustering Models and Applications. Springer, Heidelberg (1997)
14. Sato-Ilic, M., Jain, L.C.: Innovations in Fuzzy Clustering. Springer, Heidelberg (2006)
15. Sato-Ilic, M.: Fuzzy Blocking Regression Analysis. Intelligent Engineering Systems through Artificial Neural Networks 16, 767–772 (2006)
16. Sato-Ilic, M.: Weighted Fuzzy Clustering on Subsets of Variables. In: International Symposium on Signal Processing and its Applications with the International Conference on Information Sciences, Signal Processing and its Applications (2007)
17. Sato-Ilic, M.: Regression Analysis Considering Fuzzy Block Intercepts. In: International Symposium Management Engineering, T03 (2007)
18. Sato-Ilic, M.: Variable Based Fuzzy Blocking Regression Model. In: Knowledge-Based Intelligent Information and Engineering Systems, pp. 525–532 (2007)

Support Vector Machines and Features for Environment Perception in Mobile Robotics

Rui Araújo, Urbano Nunes, Luciano Oliveira, Pedro Sousa, and Paulo Peixoto

ISR-Institute of Systems and Robotics, and Department of Electrical and Computer Engineering, University of Coimbra
Pólo II, 3030-290, Coimbra, Portugal
{rui,urbano,lreboucas,pasousa,peixoto}@isr.uc.pt

Abstract. Environment perception is one of the most challenging and underlying task which allows a mobile robot to perceive obstacles, landmarks and extract useful information to navigate safely. In this sense, classification techniques applied to sensor data may enhance the way mobile robots sense their surroundings. Amongst several techniques to classify data and to extract relevant information from the environment, Support Vector Machines (SVM) have demonstrated promising results, being used in several practical approaches. This chapter presents the core theory of SVM, and applications in two different scopes: using Lidar (Light Detection and Ranging) to label specific places, and vision-based human detection aided by Lidar.

1 Introduction

Learning models of the environment [2] is required in tasks such as localization, path-planning, and human-robot interaction. Other examples of application of learning systems to environment perception are the following:

Industry Product Manufacturing
: Learning algorithms are important in a production line to check if a product contains some error. The application of these methods increases the productivity and guarantees higher quality of the product.

Security
: Important security systems are based on learning methods. Two example applications are burglar detection in banks and casinos systems, and terrorist detection in airports. These systems try to detect dangerous people by automatically identifying their face.

Biological Sciences
: For example, learning systems are intensively used on classification of protein types, based on the DNA sequence from which they are generated. Research is also being carried out for the application of learning systems for understanding how the human brain works.

The process of constructing models by explicitly considering and coding aspects of the environment is not trivial. Thus, methodologies which aim at creating automatic perception and system control are necessary.

L.C. Jain et al. (Eds.): Comp. Intel. Para.: Innov. Applications, SCI 137, pp. 219–250, 2008.
springerlink.com

The problem of learning to perceive the environment from sensor data is an active research topic and it is usually frequent that certain systems may be able to perceive or comprehend the environment in a way similar to human beings. If this is achieved, a new set of applications can be proposed, *e.g.*, intelligent systems may interact with users using a new perspective. In robotics, the robot would be able to build models and to react to the world using new approaches.

Machine learning techniques have been used to tackle the particular problem of building generic models from the input space in order to recognize patterns which are implicitly present within data. Amongst the various approaches available on the literature, SVM has been receiving attention in the last few years. These machines represent learning structures that, after being trained to accomplish classification tasks (pattern recognition) or approximation of real valued functions (regression, estimation), usually show good generalization results. SVM are based on statistical learning theory, and were originally proposed by Vapnik [51] as a binary classifier. They embody the Structural Risk Minimization (SRM), Vapnik-Chervonenski (VC) dimension theory and optimization theory [20].

The application of SVM have permitted the attainment of good results in practical applications: sufficiently complex models can be built to meet the requirements of real-world applications. SVM can be used for nonlinear classification, function approximation, and feature extraction [8], [45], [51], [15], yet the mathematical analysis of SVM is reasonably simple: SVM correspond to linear techniques in a high-dimensional feature space. A nonlinear mapping relates the input space and the feature space. However, by the use of kernels, all the required computations are directly made in the lower-dimentional input space. SVM contain other structures as particular cases. Radial basis functions (RBF) neural networks, and polynomial classifiers are included as special cases of the SVM approach.

Artificial Neural Networks (ANNs) are an alternative to SVM for dealing with problems of supervised learning. However, ANN normally present the following drawbacks with respect to SVM [20], [26], [4]:

- **Problems with local minima of the error function, *e.g.*, backpropagation training algorithm.** Local minima are undesirable points that a learning error function may attain in the process of training. It means that the algorithm may be stuck before the maximum separation of data is found. For a discussion about the presence of local minima in backpropagation learning algorithm, please refer to [26]. An approach to tackle this problem in Recurrent ANNs can be found in [4].
- **Tendency to overfit.** With a large number of weights, ANNs are prone to overfit, i.e., they present a poor performance for new samples in the test stage [16].
- **There is no mathematical evidence to help in building an ANN structure in a particular problem.** An ANN usually owns structural parameters like number of neurons, synapses function and number of hidden layers, which it is necessary to be found empirically, since there is no mathematical proof to guide in finding the best structure to a particular problem

(the Kolmogorov theorem states some useful points in order to define the number of hidden layers and neurons for continuous function, but it is restrictive to use in any ANN [20]).
- **The order of presentation of the training data affects the learning results.** With the same training data, it is not guaranteed to have the same function to split the data.

Hence, throughout the years, there has been a growing use of SVM. The main goal of this chapter is to bring a discussion of practical aspects of SVM construction, illustrated with two applications. Moreover, the chapter also shows some highlights about preprocessing of input space in order to extract relevant features that help SVM to better perform the concerned classification task, particularly in robot environment perception using lasers and vision cameras.

The chapter is organized as follows. In Section 2, SVM learning theory is presented. The main aspects of binary SVM are discussed, as well as strategies to implement SVM as a multiclassifier. The core of SVM is formulated in order to enclose all the structures that embody its theory, i.e. Quadratic Programming (QP) optimization, VC dimension and Statistical Learning Theory.

In Section 3, features commonly used for learning models of the environment are presented. This section is decomposed into two subsections: subsection 3.1 describes the features employed for learning to label places in indoor environments, and subsection 3.2 presents a brief survey about features used for image pattern recognition along with SVM.

Section 4 presents applications of the methods discussed during the theory sections. Particularly, two applications are highlighted: indoor place labeling (4.1) and vision-based object classification aided by a Lidar (4.2).

2 Support Vector Machines

SVM was originally proposed by Vapnik [51] as a binary classifier, and embodies the Structural Risk Minimization (SRM), Vapnik-Chervonenski (VC) dimension theory and Optimization Theory [20], [8], [45], [6]. The SVM methodology can be employed for classification and regression problems with good generalization results. This chapter focuses in the application of SVM to classification problems.

SVM is a supervised learning technique, so there are two stages in order to create and apply a SVM classifier: training and prediction. In the training stage, or learning stage, pairs $(\mathbf{x}_i, y_i)$ of input feature vectors, $\mathbf{x}_i = (x_1, x_2, \ldots, x_n)$, and desired outputs, y_i, are given in order to design support vectors which are used to predict desired outputs. These support vectors constitute a model, the *prediction model*. Later, after learning, the prediction model is applied in the *prediction phase* to predict outputs, y_i, for previously unseen input feature vectors, $\mathbf{x}_i$.

Let $\mathbf{X}$ represent the input space, and Y the output space (the space of classes in a classification task or real numbers in a regression task). Let a set S of training examples be denoted by $S = ((\mathbf{x}_1, y_1), \ldots, (\mathbf{x}_l, y_l)) \subseteq (\mathbf{X} \times Y)^l$, where l is the number of instances. Let us assume that there are only two classes for classification. The extension for multiclass problems will be later discussed. Let $(\mathbf{w}, b)$

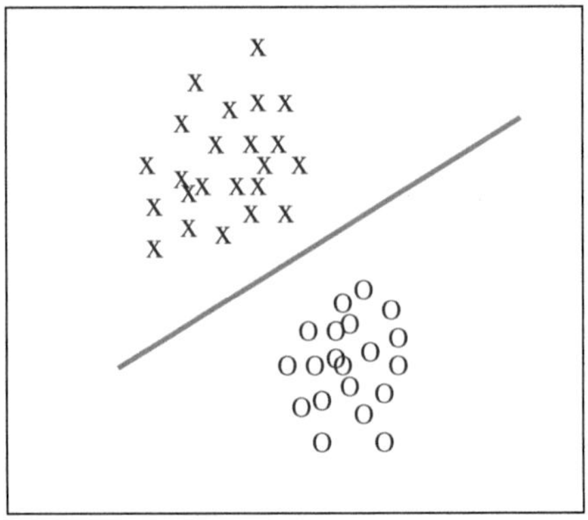

Fig. 1. Representation of two linearly separated classes

denote the weight vector, $\mathbf{w} = (w_1, w_2, \ldots, w_n)$, and bias, b, of the hyperplane which splits data from both classes. At the *training phase*, the objective is to determine the separating hyperplane, later used to classify unseen data.

Let us assume that data is linearly separable. That is, the data extracted from the process is considered to be grouped into two distinct areas, which can be split by a hyperplane (see Fig. 1). This assumption will be later relaxed. SVM classifiers are based on a class of hyperplane functions

$$f(\mathbf{x}) = <\mathbf{w} \cdot \mathbf{x}> + b = \sum_{i=1}^{n} w_i \cdot x_i + b \; . \tag{1}$$

From the above assumption of linear separability, the following equation can be used to classify the newly input data. An input vector belongs to one class if $\operatorname{sign}(f(\mathbf{x}))$ is positive; and belongs to the other class, otherwise, where

$$\operatorname{sign}(f(\mathbf{x})) = \begin{cases} 1, & \texttt{if } f(\mathbf{x}) \geq 0, \\ -1, & \text{otherwise.} \end{cases} \tag{2}$$

The data is guaranteed to be separable if there exists $(\mathbf{w}, b)$ such that for $i = 1, \ldots, l$:

$$<\mathbf{w} \cdot \mathbf{x}_i> + b \geq 1 \qquad \texttt{if} \qquad y_i = 1, \tag{3}$$

$$<\mathbf{w} \cdot \mathbf{x}_i> + b < -1 \qquad \texttt{if} \qquad y_i = -1. \tag{4}$$

The above equations (3), and (4), can be combined as

$$y_i \left(<\mathbf{w} \cdot \mathbf{x}_i> + b\right) \geq 1, \quad i = 1, 2, \ldots, l. \tag{5}$$

In order to obtain an optimal separating hyperplane, defined as the one with the maximal margin of separation between the two classes, let us consider the following constrained *optimization problem*:

$$\begin{aligned} & minimize_{\mathbf{w},b} && <\mathbf{w}\cdot\mathbf{w}>, && (6)\\ & subject\ to && y_i(<\mathbf{w}\cdot\mathbf{x}_i>+b)\geq 1, \quad i=1,2,\ldots,l. && (7) \end{aligned}$$

In Statistical Learning Theory [51], this optimization problem is quadratic and convex and has only one *global minimum*. Thus, by finding a solution to this problem we are also finding the optimal solution. This makes SVM a practical method.

If the dataset for classification is linearly separable, then the solution hyperplane of optimization problem (6)-(7) separates all the samples of the two classes. Otherwise some data points of the training set will be misclassified by the solution hyperplane. Such samples are called *outliers*. Also, in general, during the prediction phase, the SVM classifier can misclassify some input data. In this context, an *optimal separating hyperplane* is defined as the hyperplane which minimizes the probability that randomly generated examples are misclassified [51].

The optimization problem (6)-(7) has a solution only if the problem is linearly separable, i.e. only if all the conditions (7) can be simultaneously met. In order to deal with the possibility that (7) is not met, *slack variables* ξ_i $(i=1,\ldots,n)$ are introduced into the formulation and subsequent solution process of the optimization problem. For outlier samples, slack variables are positive, and represent the distance measured along an axis perpendicular to the separating hyperplane, indicating the amount of "degradation" in the classification of the sample. Slack variables are zero for non-outlier samples. With the introduction of slack variables, the following minimization problem is obtained:

$$\begin{aligned} & minimize_{\mathbf{w},b} && <\mathbf{w}\cdot\mathbf{w}>+C\sum_{i=1}^{l}\xi_i, \\ & subject\quad to && y_i(<\mathbf{w}\cdot\mathbf{x}_i>+b)\geq 1-\xi_i, \\ & && \xi_i\geq 0, \end{aligned} \tag{8}$$

where C is a large positive constant.

In the optimization problem (8), the objective function can be combined with the restrictions. For this purpose, the problem is reformulated such that the following Lagrangian function should be minimized:

$$\begin{aligned} L(\mathbf{w},b,\xi,\alpha,\mathbf{r}) = & \frac{1}{2}<\mathbf{w}\cdot\mathbf{w}>+C\sum_{i=1}^{l}\xi_i \\ & -\sum_{i=1}^{l}\alpha_i\left\{\xi_i+y_i\left[(\mathbf{w}\cdot\mathbf{x}_i)+b\right]-1\right\}-\sum_{i=1}^{l}r_i\xi_i && (9)\\ subject\ to\quad & \alpha_i\geq 0, r_i\geq 0. && (10) \end{aligned}$$

In (9)-(10) $\alpha=(\alpha_1,\alpha_2,\ldots,\alpha_n)$ is a vector of Lagrange multipliers, $\mathbf{r}=(r_1,r_2,\ldots,r_n)$, and $r_i=C-\xi_i$ $(i=1,\ldots,n)$. This is called the primal Lagrangian formulation. The minimum solution to this primal minimization problem must satisfy the Karush-Kuhn-Tucker conditions [8].

The solution of the Lagrangian using the primal form is difficult, because it contains two inequality constraints. An alternative form, the dual form, can be constructed by setting to zero the derivatives of the primal variables, and substituting the relations in the initial Lagrangian, removing the primal variables from the expressions.

For finding the solution of the dual formulation of the primal (9)-(10), the zeros of the partial derivatives of $L(\mathbf{w}, b, \xi, \alpha, \mathbf{r})$, with respect to the primal variables $\mathbf{w}$, b, and ξ_i $(i = 1, \ldots, n)$ must be obtained:

$$\frac{\partial L(\mathbf{w}, b, \xi, \alpha, \mathbf{r})}{\partial \mathbf{w}} = \mathbf{w} - \sum_{i=0}^{l} y_i \alpha_i \mathbf{x}_i = 0, \tag{11}$$

$$\frac{\partial L(\mathbf{w}, b, \xi, \alpha, \mathbf{r})}{\partial \xi_i} = C - \alpha_i - r_i = 0, \quad i = 1, \ldots, n, \tag{12}$$

$$\frac{\partial L(\mathbf{w}, b, \xi, \alpha, \mathbf{r})}{\partial b} = \sum_{i=0}^{l} y_i \alpha_i. \tag{13}$$

Substituting the solutions $\mathbf{w}$, ξ_i and b of (11)-(13) in the expression of the primal objective function (9), the following dual objective function is obtained:

$$L(\alpha, \mathbf{r}) = \sum_{i=0}^{l} \alpha_i - \frac{1}{2} \sum_{i,j=0}^{l} y_i y_j \alpha_i \alpha_j < \mathbf{x}_i \cdot \mathbf{x}_j > . \tag{14}$$

In this case the dual objective function (14) is constrained by the following conditions:

$$\alpha_i \left[y_i \left(< \mathbf{w} \cdot \mathbf{x} > + b \right) - 1 + \xi \right] = 0, \quad i = 1, \ldots, l, \tag{15}$$

$$\xi_i \left(\alpha_i - C \right) = 0, \quad i = 1, \ldots, l. \tag{16}$$

The optimization of the dual problem (14)-(16) yields the optimal values of α, and $\mathbf{r}$. Then, the optimal value of $\mathbf{w}$ for the primal problem can be found from (11) - let us call it $\mathbf{w}^*$. Since b does not appear in the dual problem, its optimal value, b^* must be found from the primal constraints:

$$b^* = \frac{\max_{y_i=-1}(< \mathbf{w}^* \cdot \mathbf{x}_i >) + \min_{y_i=1}(< \mathbf{w}^* \cdot \mathbf{x}_i >)}{2}. \tag{17}$$

The quadratic optimization problem formulated in (14)-(16), can be solved using the Sequential Minimal Optimization algorithm (SMO) [41], or the algorithm presented in [19].

The previous formulation can be used when (training and prediction) data is assumed to be linearly separable. This is attained if, after training, a low error rate is obtained. If this does not hold, the above formulation is extended as described below. Let us note an important property of the algorithm discussed thus far: both the quadratic programming problem that is solved and the decision function (1)-(2) only depend on inner products between vectors. This fact allows us to generalize to the nonlinear case.

The extension of the previous method is done by mapping the input features ($\mathbf{x}_i$) using a non-linear function. This mapping is done by a kernel function $\Phi : \mathbb{R}^n \rightarrow F$, which maps the input features into a higher dimensional space, named *feature space*, F, such that data is linearly separable after mapping. Then, the linear algorithm is performed in F.

The non-linear mapping is performed in the dual Lagrangian formulation (14). This mapping is implicitly performed simply by substituting in equation (14) the inner product $< \mathbf{x}_i \cdot \mathbf{x}_j >$ by $K(\mathbf{x}_i, \mathbf{x}_j)$, where

$$K(\mathbf{x}_i, \mathbf{x}_j) =< \Phi(\mathbf{x}_i) \cdot \Phi(\mathbf{x}_j) >, \tag{18}$$

and $\Phi(\mathbf{x})$ is a mapping function from the input space into a (inner product) feature space. As a consequence of this mapping approach no additional change is required in the training and prediction phases; The solution of the problem remains the same as described above. This has another desirable consequence: the computation effort of the method is not significantly affected by the kernel formulation. If F is high dimentional, the computation of the right hand side of (18) may be expensive. There are however, simple kernel functions that can be computed efficiently. Some examples are given below.

A function may be considered a *kernel* if it satisfies Mercer's condition, i.e., the matrix $\mathbf{K} = (K(\mathbf{x}_i, \mathbf{x}_j))_{i,j=1}^n$, that represents the kernel function, is positive semidefinite. There are several *kernel* functions proposed in the literature. The most usual kernels are the following:

1. Linear: $K(\mathbf{x}_i, \mathbf{x}_j) = \mathbf{x}_i^T \mathbf{x}_j$ (identity map).
2. Polynomial: $K(\mathbf{x}_i, \mathbf{x}_j) = (\gamma \mathbf{x}_i^T \mathbf{x}_j + h)^d, \quad \gamma > 0$.
3. Gaussian RBF: $K(\mathbf{x}_i, \mathbf{x}_j) = \exp(-\gamma \| \mathbf{x}_i - \mathbf{x}_j \|^2), \quad \gamma > 0$.
4. Sigmoid: $K(\mathbf{x}_i, \mathbf{x}_j) = \tanh(\gamma \mathbf{x}_i^T \mathbf{x}_j + h)$.

where γ, h and d are *kernel* parameters.

The difficulty of the learning task depends on the way the function to be learned is represented. A good representation of the specific learning problem should be chosen. The application of the kernel mapping would be equivalent to the substitution of the input feature vectors $\mathbf{x}$ by $\Phi(\mathbf{x})$, i.e. the referred *kernel* mapping can be seen as a preprocessing step that implements a change in the representation of data $\mathbf{x} = (x_1, \ldots, x_n) \longmapsto \Phi(\mathbf{x}) = \phi(\mathbf{x}) = (\phi_1(\mathbf{x}), \ldots, \phi_N(\mathbf{x}))$, where N is the dimensionality of F. The new set of input features may now be defined as $F = \{\phi(\mathbf{x}) \mid \exists y \in Y : (\mathbf{x}, y) \in S\}$. To take into account the feature set F, an optimization strategy is now applied as if it was a linearly separable space. This linear SVM has the following modified form:

$$f(\mathbf{x}) = \sum_{i=1}^{N} w_i \phi_i(\mathbf{x}) + b\,. \tag{19}$$

Fig. 2 illustrates the SVM operation with the kernel mapping.

Fig. 3 shows a general SVM framework. When the input space is linearly separable, an optimal separating hyperplane is obtained, casting an optimization problem and solving it. When the input space is non-linearly separable, a

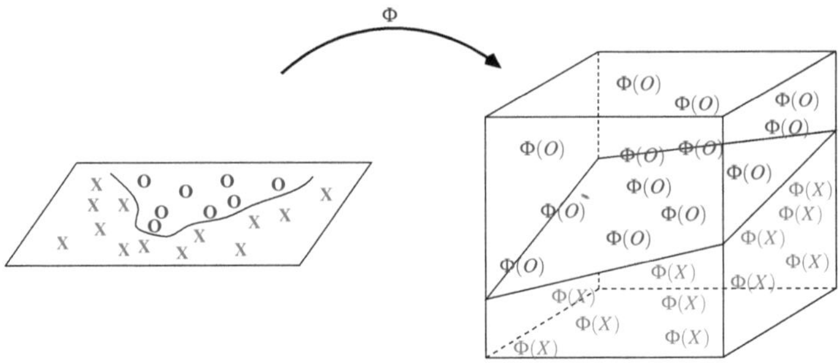

Fig. 2. Mapping to a feature space

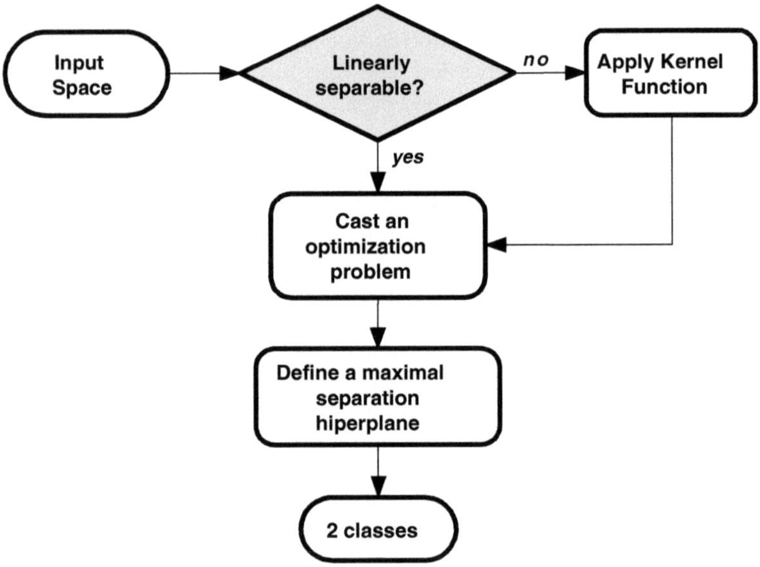

Fig. 3. SVM framework

non-linear function (*kernel*) is required in order to firstly map it into a higher dimensional space, where data will be linearly separable.

To decide if the input space is linearly separable, initially, a SVM with linear kernel is tested to see if the resulting separating hyperplane separates the data. If the training error is high or there is no convergence in the solution of the problem, then a different *kernel* function is applied in order to map the input space into a higher dimensional space. After the kernel mapping, the SVM problem is solved as described earlier.

Other considerations are important in this context. Sometimes, there may be a subset of *irrelevant features*, which are defined as follows: the removal of such features from the problem does not affect the performance of the classifier. Thus, it is often important to determine the smallest subset of features that is sufficient to correctly represent the problem, and the corresponding relevant information that is present in the original set of features or attributes. This process is known as *dimensionality reduction*.

Selecting different kernels yields different special classes of SVM. In particular, SVM contain other structures as special cases. For example, choosing polynomial kernels we may get polynomial classifiers; With a Gaussian kernel an RBF classifier is obtained; Using a sigmoid kernel, a three-layer neural network is attained.

2.1 VC Dimension

VC dimension is a core concept in Vapnik-Chervonenkis theory and may be defined as *the largest number of points that can be shattered in all possible ways* [20]. For binary classification, a set of l points can be labeled in 2^l possible ways.

Now, we are prepared to define the core concept of support vectors which is the subset of training patterns that form the maximal margin that separates two classes. In Fig. 4, the letter "o" represents the support vectors. The optimal separating hyperplane that results from the solution of the previously described optimization problem can be characterized by $\mathbf{w}$ which has a solution of the form $\mathbf{w} = \sum_i v_i \mathbf{x}_i$ in terms of the support vectors. Thus, these training vectors have all relevant information about the classification problem.

The VC dimension of a set of l points represents the number of support vectors and, in this case, the number of functions that defines the hyperplane inside of the smallest hypothesis space. The connection between the number of support vectors and the generalization ability, $E_l[P(error)]$, may be expressed by:

$$E_l[P(error)] \leq \frac{E[\text{number of support vectors}]}{l} . \tag{20}$$

where E_l denotes expectation over all training data sets of size l, and $P(error)$ is the misclassification probability. The value of this bound on the expected

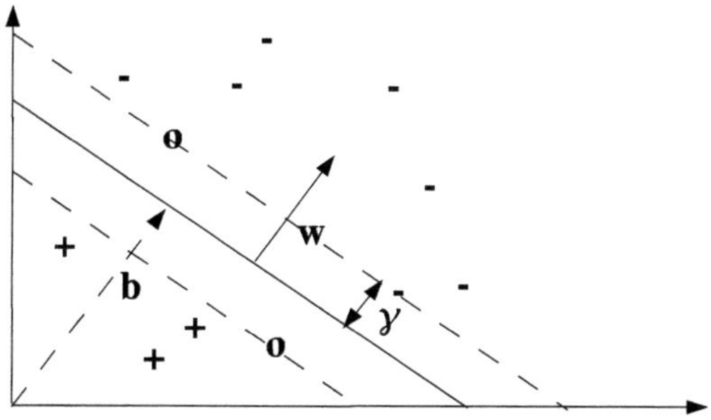

Fig. 4. Maximal margin hyperplane with its support vectors (o)

misclassification rate is independent of the dimensionality of the input space. Additionally, an SVM which has a small number of vectors will have good generalization ability and, consequently, will classify the input space faster.

2.2 Multiclassification with SVM

Although SVM is a binary classifier, strategies were formulated in order to obtain a multiclassification approach. The most used ones are: one-against-one, one-against-all and DAGSVM [18], [9], [23].

In a multiclassification problem, for each of the m classes of the output space $\mathrm{Y} = \{1, 2, \ldots, m\}$ it is associated an input vector and a bias, $(\mathbf{w}_i, \mathbf{b}_i)$, with $i \in \{1, \ldots, m\}$, and the decision function (1) is rewritten as:

$$c(\mathbf{x}) = \arg \max_{1 \leq i \leq m} \left(< \mathbf{w}_i \cdot \mathbf{x} > + \mathbf{b}_i\right) . \tag{21}$$

Geometrically, this is equivalent to associate an hyperplane for each class.

2.3 Practical Considerations

Nowadays, there are several resources and libraries that support the work with SVM (e.g. [34], [43]). In this section, we discuss the main features of some libraries, while detailing some practical aspects that might help one in dealing with SVM classifiers. A list with many SVM softwares available in the internet can be found in [33]. In this list, two libraries can be highlighted: LightSVM [31] and LibSVM [32]. Both LightSVM and LibSVM work in the same manner from the user point of view, and all aspects herein analyzed, will apply to both of these libraries:

1. **Supervised training** – as stated in the previous sections, SVM are supervised machines. Hence, there are always two stages to be carried out in order to classify a given pattern: training and classification. On all situations where a supervised training takes place, it will always be necessary to firstly gather training examples in a sense that it may contribute to enrich the way that the SVM model can be created. Thus, one ought to give an input file in some format. After that, the given examples will be scanned and a QP problem solved in order to obtain the support vectors and weights (SVM model).

2. **Input space normalization** – data are needed to be normalized before being given to an SVM training algorithm. Two methods usually applied for normalization are: L1-norm $\frac{X}{\|X\|}$ and L2-norm $\frac{X}{\|X\|^2}$. An interesting discussion about normalization and its effects in SVM is found in [3].

3. **Input space compactness** – sometimes, the higher the input space dimension, the higher the number of support vectors in the final SVM model. The speed of classification is directly proportional to the number of support

vectors. Hence, if one might reduce the input space by using just the "good" features, it would be possible to decrease the classification time keeping a certain level of classification performance. The compactness of the input space is achieved by applying feature selection strategies. The Recursive Feature Elimination (RFE) has drawn attention in the last few years, showing good results along with SVM. An interesting and practical discussion about feature selection, for SVM, can be found in [14].

4. **Choice of the kernel** – the choice of the kernel (and its parameters) which will be used in the SVM model should be made carefully, since it will affect the results. The main method to deal with this problem is cross validation, that is, to split the input space into v folds and use systematically one of them in each interaction as a test sample and the others as training samples. At the end, the chosen kernel should be the one with the lowest percentage of errors. This process must be made in the training stage and it gives a general idea of the classification performance.

The previous list is not intended to be exhaustive but includes the more important items to be followed in order to obtain a "good" classifier.

One important issue, that should be better analyzed by the SVM library community, is the disk space management, since when the number of examples and the dimensionality of the input space is high, the input file can get big enough to not fit in the Operating System (OS) environment. Thus, one might think of alternatives to compact the input file so that also the reading of the input data could be made faster than in text format. Some implementations already deal with binary files in a dense way, compacting the input file up to one third of the size of the text file format usually used.

3 Features

Sensor data, normally, presents a high dimensionality. Thereat, it is normally prohibitive to directly process this complex data for classification purposes. Twofold problems normally arise, in this context:

1. For analyzing a large number of variables, it is required a large amount of memory and computational resources;
2. In classification systems, the usage of large amount of data, normally, leads to overfitting of the training samples, and consequently poor generalization.

An alternative approach is to define features[1] as variables constructed from the input space [14], [11]. Extracting relevant information (features), from the input space, may lead to a more powerful and effective classification. Then,

[1] It is necessary to distinguish between the feature space, relative to data, and the feature space that comes from a "kernelization" of the input space, applied in SVM, to achieve a higher dimensional inner product space.

before the application of SVM, one may think of analysing the input vector to find a higher level representation of variables that would "drive" the classifier to more accurate classification results. On the design phase, there are difficulties in choosing a set of features that permits the attainment of good classification results for the particular problem being solved, and the choice of wrong features may compromise the performance of the classifier.

In the next sections, features extracted from Lidar and image sensor data are presented. These features are used in two SVM-based classification tasks: learning to label places in an indoor environment using Lidar data, and vision-based human detection.

3.1 Features in Lidar Space

Range-bearing data, provided by Lidar, is commonly used in mobile robotics to perform tasks such as map learning, localization, and obstacle detection [49], [2], [1], [7], [42]. Each measurement collected from a range sensor such as Lidar represents the distance from the sensor to an object[2] at a certain range and bearing.

The features described in this subsection, were designed for application with 2D data, although some of them can be adapted for application to higher dimentional data. The idea behind the method is to extract simple and computationally non-intensive features, which by themselves do not distinguish between the various classes, but in group perform the expected task. They are geometric features adapted from computer vision methods [38], [52], [24], [13], [28]. These features were selected because they are invariant to rotation, scale and translation. Later, in Sec. 4.1, these features will be employed for learning to semantically classify places with SVM.

Data collected with a range sensor is composed by a set of measures

$$Z = \{b_0, \ldots, b_{l-1}\}, \tag{22}$$

where b_i is a tuple (ϕ_i, d_i), and ϕ_i and d_i are the bearing and range measures respectively (e.g. Fig. 5). Let $\mathcal{Z}$ define the set of all possible observations. Features are directly extracted from the gathered sensor data, where each data set Z is transformed into a real value $f_i(Z) : \mathcal{Z} \rightarrow \mathbb{R}$, which defines a feature that is used as a component of an input vector employed for classification purposes. Thus, each set of raw measurements Z is transformed into a vector of features that is subsequently used as input for classification.

For each set of input measurements, Z, the extraction of features broadly occurs using two different representations and processes. In the first representation, raw measurements (the elements of Z, e.g. Fig. 5) are directly used in the computation of features. These are called raw features. In the second representation, features are extracted from an intermediate polygonal representation (e.g. Fig. 6) of the raw data. These are called polygonal features.

[2] Every sensor presents a maximum and minimum value. The objects must be contained inside this interval.

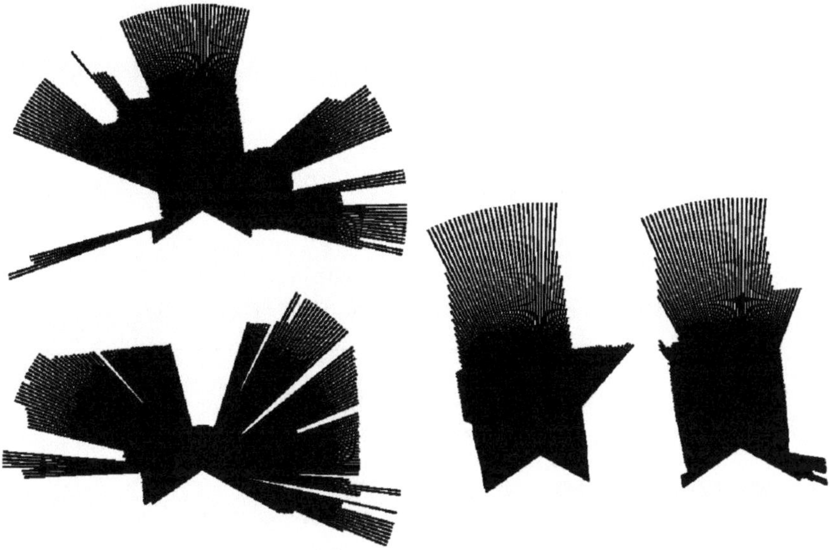

Fig. 5. Examples of data extracted from Lidar sensor

The following features are directly computed from raw sensor data:

1. **Average of the length difference of consecutive measures**
   This feature represents the average difference between the length of consecutive range measures:

$$f_{avr-dif} = \frac{1}{l}\sum_{i=0}^{l-2}(d_i - d_{i+1})\,. \tag{23}$$

2. **Standard deviation of the length difference of consecutive measures**
   This feature represents the standard deviation of the difference between the length of consecutive range measures. It is defined as follows:

$$f_{std-avr} = \sqrt{\frac{1}{l}\sum_{i=0}^{l-2}((d_i - d_{i+1}) - f_{avr-dif})^2}\,. \tag{24}$$

3. **Average of the length difference of consecutive measures using a maximal range**
   The processing of this feature is performed in two stages. First, each measurement is limited to a maximum threshold value, as follows:

$$d_\theta(i) = \begin{cases} d_i, & \text{if } d_i < \theta, \\ \theta, & \text{otherwise.} \end{cases} \tag{25}$$

After transforming all collected values, the feature is determined as follows:

$$f_{avr-d_\theta} = \frac{1}{l}\sum_{i=0}^{l-2}\left[d_\theta(i) - d_\theta(i+1)\right]. \tag{26}$$

4. **Standard deviation of the length difference of consecutive measures using a maximal range**
The calculus of this feature is also performed in two stages as for feature 3. First, a limit value is applied to all measurements using expression (25). After this pre-processing stage, the feature is determined by:

$$f_{std-d_\theta} = \sqrt{\frac{1}{l}\sum_{i=0}^{l-2}\left\{\left[d_\theta(i) - d_\theta(i+1)\right] - f_{avr-d_\theta}\right\}^2}. \tag{27}$$

5. **Average range length**
This feature represents the average of the measures gathered from range sensor. It is defined as:

$$f_{avr-length} = \frac{1}{l}\sum_{i=0}^{l-1} d_i. \tag{28}$$

6. **Standard deviation of the range length**
The standard deviation of the range length is defined as, where $f_{avr-length}$ is defined in (28):

$$f_{std-length} = \sqrt{\frac{1}{l}\sum_{i=0}^{l-1}\left(d_i - f_{avr-length}\right)^2}. \tag{29}$$

7. **Number of gaps in a scan between consecutive measures**
If two consecutive scan measurement values have a range difference greater than a given threshold (θ_g), it is considered that there exists a gap between these two measurements. A gap is defined as:

$$gap_{\theta_g}(d_i, d_{i+1}) = \begin{cases} 1, & \text{if } |d_i - d_{i+1}| > \theta_g, \\ 0, & \text{otherwise.} \end{cases} \tag{30}$$

The total number of gaps is defined as:

$$f_{gaps,\theta} = \sum_{i=0}^{l-2} gap_{\theta_g}(d_i, d_{i+1}). \tag{31}$$

In addition to the raw features defined above, a second set of features, the polygonal features, was employed for learning to label places. Define $b_l \equiv b_0$. Differently from the previous features, polygonal features are computed from an intermediate polygonal representation (e.g. Fig. 6) of the input raw sensor data

Fig. 6. Polygon representation of data extracted from Lidar range sensor

set. In these features, each data is seen as being the vertices of a polygon, $\mathbf{P}(z)$, defined by the corresponding raw sensor measurements

$$\mathbf{P}(z) = \{v_0, v_1, \ldots, v_{l-1}\} \tag{32}$$

where $v_i = (x_i, y_i)$ is a Cartesian representation of the range-bearing measure b_i, with $x_i = d_i \cos(\phi_i)$ and $y_i = d_i \sin(\phi_i)$. Below, the set of polygonal features used for place labeling with SVM are defined.

1. **Area of $\mathbf{P}(z)$**
   The area of $\mathbf{P}(z)$ is defined as

$$Area_{\mathbf{P}(z)} = \frac{1}{2}\sum_{i=0}^{l-1}(x_i y_{i+1} - x_{i+1} y_i). \tag{33}$$

2. **Perimeter of $\mathbf{P}(z)$**
   The perimeter of $\mathbf{P}(z)$ is computed as follows:

$$Perimeter_{\mathbf{P}(z)} = \sum_{i=0}^{l-1} dist(v_i, v_{i+1}), \tag{34}$$

   where function $dist(v_i, v_{i+1})$ is defined as

$$dist(v_i, v_{i+1}) = \sqrt{(x_i - x_{i+1})^2 + (y_i - y_{i+1})^2}. \tag{35}$$

3. **Mean distance between the centroid and the vertices of the polygon**
The central point (centroid), $c = (c_x, c_y)$, of $\mathbf{P}(z)$ is defined as:

$$c_x = \frac{1}{6 \cdot Area_{\mathbf{P}(z)}} \sum_{i=0}^{l-1} (x_i + x_{i+1})(x_i y_{i+1} - x_{i+1} y_i) \tag{36}$$

$$c_y = \frac{1}{6 \cdot Area_{\mathbf{P}(z)}} \sum_{i=0}^{l-1} (y_i + y_{i+1})(x_i y_{i+1} - x_{i+1} y_i) \tag{37}$$

The mean distance between the centroid and the vertices is then determined as follows:

$$f_{mean-centroid} = \frac{1}{l} \sum_{i=0}^{l-1} dist(v_i, c) \tag{38}$$

where

$$dist(v_i, c) = \sqrt{(x_i - c_x)^2 + (y_i - c_y)^2}. \tag{39}$$

4. **Standard deviation of the distances between the centroid and polygon vertices**
The standard deviations of the distances from the centroid to the polygon vertices is calculated as follows:

$$f_{std-centroid} = \sqrt{\frac{1}{l} \sum_{i=0}^{l-1} \left(dist(v_i, c) - f_{mean-centroid}\right)^2}, \tag{40}$$

where $f_{mean-centroid}$ is the value defined at (38).
5. **Invariants from the central moments of $\mathbf{P}(z)$**
For a bidimentional continuous function $f(x, y)$, the moment of order $p + q$ is defined as

$$m_{pq} = \int_{-\infty}^{+\infty} \int_{-\infty}^{+\infty} x^p y^q f(x, y) dx dy, \quad \text{for } p, q = 0,\ 1,\ 2,\ \ldots \tag{41}$$

It can be shown [13] that if $f(x, y)$ is piecewise continuous and has nonzero values only in a finite part of the xy plane, moments of all order exist and the moment sequence $\{m_{pq}\}$ is uniquely determined by $f(x, y)$. Conversely, the sequence $\{m_{pq}\}$ uniquely determines $f(x, y)$. The central moments are defined as

$$\mu_{pq} = \int_{-\infty}^{+\infty} \int_{-\infty}^{+\infty} (x - \bar{x})^p (y - \bar{y})^q f(x, y) dx dy \tag{42}$$

where

$$\bar{x} = \frac{m_{10}}{m_{00}},$$

and

$$\bar{y} = \frac{m_{01}}{m_{00}}.$$

For discrete functions, expression (42) becomes:

$$\mu_{pq} = \sum_{x}\sum_{y}(x-\bar{x})^p(y-\bar{y})^q f(x,y) \tag{43}$$

In our case, the computation of expression (43) can be significantly simplified since $f(x, y)$ is defined as:

$$f(x,y) = \begin{cases} 1, \text{ if}(x,y) = v_i, \; i = 0,..,l-1 \\ 0, \text{ otherwise} \end{cases} \tag{44}$$

Then, the following normalized central moments (denoted by η_{pq}) are computed:

$$\eta_{pq} = \frac{\mu_{pq}}{\mu_{pq}^{\gamma}}, \text{for} \quad p+q = 2,\ 3,\ \ldots, \tag{45}$$

where

$$\gamma = \frac{p+q}{2} + 1, \text{for} \quad p+q = 2,\ 3,\ \ldots. \tag{46}$$

Finally, the following seven invariant moments are used as features for the classification process:

$$\phi_1 = \eta_{20} + \eta_{02} \tag{47}$$

$$\phi_2 = (\eta_{20} - \eta_{02})^2 + 4\eta_{11}^2 \tag{48}$$

$$\phi_3 = (\eta_{30} - 3\eta_{12})^2 + (3\eta_{21} - \eta_{03})^2 \tag{49}$$

$$\phi_4 = (\eta_{30} + 3\eta_{12})^2 + (\eta_{21} - \eta_{03})^2 \tag{50}$$

$$\begin{aligned}\phi_5 = {} & (\eta_{30} - 3\eta_{12})(\eta_{30} + 3\eta_{12})\left[(\eta_{30} + 3\eta_{12})^2 - 3(\eta_{21} + \eta_{03})^2\right] \\ & + (3\eta_{21} - \eta_{03})(\eta_{21} + \eta_{03})\left[3(\eta_{30} + 3\eta_{12})^2 - (\eta_{21} + \eta_{03})^2\right]\end{aligned} \tag{51}$$

$$\begin{aligned}\phi_6 = {} & (\eta_{20} - \eta_{02})\left[(\eta_{30} + \eta_{12})^2 - (\eta_{21} + \eta_{03})^2\right] \\ & +4\eta_{11}(\eta_{30} + \eta_{12})(\eta_{21} + \eta_{12})\end{aligned} \tag{52}$$

$$\begin{aligned}\phi_7 = {} & (3\eta_{21} - \eta_{03})(\eta_{30} + \eta_{12})^2\left[(\eta_{30} + \eta_{12})^2 - 3(\eta_{21} + \eta_{03})\right] \\ & + (3\eta_{12} - \eta_{30})(\eta_{21} + \eta_{03})\left[3(\eta_{30} + \eta_{12})^2 - (\eta_{21} + \eta_{03})\right]\end{aligned} \tag{53}$$

These moments are invariant to scale, translation and rotation.

6. **Normalized compactness of $\mathbf{P}(z)$**
The normalized compactness of $\mathbf{P}(z)$ is defined as:

$$M_{compactness} = \frac{Area_{P(z)}}{\mu_{20} + \mu_{02}},\ 0 \le M_{compactness} \le 1, \tag{54}$$

where μ_{20}, and μ_{02} are cental moments of second order determined from (43).

7. **Normalized eccentricity of $\mathbf{P}(z)$**
   The normalized eccentricity of $\mathbf{P}(z)$ is defined as:

$$M_{eccentricity} = \frac{\sqrt{(\mu_{20} + \mu_{02})^2 + 4\mu_{11}^2}}{\mu_{20} + \mu_{02}}, \ 0 \leq M_{eccentricity} \leq 1, \qquad (55)$$

   where μ_{20}, μ_{02}, and μ_{11} are the central moments of second order computed from (43).
8. **Form factor of $\mathbf{P}(z)$**
   The form factor of $\mathbf{P}(z)$ is defined as:

$$M_{factor-form} = \frac{4\pi Area_{\mathbf{P}(z)}}{\sqrt{Perimeter_{\mathbf{P}(z)}}} \qquad (56)$$

3.2 Features in Camera Space: A Brief Survey

A digital image is a two dimensional matrix with color information at each pixel. Regardless of its density, i.e., the dimensionality of its color space (if it is a gray color image or a 24-bit per pixel one), one may try to extract statistical representations of the pixels in order to find, as much as possible, objects over the image. These representations are guided to an underlying characteristic of a good image classification algorithm that is the discriminability. Hence, the main property of a feature representation is the degree of discrimination to represent a part or the whole object.

When we think about pattern recognition based on vision, it is worth noting how discriminative features are, since the way to acquire an image is mainly influenced by external factors, such as illumination and noise, that usually modify the way how the image is perceived by the classifier. Thus, to design a robust image classification algorithm it is necessary not only a robust representation (features) of this object in the image, but also a robust classifier that is able to share these features into object and non-object accurate classes.

In this section, a brief survey of some features usually applied in object recognition, and classified by SVM, is presented. Table 1 summarizes the contents of the rest of the section.

In Section 4.2, particular implementation details of a HOG/SVM based system are presented. This vision-based system, along with a Lidar sensor, is able to give spatial information about pedestrians in front of a vehicle.

Histogram of Oriented Gradients. This feature has reached a maturity version for object matching in [25] and it was further modified by [10] to recognize pedestrians sequences.

In Fig. 7, the feature is shown. It is composed by block descriptors, which, in turn, are formed by a set of cells, where the oriented gradients are accumulated. The size of the blocks and cells will determine how discriminative the recognition will be.

In [25], the block descriptor is a set of 16×16 pixels. The oriented gradients, then, are obtained in a region of 4×4 pixels, which form the cell of the block

Table 1. Brief survey of features for object recognition systems

Feature	SVM kernel type	Application	Reference
Histogram of Oriented Gradients	Linear	Pedestrian recognition	[10]
Haar Wavelet templates	Polynomial	Pedestrian recognition	[39]
C2	Linear	Recognition of a set of objects	[44]
Local Receptive Fields		Pedestrian recognition	[29], [36]

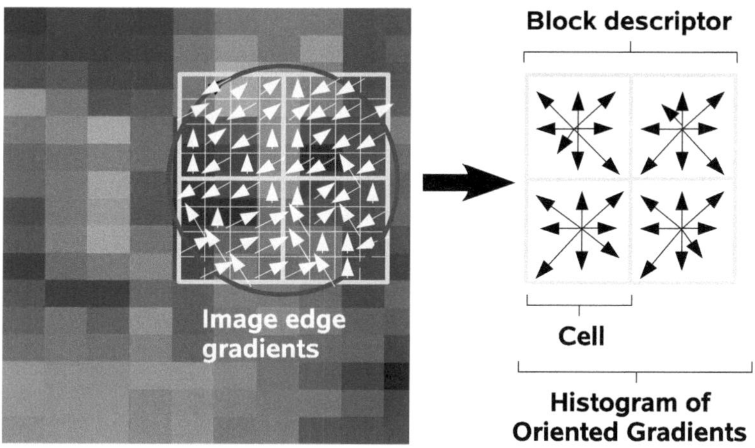

Fig. 7. Histogram of Oriented Gradients (HOG) is created by computing the values of the gradients (magnitude and orientation), as shown on left side of the picture. These values are accumulated into bins to form the histogram, right picture. The size of the histogram (size of the blocks and cells) will determine how discriminative the recognition will be.

descriptor. For each histogram, in each one of the 4 regions, there are 8 bins, computed around a full circle, i.e., each 45 degrees of orientation accumulates a value of the gradients. The circle, in Fig. 7, stands for a Gaussian function to weight each block descriptor, according to:

$$G(x, y, \sigma) = \frac{1}{2\pi\sigma^2} e^{-(x^2+y^2)/2\sigma^2} . \qquad (57)$$

where x and y are the coordinates of the centered pixel in the block descriptor window and σ is equal to one half of the width of the block descriptor window.

The computation of the gradient at each point in the image $I(x, y)$ is performed by calculating its magnitude $m(x, y)$ and orientation $\theta(x, y)$, according to:

$$m(x,y) = \sqrt{[I(x+1,y) - I(x-1,y)]^2 + [I(x,y+1) - I(x,y-1)]^2}\,. \tag{58}$$

$$\sigma(x,y) = \tan^{-1}\left[\frac{I(x,y+1) - I(x,y-1)}{I(x+1,y) - I(x-1,y)}\right]\,. \tag{59}$$

After the histograms are found, they are normalized by the *L2-Hys* norm, which corresponds to *L2-norm*, ($v \leftarrow v/\sqrt{||v||_2^2 + \epsilon^2}$), followed by limiting the maximum values of v to 0.2, and renormalizing [25].

In [25], HOGs are just computed at stable points, i.e., partial rotation, translation and scale invariant points which are achieved by means of a pyramid of Gaussians. On the other hand, in [10], HOGs are computed over all image, densely. In the latter approach, HOGs are gathered through a normalized image window of predetermined size. Thereat, the input vector passed to a linear SVM has a constant size and represents an object/non-object decision.

Haar Wavelet Templates. This kind of feature aims to describe regions with difference of color intensity. The name 'Haar Wavelet' indicates a particular discrete case of the wavelet transform. One of the first approaches using these features along with SVM can be found in [37] and further improved in [39]. These features have been represented by the following templates [39]:

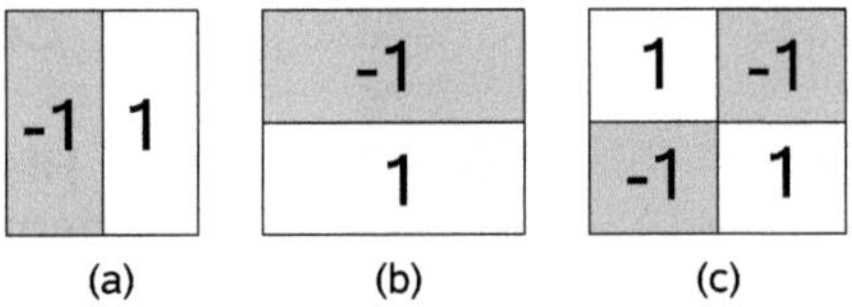

Fig. 8. Wavelets templates: (a) vertical, (b) horizontal and (c) corner

The templates shown in Fig. 8 come from the following equation:

$$\gamma(t) = \begin{cases} 1, & \text{if } 0 \leq g < a, \\ -1, & \text{if } a \leq g < b, \\ 0, & \text{otherwise,} \end{cases} \tag{60}$$

where g is a gray level intensity, and a and b correspond to the desired threshold.

In [39], a set of wavelet templates have been collected over a 64×128 normalized image window for pedestrian detection, and submitted to an SVM of second degree polynomial kernel, according to:

$$f(x) = \phi\left(\sum_{i=1}^{N} \alpha_i y_i (x.x_i + 1)^2 + b\right) \tag{61}$$

where N is the number of support vectors and α_i are Lagrange parameters.

It is worthing notting that a number of new Haar wavelets features has been proposed and further utilized in [50] along with an Adaboost classifier.

C2. Its creation was motivated by researches about how the brain of primate primary visual cortex works, which contains cells with the same name [44].

To extract the features, four stages are necessary (S1, C1, S2, C2):

- **S1:** Application of a set of Gabor filters and elimination of those which are incompatible to biological cells. At the end, 8 bands are given to the next stage. This stage corresponds to simple cells in primate visual cortex;
- **C1:** Extraction of patches from the bands S1;
- **S2:** Obtaining S2 maps according to Eq. (63) below;
- **C2:** Computation of the maximum over all positions and scales for S2 maps. The output of this stage gives C2 maps which are scale invariants.

In order to compute the bands for S1, a set of Gabor filters $G(x, y, \lambda, \theta, \psi, \sigma, \gamma)$ are applied to the input image:

$$G(x, y, \lambda, \theta, \psi, \sigma, \gamma) = \exp\left(-\frac{x'^2 + \gamma y'^2}{2 * \sigma^2}\right) \cos\left(2\pi \frac{x'}{\lambda} + \psi\right) . \quad (62)$$

where $x' = x\cos\theta + y\sin\theta$, $y' = -x\sin\theta + y\cos\theta$, θ represents the orientation of the normal to the parallel stripes function, ψ represents the offset of the phase, and γ specifies the shape of the Gabor function support. In stage C1, after computing the final bands in the previous stage, patches are extracted in order to compose a set of 'interesting points' or 'keypoints'. For all image patches (X) and each patch (P) learned during training, the following equation is applied so that S2 maps are obtained:

$$Y = \exp(-\gamma ||X - P_i||^2) . \quad (63)$$

In the last stage, $\max(S2)$ over all positions and scales is computed and C2 features, which are scale invariants, are achieved. A study of C2 features is made in [44], where a linear SVM and a Gentle Adaboost are applied to classify the features, with similar classification results being reported for both classifiers.

Local Receptive Fields. The structure of these features follows the same roots as C2. The main difference is that Local Receptive Fields (LRF) are trainable types of features.

LRF were firstly proposed by [22] and have shown excellent performance in many fields [22], [21], [12]. Recently, [29] demonstrated that LRF can achieve hit rates comparable to state-of-art features for pedestrian recognition and, in [36], an ensemble method, using LRF, achieved the highest performance in NiSIS competition [35], in which the main objective was to recognize pedestrians, using Daimler Chrysler dataset. The challenge of the task was to find a discriminant model between pedestrians and non-pedestrians using noisy, low quality images, where sometimes pedestrians are partially occluded.

Since LRF are trainable features, one should provide a complete feature extraction method along with a classifier, at the same time, extracting the features and training the classifier. The best choice for this is a Multi Layer Perceptron (MLP) with a backpropagation training algorithm to find the best weights to extract the features. After this training procedure, features can be fed into any other classifier. In [36], the proposed method, after extracting LRF features, these are used along with an SVM to improve the lexicon of classification.

4 Applications

4.1 Learning to Label Environment Places

This subsection describes a method for learning to classify places in indoor environments using SVM. The features presented in Sec. 3.1 are employed to the SVM classifier. The application method was performed on a embedded PC, located on-board of a robotic wheelchair (Fig. 9) [40], [30], [47].

The results of the method are being used in other developments such as methods for topological mapping of the environment, localization, and human-robot interaction.

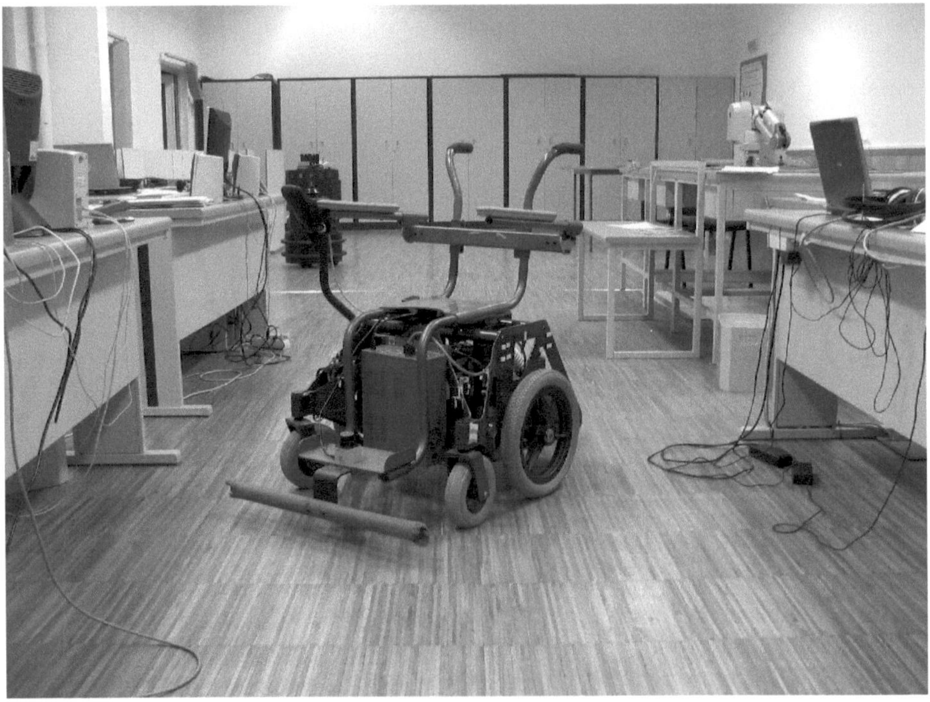

Fig. 9. Picture of the testbed robot (Robchair)

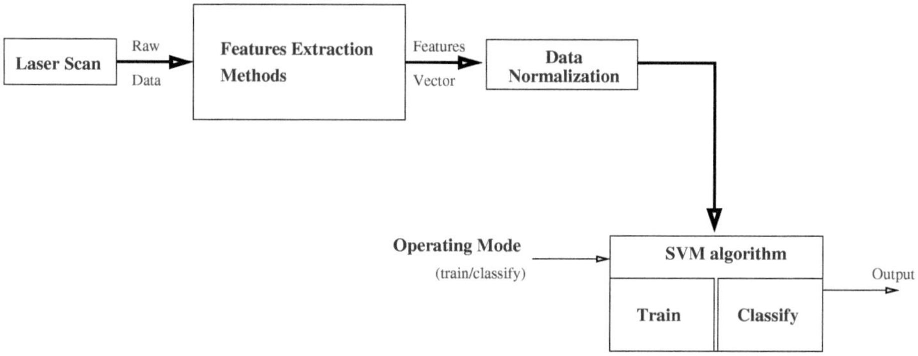

Fig. 10. Overall structure of the place labeling method

The global architecture of the method is presented in Fig. 10. Three distinct phases are involved in the operation of the method. In a first stage, data is gathered from the environment through a Lidar sensor. At a second stage, the raw Lidar data is processed into multiple features. Finally, the computed features are feed as inputs to the SVM classifier. The SVM classifier has two operation modes: the learning phase and the prediction phase (see section 2).

Range data is collected using one Hokuyo URG-04LX Lidar [17] connected to the embedded computer through an Universal Serial Bus (USB). Range measurement scans are periodically acquired and post-processed into a collection of features for classification. The Hokuyo has a scanning range of 240 degrees. Each scan is composed of 632 range-bearing readings which are radially equally spaced by 0.36 degrees, i.e., each raw sensor observation is $\mathbf{z} = \{b_1, \ldots, b_{632}\}$. Each raw sensor measure is used as input to the 15 feature extraction algorithms explained in Sec. 3.1. The outputs of the feature extraction algorithms are gathered into a fixed size vector. Thus, the feature mapping function is $\mathbf{f} : \mathbb{R}^{632} \rightarrow \mathbb{R}^{15}$.

The next phase is the normalization of the feature vectors, with the objective of increasing the numerical stability of the SVM algorithm.

The sensor data sets were collected in the office-like building of ISR-UC (illustrated in Fig. 11). The design of the method enables it to operate independently of specific environment characteristics (both in learning and prediction phases), i.e. the method is prepared to operate in various environments and to be robust to environment changes. In this context, five data sets were collected in several corridors and rooms. The first data set was used for training the classifier, and the other data sets were used to test the classification system with data, representing new observation situations not present in the training data. The training data set was composed of 527 sensor observations $\mathbf{z}_i$ and the corresponding place classifications v_i. This is a relatively small training set. One of the four testing data sets corresponds to a corridor. The other three testing data sets were obtained in different rooms.

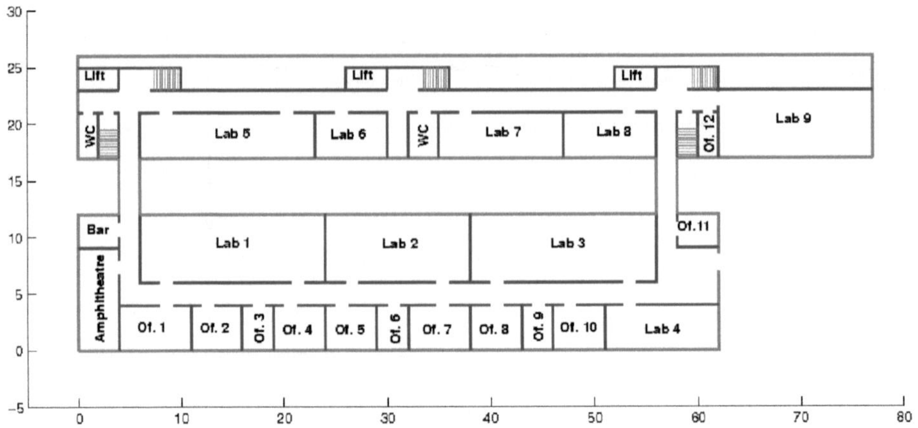

Fig. 11. Map of the environment used in the experiments

Table 2. Classification results for different indoor areas and using a sigmoid kernel

	Total samples	Correctly classified	Correct rate
Corridor	48	36	75.00%
Lab. 7	112	84	75.00%
Lab. 4	27	18	66.67%
Lab. 5	57	43	75.44%

Table 3. Classification results for different indoor areas and using a polynomial of degree $d = 5$ kernel

	Total samples	Correctly classified	Correct rate
Corridor	48	38	79.16%
Lab. 7	112	88	78.57%
Lab. 4	27	21	77.78%
Lab. 5	57	43	75.44%

In different training sessions, the SVM classifier was tested with three different kernels: sigmoid, polynomial of degree $d = 5$, and Gaussian RBF. After obtaining the SVM model, the classifier was tested with different data sets. Tables 2, 3, and 4 present the results obtained with the three different kernels. The best results were obtained with the Gaussian RBF kernel, where the hit ratio was always above 80%. These are promising results obtained with the relatively small data set used to train the SVM-based classifier.

Table 4. Classification results for different indoor areas and using a Gaussian RBF kernel

	Total samples	**Correctly classified**	**Correct rate**
Corridor	48	41	85.42%
Lab. 7	112	92	82.14%
Lab. 4	27	22	81.48%
Lab. 5	57	44	77.19%

The results show that place classification with SVM is possible with a high degree of confidence. The method can be extended by the introduction and testing of other kernels and additional features.

4.2 Recognizing Objects in Images

As mentioned in Section 3, it does not usually suffice to use a raw input vector directly into a classifier, since this vector is not discriminative enough to lead to a separation of objects and non-objects in a consistent way. Hence, one usually need to find robust features to do the job.

Motivated by [10], in this section, we explain the implementation details of our system to recognize and locate pedestrians in front of a vehicle. In Fig. 12, the framework of the system is illustrated. Firstly, the sensor data are acquired separately by using two different threads, one for the camera and the other for the Lidar. These threads communicate and are synchronized using shared memory and semaphores.

The module in charge of the Lidar scanner processing, then, segments the Lidar scanner points [5], as they arrive, and tries to cluster them, following the most probable hypothesis to be an object in the scene. Finally, this module transforms the coordinates of the world into image coordinates, by means of a rigid transformation matrix (see Fig. 13).

Concerning the image pattern recognition module, the algorithm proposed by [10] was implemented from scratch, with some slight modifications in order to speed up the classification process. The overall sketch of this method is described in Algorithm 1.

Firstly, a set of HOG features is extracted densely across the image. After feature normalization, feature vectors are passed to a linear SVM in order to obtain a support vector model. In case of negative features, which are extracted randomly across non-object images, the model is retrained with an augmented set of hard examples to improve performance (bootingstrap strategy).

The classification stage scans the input image over several orientations and scales through an image pyramid (see Fig. 14), all bounding boxes that overlap are clustered. For all stages, several parameters may be used and there exists a set of default optimal values [10].

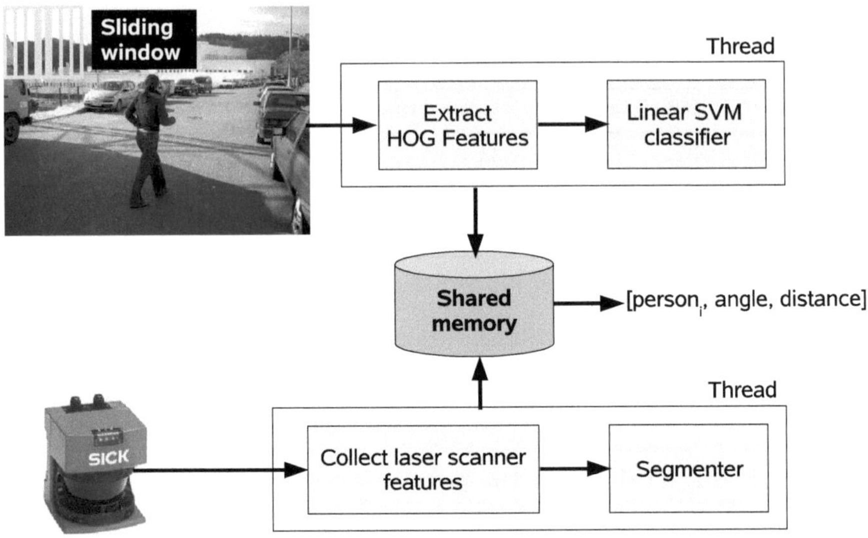

Fig. 12. Two threads are performed in the system: one responsible to the vision module and another one, responsible to Lidar segmentation. A rigid transformation is applied in order to map Lidar scanner space points to camera space points. At the end, the complete spatial location of the person in the image is given.

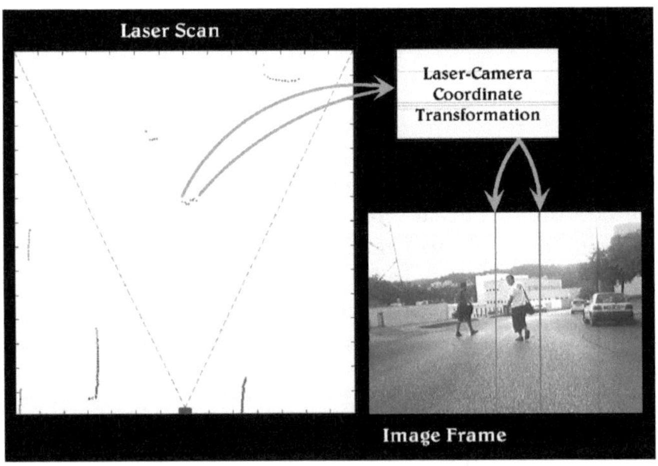

Fig. 13. A rigid transformation matrix is applied to transform world coordinates into image coordinates in order to give spatial information of the objects in the scene

Fig. 14 illustrates the operation of Algorithm 1. Each level of the image pyramid is scanned by means of a sliding window. Afterwards, all the bounding boxes (red rectangle) are clustered with a non-maximum suppression algorithm (the

Algorithm 1. Training and Classification of HOG / SVM

Training stage
- Extract feature vectors of normalized positive samples and random negative samples;
- Normalize vectors;
- Run classifier against negative samples, thoroughly;
- Retrain hard examples;
- Create final linear SVM model;

Classification Stage
- Scan images at different location and scale;
- Predict classes with linear SVM model;
- Fuse multiple detections;
- Give object bounding boxes in image;

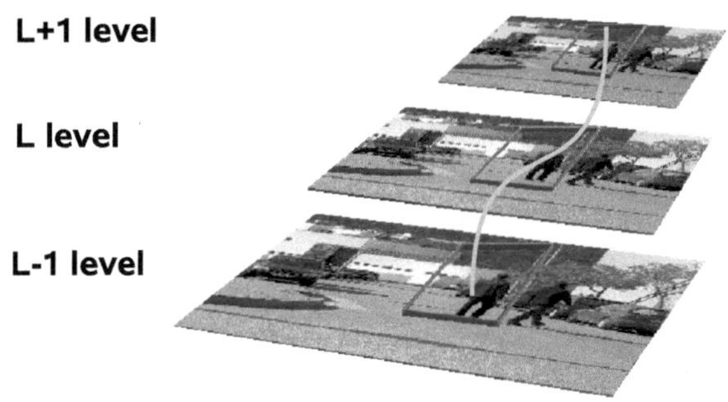

Fig. 14. The input image is scanned over several orientations and scales through the image pyramid

blue path shows that the criterion to cluster the bounding boxes should be based on overlapping situations over the pyramid).

Below, some more details are given concerning the classification tasks and sensor fusion, as well.

Implementation Details. In our HOG/SVM implementation, the steps in Algorithm 1 were followed, except for two points:

– the final bounding boxes were clustered not by means of a mean shift algorithm, but according to an area overlapping criterion:

$$OA = \frac{A_{L-1} \cap A_L}{A_L \cup A_{L-1}} \geq 0.5 \,, \tag{64}$$

where A_{L-1} and A_L represent the bounding boxes of the object in level $L-1$ and L, respectively.

– The maximum number of pyramid levels (S_n) (see Fig. 14) is given by

$$S_n = \text{floor}\left(\frac{\log(S_e/S_s)}{\log(S_r)} + 1\right), \tag{65}$$

where 'floor' is the largest integer value not greater than argument, S_r is the step constant, S_s is the initial step, equal to 1, and $S_e = \min(W_i/W_n, H_i/H_n)$, where W_i, H_i and W_n, H_n are the image size and positive normalized window size, respectively. However, S_n has been limited to 10 in order to decrease the computational cost.

Additionally, in order to obtain an improved classification speed, another implementation detail is worth to be realized. Since the algorithm proposed by Dalal and Triggs [10] presents high computational cost, we have decided to limit the classification region in the image by the Lidar Region of Interest (ROI), according to Fig. 13. In other words, the information that comes from the Lidar is used to establish as a focus of attention, narrowing the areas of the image that show more hypothesis to have a pedestrian.

In Fig. 15, a snapshot illustrates results of the system. The output of the system is depicted in the image as different bounding box colors with the distance from the Lidar scanner, shown on the top of the bounding box. If the image pattern recognition module does not recognize the object as a pedestrian, only the Lidar works, giving the distance to the target. The system runs at 4 fps in a pentium M laptop, using linux.

As the system has been proposed to work object classification and collision detection system, it was necessary to illustrate different distances from the object in front of the vehicle. Then, different colors highlight different levels of attention.

Fig. 15. The output of the system is depicted in the image as different bounding boxes from the Lidar. Between 0 and 2 meters, the bounding box of the related object is red; between 2 and 4 meters, it is yellow, and beyond 4 meters, it is light blue.

Linear SVM. Hitherto, nothing is known about the choice of linear SVM (Eq. (1)) along with HOG features. In [10], a complete study shows the reason to use this type of SVM. After evaluating some kernels (polynomial and Gaussian), the linear SVM was chosen, since it was faster than the others kernelized SVM and has shown a better tradeoff between speed and classification performance. Although Gaussian kernel has shown the best classification curve.

It is worth remembering that the generalization ability of SVM mainly relies on Eq. (20), since the smaller the number of support vectors concerning to the same number of training data sets, the better the generalization ability of the SVM.

5 Conclusion

This chapter has presented SVM theory for learning classification tasks with some highlights of different features prone to work well with SVM. It has been also shown that it is possible to apply SVM for learning to semantically classify places in indoor environments, and to classify objects with good results. The proposed techniques are based on training an SVM using a supervised learning approach, and features extracted from sensor data. The object recognition method was designed to improve the computational costs enabling the speed-up in operation which is an important factor for the application in real domains. The classification tasks employed a set of features extracted from Lidar and a vision camera. The good results attained with SVM encourage the development of new applications.

References

1. Almeida, J., Almeida, A., Araújo, R.: Tracking Multiple Moving Objects for Mobile Robotics Navigation. In: Proc. 10th IEEE International Conference on Emerging Technologies and Factory Automation (ETFA 2005) (2005)
2. Araújo, R.: Prune-Able Fuzzy ART Neural Architecture for Robot Map Learning and Navigation in Dynamic Environments. IEEE Transactions on Neural Networks 17(5), 1235–1249 (2006)
3. Arnulf, G., Silvio, B.: Normalization in Support Vector Machines. In: Radig, B., Florczyk, S. (eds.) DAGM 2001. LNCS, vol. 2191, pp. 277–282. Springer, Heidelberg (2001)
4. Bianchini, M., Gori, M., Maggini, M.: On the Problem of Local Minima in Recurrent Neural Networks. IEEE Transaction on Neural Networks, Special Issue on Dynamic Recurrent Neural Networks, 167–177 (1994)
5. Borges, G., Aldon, M.: Line Extraction in 2D Range Images for Mobile Robotics. Journal of Intelligent and Robotic Systems 40(3), 267–297 (2004)
6. Burges, C.J.: A Tutorial on Support Vector Machines for Pattern. Data Mining and Knowledge Discovery 2, 121–167 (1998)
7. Costa, J., Dias, F., Araújo, R.: Simultaneous Localization and Map Building by Integrating a Cache of Features. In: Proc. 11th IEEE International Conference on Emerging Technologies and Factory Automation (ETFA 2006) (2006)

8. Cristianini, N., Shawe-Taylor, J.: An Introduction to Support Vector Machine and Other Kernel-based Learning Methods. Cambridge University Press, Cambridge (2000)
9. Duanl, K.-B., Sathiya Keerthi, S.: Which Is the Best Multiclass SVM Method? An Empirical Study. In: Oza, N.C., Polikar, R., Kittler, J., Roli, F. (eds.) MCS 2005. LNCS, vol. 3541, pp. 278–285. Springer, Heidelberg (2005)
10. Dalal, N., Triggs, B.: Histograms of Oriented Gradients for Human Detection. In: International Conference on Computer Vision and Pattern Recognition, pp. 886–893 (2005)
11. Duda, R., Hart, P., Stork, D.: Pattern Classification. Book News Inc. (2000)
12. Garcia, C., Delakis, M.: Convolutional Face Finder: A Neural Architecture for Fast and Robust Face Detection. Proceedings of the IEEE 26(11), 1408–1423 (2004)
13. Gonzalez, R., Woods, R.: Digital Image Processing. Addison-Wesley, Reading (1993)
14. Guyon, I., Gunn, S., Nikravesh, M., Zadeh, L.: Feature Extraction: Foundations and Applications (Studies in Fuzziness and Soft Computing). Springer, Heidelberg (2006)
15. Hearst, M.A., Dumais, S.T., Osman, E., Platt, J., Scholkopf, B.: Support Vector Machines. IEEE Intelligent Systems 13(4), 18–28 (1998)
16. Hippert, H.S., Pedreira, C.E., Souza, R.C.: Neural Networks for Short-term Load Forecasting: a Review and Evaluation. IEEE Trans. on Power Systems 16(1), 44–55 (2001)
17. Hokuyo: Range-Finder Type Laser Scanner URG-04LX Specifications, Hokuyo Automatic Co. (2005)
18. Hsu, C., Lin, C.: A Comparison Methods for Multi-Class Support Vector Machine. IEEE Transactions on Neural Networks 13, 415–425 (2002)
19. Joachims, T.: Optimizing Search Engines Using Clickthrough Data. In: Proceedings of the ACM Conference on Knowledge Discovery and Data Mining (KDD), ACM, New York (2002)
20. Kecman, V.: Learning and Soft Computing. MIT Press, Cambridge (2001)
21. Lawrence, S., Giles, L., Tsoi, A., Back, A.: Face Recognition: A Convolutional Neural Network Approach. IEEE Transactions on Neural Networks, Special Issue on Neural Network and Pattern Recognition 8(1), 98–113 (1997)
22. LeCunn, Y., Bengio, L., Haffner, P.: Gradient-based Learning Applied to Document Recognition. Journal of Neurophysiology, Proceedings of the IEEE 86(11), 2278–2324 (1998)
23. Li, H., Qi, F., Wang, S.: A Comparison of Model Selection Methods for Multi-class Support Vector Machines. In: Gervasi, O., Gavrilova, M.L., Kumar, V., Laganá, A., Lee, H.P., Mun, Y., Taniar, D., Tan, C.J.K. (eds.) ICCSA 2005. LNCS, vol. 3483, pp. 1140–1148. Springer, Heidelberg (2005)
24. Loncaric, S.: A Survey of Shape Analysis Techniques. Pattern Recognition Journal (1998)
25. Lowe, D.: Distinctive Image Features From Scale-invariant Keypoints. International Journal of Computer Vision 60, 91–110 (2004)
26. McInerney, J., Haines, K., Biafore, S., Hecht-Nielsen: Back Propagation Error Surfaces Can Have Local Minima. In: International Joint Conference on Neural Networks (IJCNN) (1989)
27. Mozos, O.M., Stachniss, C., Rottmann, A., Burgard, W.: Using AdaBoost for Place Labeling and Topological Map Building. In: Robotics Research: Results of the 12th International Symposium ISRR, pp. 453–472 (2007)

28. Mozos, O.M., Stachniss, C., Burgard, W.: Supervised Learning of Places from Range Data Using Adaboost. In: IEEE International Conference Robotics and Automation, April 2005, pp. 1742–1747 (2005)
29. Munder, S., Gavrila, M.: An Experimental Study on Pedestrian Classification. IEEE Transactions on Pattern Analysis and Machine Intelligence 28, 1863–1868 (2006)
30. Nunes, U., Fonseca, J.A., Almeida, L., Araújo, R., Maia, R.: Using Distributed Systems in Real-Time Control of Autonomous Vehicles. In: ROBOTICA, May-June 2003, vol. 21(3), pp. 271–281. Cambridge University Press, Cambridge (2003)
31. Online. Acessed in: `http://svmlight.joachims.org/`
32. Online. Acessed in: `http://www.csie.ntu.edu.tw/~cjlin/libsvm/`
33. Online. Acessed in: `http://www.support-vector-machines.org/SVM_soft.html`
34. Online. Acessed in: `http://www.kernel-machines.org/`
35. `http://www.nisis.risk-technologies.com`
36. Oliveira, L., Nunes, U., Peixoto, P.: A Nature-inspired Model for Partially Occluded Pedestrian Recognition with High Accuracy. IEEE Transactions on Intelligent Transportation Systems (submitted, 2007)
37. Oren, M., Papageorgiou, C., Shina, P., Poggio, T.: Pedestrian Detection Using Wavelet Templates. In: IEEE Conference on Computer Vision and Pattern Recognition, pp. 193–199 (1997)
38. O'Rourke, J.: Computational Geometry in C, 2nd edn. Cambridge University Press, Cambridge (1998)
39. Papageorgiou, C., Poggio, T.: Trainable Pedestrian Detection. In: International Conference on Image Processing, vol. 4, pp. 35–39 (1999)
40. Pires, G., Nunes, U.: A Wheelchair Steered through Voice Commands and Assisted by a Reactive Fuzzy-Logic Controller. Journal of Intelligent and Robotic Systems 34(3), 301–314 (2002)
41. Platt, J.: Using sparseness and analytic qp to speed training of support vector machines, Neural Information Processing Systems (1999), `http://research.microsoft.com/users/jplatt/smo.html`
42. Premebida, C., Nunes, U.: A Multi-Target Tracking and GMM-Classifier for Intelligent Vehicles. In: Proc. 9th IEEE Int. Conf. on Intelligent Transportation Systems, Toronto, Canada (2006)
43. Press, W.H., Teukolsky, S.A., Vetterling, W.T., Flannery, B.P.: Numerical Recipies, 3rd edn., September 2007. Cambridge University Press, Cambridge (2007)
44. Serre, T., Wolf, L., Poggio, T.: Object Recognition with Features Inspired by Visual Cortex. In: IEEE Conference on Computer Vision and Pattern Recognition, pp. 994–1000 (2005)
45. Shawe-Taylor, J., Cristianini, N.: Kernel Methods for Pattern Analysis. Cambridge University Press, Cambridge (2004)
46. Smola, A., Schölkopf, B.: A Tutorial on Support Vector Regression, NeuroCOLT2 Technical Report NC2-TR, pp. 1998–2030 (1998)
47. Sousa, P., Araújo, R., Nunes, U., Alves, L., Lopes, A.C.: Real-Time Architecture for Mobile Assistant Robots. In: Proc. IEEE International Conference on Emerging Technologies and Factory Automation (ETFA 2007), Patras, Greece, September 25-28, 2007, pp. 965–972 (2007)
48. Stachniss, C., Mozos, O.M., Burgard, W.: Speeding-Up Multi-Robot Exploration by Considering Semantic Place Information. In: Proc. the IEEE Int.Conf.on Robotics & Automation (ICRA), Orlando, USA (2006)
49. Thrun, S., Burgard, W., Fox, D.: Probabilistic Robotics. The MIT Press, Cambridge (2005)

50. Viola, P., Jones, M.: Rapid Object Detection using a Boosted Cascade of Simple Features. In: IEEE Conference on Computer Vision and Pattern Recognition, pp. 511–518 (2001)
51. Vapnik, V.: The Nature of Statistical Learning Theory. Springer, Heidelberg (1995)
52. Young, A.: Handbook of Pattern Recognition and Image Processing. Academic Press, London (1986)

Linkage Analysis in Genetic Algorithms

Miwako Tsuji and Masaharu Munetomo

Information Initiative Center, Hokkaido University, North 11, West 5,
Sapporo, 060-0811, Japan

Abstract. A series of advanced techniques in genetic and evolutionary computation have been proposed that analyze gene linkage to realize *competent genetic algorithms*. Although it is important to encode linked variables tightly for simple GAs, it is sometimes difficult because it requires enough knowledge of problems to be solved. In order to solve real-world problems effectively even if the knowledge is not available, we need to analyze gene linkage.

We review algorithms which have been proposed that identify linkage by applying perturbations, by building a probabilistic model of promising strings, and a recombination of the both of the above.

We also introduce a context-dependent crossover that can utilize overlapping linkage information in a sophisticated manner. By employing linkage identification techniques with context dependent crossover, we can solve practical real-world application problems that usually have complex problem structures without knowing them before optimization.

1 Introduction

In this chapter, we present advanced techniques in genetic and evolutionary computation that analyze gene linkage to realize competent genetic algorithms (GAs). In genetic algorithms, it is essential to process building blocks (BBs) effectively through genetic recombination such as crossovers. To apply simple GAs, tight linkage of BBs is necessary in encoding strings because simple genetic operators such as one-point crossovers should disrupt BBs when they are encoded sparsely over a string. It is, however, sometimes difficult to preserve tightness of linkage in solving practical problems because it is necessary to obtain prior information of the target problems. In order to solve practical application problems without enough knowledge of them, we need to analyze gene linkage before/along genetic optimization, which is employed to perform recombination without disrupting BBs.

A series of algorithms have been proposed that identify linkage by applying perturbations. Linkage identification by nonlinearity check (LINC) applies bitwise perturbations for each pair of loci to detect nonlinear interactions between them as their linkage information. Linkage identification with non-monotonicity detection (LIMD) detects non-monotonicity by perturbations instead of nonlinearity to detect more general information of linkages. We have also developed more general framework of linkage information based on epistasis measures defined based on the conditions of the LINC and the LIMD.

L.C. Jain et al. (Eds.): Comp. Intel. Para.: Innov. Applications, SCI 137, pp. 251–279, 2008.
springerlink.com

This chapter also discusses briefly linkage analysis based on probabilistic models. The methods are called Estimation of Distribution Algorithms (EDAs) The EDAs generates probabilistic models from a set of promising solutions, which are employed to generate offspring of the next generation.

We have developed a recombination of the both of the above to realize most effective way to identify linkage; the resultant algorithm is called Dependency Detection for Distribution Derived from fitness Differences (D^5), which builds probabilistic models from information on fitness differences by bit-wise perturbations. The D^5 can identify genetic linkage effectively compared with those of the above. By combining a context-dependent crossover that can utilize overlapping linkage information in a sophisticated manner, we can solve practical problems with complex problem structures which we cannot obtain before solving them.

2 Linkage Identification in Genetic Algorithms

From the earlier history of genetic and/or evolutionary algorithms, the importance of *tight linkage* has been recognized; that is, a set of loci that form a BB of an optimal solution should be tightly encoded on a string. Otherwise, simple GAs employing one-point crossover, etc. can easily disrupt them. A classical approach to ensure tight linkage is simply encode strings that preserve tightness of building blocks; however, it is often difficult to encode strings ensuring tight linkage in advance.

There are several approaches have been developed to generate tight linkage *online* along genetic optimization process. *Inversion* operator was proposed by Holland in his famous book that open the field of genetic and evolutionary computation[6]. The inversion operator *inverts* the order of loci that belong to

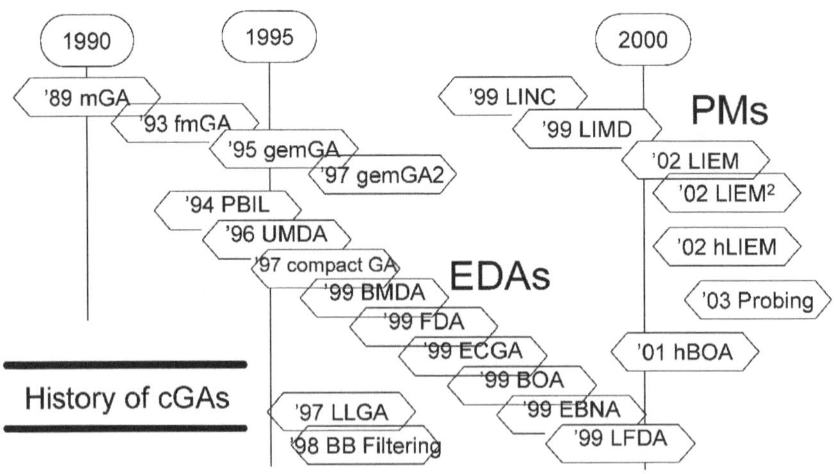

Fig. 1. History of competent Genetic Algorithms (cGAs)

```
P = initialize n strings

while !terminated
   select promising solutions P' from P
   construct a model M based on the P'
   generate new solutions O using M
   replace some of P with O
end
```

Fig. 2. The Procedure of General EDAs

a substring. By employing the operator with selection, we expect strings that have tight linkage should survive along generations; however, the convergence is actually too slow to generate good linkage. In order to generate tight linkage more effectively, a series of advanced techniques on *competent genetic algorithms (cGAs)* have been proposed, which is illustrated in Fig. 1.

The messy GA proposed by Goldberg et al. [3] in 1989 pioneers the field of cGAs, which performs genetic operators directly on schemata to generate building blocks. It employs variable-length strings to represent schemata. Each element of the strings in the mGA is tagged with its name (bit-position) and value. For example, ((1 0) (4 1) (3 0) (3 1)) means the value of the 1st position is 1 and that of the 4th is 1 and so on. As this example shown, the values of a position can be ether under- or over-specified. The under-specified positions are filled with corresponding values of a certain string called template. The over-specified positions are interpreted using the left-to-right rule. The process of the mGA consists of two phases — primordial phase and juxtapositional phase. In the primordial phase, the mGA generates all schemata explicitly of a user-specified order, and performs selections over them to generate candidates of BBs. In the juxtapositional phase, the mGA combines them to find better strings. Fast Messy GA (fmGA) [2] reduces computational cost. Unlike the mGA, it does not generate all possible schemata. Instead, it generates higher-order schemata than them in the mGA and filters building blocks from them.

Estimation of Distribution Algorithms (EDAs) [10, 17], which are also called probabilistic model-building GAs (PMBGAs), construct linkage information statistically. Instead of combining or mutating strings directly, EDAs learn and sample from the probability distribution of promising strings. Fig. 2 shows the algorithm of general EDAs. The model M characterizes each EDA. For example, Bayesian Optimization Algorithm (BOA) [16], Estimation of Bayesian Networks Algorithm (EBNA) [1] and Learning Factorized Distribution Algorithm (LFDA) [12] construct Bayesian network to represent linkages between variables. First, they construct Bayesian networks which fit the promising strings and calculate the frequencies of variables over the networks. Then, new strings are generated from the Bayesian networks.

EDAs are sensitive to the scaling of building blocks. If all building blocks contribute fitness uniformly, EDAs can build their probabilistic models for all building blocks simultaneously. However, if there are building blocks which are not under selection pressure due to less contribution to overall fitness, EDAs cannot learn correct models. Before modeling them, EDAs should wait the convergence of other weighty building blocks. This phenomenon is sometimes seen in simple GAs and called "domino convergence"[11, 18].

There are another series of researches that identify *linkage sets* explicitly. The linkage identification techniques which examine fitness differences by perturbing variables — change values of the variables $0 \rightarrow 1$ or $1 \rightarrow 0$ — are also proposed to detect building blocks. Even these approaches require additional fitness evaluations to compute fitness differences, they can detect the building blocks with small fitness contribution. In the remainder of this chapter, we review this sort of linkage identification techniques described as Perturbation Methods (PMs).

3 Problem Decomposability

Before reviewing the perturbation methods, we describe notations used in this chapter. It is considered that most of real-world problems are loosely decomposable. Then, they can be defined by the sum of sub-functions with some overlaps as follows:

$$f(\boldsymbol{s}) = \sum_{j=1}^{m} f_j(\boldsymbol{s}_{\boldsymbol{v}_j}), \tag{1}$$

where $\boldsymbol{s} = s_1 s_2 \cdots s_i \cdots s_l$ (l is string lengh) is a string, m is the number of sub-functions, $\boldsymbol{s}_{\boldsymbol{v}_j}$ is a sub-string and a vector $\boldsymbol{v}_j$ specifies which variables construct the sub-string. For example, if $\boldsymbol{v}_j = (3, 1, 7)$, $\boldsymbol{s}_{\boldsymbol{v}_j} = s_3 s_1 s_7$. Let V_j is a set of loci which consist $\boldsymbol{v}_j$. If $\boldsymbol{v}_j = (3, 1, 7)$, $V_j = \{1, 3, 7\}$. The sub-functons can be either overlaping or non-overlapping.

Sometimes, we use superscripts $\boldsymbol{s}^1, \boldsymbol{s}^2, \cdots, \boldsymbol{s}^p, \cdots$ to index particuler strings. The superscripts are omitted when it is clear from the context.

Building blocks, which are highly fit sub-strings, must be candidate sub-solutions of sub-function in (1) to solve problems by combining the building blocks. Therefore, we call the set V_j "linkage set", which we would like to find. Variables in a same linkage set are linked.

4 Linkage Identification by Perturbation

Knowing which variables must be processed together helps us optimize functions efficiently. Linkage Identification by Nonlinearity Check (LINC) [15] judges whether any two variables are independent or not by investigating the non-linearity between the two.

Fig. 3 shows the algorithm of the LINC. The LINC detects the non-linearity in a pair of variables (i, j) caused by perturbations. First, it calculates fitness differences as follows:

$$\Delta f_i(\boldsymbol{s}) = f(\cdots \overline{s}_i \cdots\cdots) - f(\cdots s_i \cdots\cdots) \tag{2}$$
$$\Delta f_j(\boldsymbol{s}) = f(\cdots\cdots \overline{s}_j \cdots) - f(\cdots\cdots s_j \cdots) \tag{3}$$
$$\Delta f_{ij}(\boldsymbol{s}) = f(\cdots \overline{s}_i \cdot \overline{s}_j \cdots) - f(\cdots s_i \cdot s_j \cdots), \tag{4}$$

where $\overline{s}_i = 1 - s_i$ and $\overline{s}_j = 1 - s_j$. If

$$\Delta f_i(\boldsymbol{s}) + \Delta f_j(\boldsymbol{s}) = \Delta f_{ij}(\boldsymbol{s}) \tag{5}$$

for all strings in a population, the LINC considers that the variables i and j are separable; otherwise, they have a non-linear relationship.

Linkage Identification by non-Monotonicity Detection (LIMD) [15] relaxes the non-linearity condition in the LINC to non-monotonicity condition, bacause the non-linearity condition is sometimes too strict to optimize real-world problems. The non-monotonicity condition is defined as follows:

$$\begin{aligned}&\text{if } (\Delta f_i(\boldsymbol{s}) > 0 \text{ and } \Delta f_j(\boldsymbol{s}) > 0)\\ &\quad\text{then } (\Delta f_{ij}(\boldsymbol{s}) > \Delta f_i(\boldsymbol{s}) \text{ and } \Delta f_{ij}(\boldsymbol{s}) > \Delta f_j(\boldsymbol{s}))\end{aligned} \tag{6}$$
$$\begin{aligned}&\text{if } (\Delta f_i(\boldsymbol{s}) < 0 \text{ and } \Delta f_j(\boldsymbol{s}) < 0)\\ &\quad\text{then } (\Delta f_{ij}(\boldsymbol{s}) < \Delta f_i(\boldsymbol{s}) \text{ and } \Delta f_{ij}(\boldsymbol{s}) < \Delta f_j(\boldsymbol{s})).\end{aligned} \tag{7}$$

Fig. 4 shows the non-linearity and non-monotonicity conditions in optimization.

Sometimes, the strict condition of the LINC, which checks whether the equation (5), are not suitable to practical applications. Therefore, Linkage Identification with Epistasis Measures (LIEM) [13] relaxes the strict condition. It identifies

```
P = initialize n strings

for each s ∈ P
  for i = 1 to l
    s' = s_1 ··· s̄_i ··· s_l /* Perturb i-th variable */
    Δf_i(s) = f(s') − f(s)
    for j = i + 1 to l
      s'' = s_1 ··· s̄_j ··· s_l /* Perturb j-th variable */
      Δf_j(s) = f(s'') − f(s)
      s''' = s_1 ··· s̄_i ··· s̄_j ··· s_l /* Perturb i and j-th variables */
      Δf_ij(s) = f(s''') − f(s)
      if |Δf_ij(s) − (Δf_i(s) + Δf_j(s))| > ε /* Non-linearity condition */
        add i to the linkage set V_j
        add j to the linkage set V_i
      end
    end
  end
end
```

Fig. 3. The Linkage Identification by Nonlinearity Check (LINC)

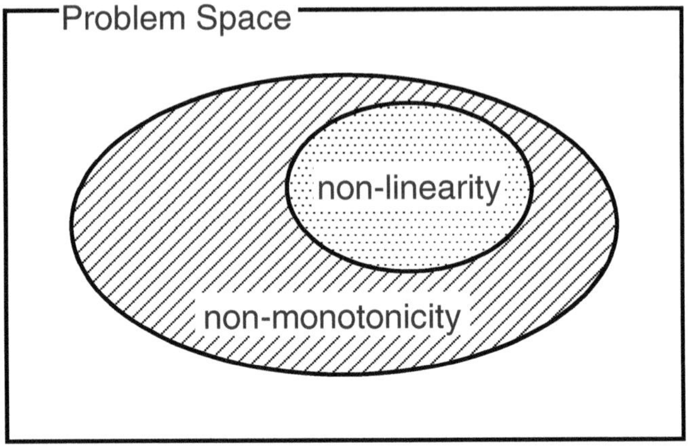

Fig. 4. The non-linearity and non-monotonicity conditions

```
P = initialize n strings

for i = 1 to l
  for j = i + 1 to l
    e_ij = 0
    for each s ∈ P
      s' = s_1 ··· s̄_i ··· s_l              /* Perturb i-th variable */
      Δf_i(s) = f(s') − f(s)
      s'' = s_1 ··· s̄_j ··· s_l             /* Perturb j-th variable */
      Δf_j(s) = f(s'') − f(s)
      s''' = s_1 ··· s̄_i ··· s̄_j ··· s_l /* Perturb i and j-th variables */
      Δf_ij(s) = f(s''') − f(s)
      e_tmp = |Δf_ij(s) − (Δf_i(s) + Δf_j(s))|
      if e_tmp > e_ij
        e_ij = e_tmp
      end
    end
  end
end

for i = 1 to l
  while |V_i| < k
    add j with the largest e_ij and j ∉ V_i to V_i
  end
end
```

Fig. 5. The Linkage Identification with Epistasis Measures (LIEM)

linkage sets based on an epistasis measure that measures strength of nonlinearity among a pair of variables. The epistasis measure in the LIEM is designed to measure the amount of non-linearity based on the LINC condition.

The epistasis for a pair of variables (i, j) is calculated as follows:

$$e_{ij} = \max_{\boldsymbol{s} \in P} |\Delta f_{ij}(\boldsymbol{s}) - (\Delta f_i(\boldsymbol{s}) + \Delta f_j(\boldsymbol{s}))\,|, \tag{8}$$

where $\Delta f_{ij}(\boldsymbol{s}), \Delta f_i(\boldsymbol{s}), \Delta f_j(\boldsymbol{s})$ are the same as those in equations (2) – (4) and P is a population.

The LIEM considers that a larger value of e_{ij} indicates tighter linkages between (i, j) and a linkage set is constructed from variables with larger e_{ij}. These process of the LIEM is shown in Fig. 5.

Similarly, the non-monotonicity condition in the LIMD is relaxed to an epistasis measure in Linkage Identification with Epistasis Measures considering Monotonicity (LIEM^2)[14]. The epistasis measure in the LIEM^2 measures the strength of non-monotonicity:

$$e_{ij} = \max_{\boldsymbol{s} \in P} g\left(\Delta f_{ij}(\boldsymbol{s}), \Delta f_i(\boldsymbol{s}), \Delta f_j(\boldsymbol{s})\right). \tag{9}$$

$$g(x, y, z) = \begin{cases} tr(y-x) + tr(z-x), & (y > 0, z > 0) \\ tr(x-y) + tr(x-z), & (y < 0, z < 0) \\ 0, & \text{otherwise} \end{cases} \tag{10}$$

$$tr(x) = \begin{cases} x, & (x \geq 0) \\ 0. & (x < 0) \end{cases} \tag{11}$$

The algorithm of the LIEM^2 is as same as the one of the LIEM (Fig. 5) except the condition used.

In addition to the LINC-variations shown in this section, there have been many linkage identification algorithms which uses fitness differences cased by perturbations in variables [5, 7, 8, 9].

While the perturbation methods can detect dependencis in the building blocks which EDAs sometimes fail to detect — those which have less contribution to fitness — they require additional fitness computation and the computational cost is generally $O(l^2)$ or more where l is string length.

5 D^5: Dependency Detection by Fitness Clustering

In PMs and EDAs, EDAs are more efficient than PMs for functions composed of uniformly-scaled sub-functions; however, they cannot solve problems with variously-scaled BBs efficiently and their computational cost increases for such problems. In contrast, PMs can detect BBs with lower fitness contribution, however, their computational cost is relatively high and virtually constant regardless of the scaling effect.

In this section, we describe another approach which combines the mechanisms of PMs and EDAs, called Dependency Detection for Distribution Derived from fitness Differences (D^5). It detects dependencies of variables by estimating the

```
P = initialize n strings

for i = 1 to l
  /* Calculate fitness differences */
  for each s ∈ P
    s' = s_1 ··· s̄_i ··· s_l /* Perturb i-th variable */
    Δf_i(s) = f(s') − f(s)
  end

  /* Classify strings according to Δf_i */
  classify s ∈ P into C_1, C_2, ··· according to Δf_i(s)

  /* Estimate each cluster */
  for each C_i
    investigate C_i to estimate the linkage set V_i
  end
end
```

Fig. 6. The Linkage Identification by D^5

distributions of strings clustered according to fitness differences. In each cluster, it explores biases of strings which are extracted not by natural selection pressure but by clustering. In the fitness differences, the contribution from independent sub-functions are canceled and we can focus on an arbitrary sub-function, even on a sub-function with small fitness contribution. Moreover, because we perform estimation instead of order-2 or more perturbations, the number of evaluations is smaller than that of PMs. Therefore, the D^5 can solve EDA-difficult problems using less computational cost than PMs.

The Dependency Detection

D^5 combines PMs and EDAs in order to obtain a distinctive distribution in a sub-population rapidly even if there are BBs which contribute to fitness smally. It is implemented by selecting strings according to fitness differences using perturbations instead of absolute fitness values. Because the fitness differences depend only on a sub-function including the perturbed variable, even sub-functions with small contributions can be detected. For each variable i, D^5 calculates fitness differences by perturbations at variable i on all strings in the population, classifies strings according to the fitness differences, and estimates the distribution of strings in each cluster. Although additional fitness evaluations are needed to investigate fitness differences, the number of evaluations is $O(l)$, which is less than that of PMs especially for large l.

Fig. 6 shows the algorithm of the D^5. For each variable, the D^5 detects variables which depend on the variable. The detection consists of three stages: (1) calculation

```
for each s ∈ P = {s^1, s^2, ···}
   C_p = {s^p}
   df(C_p) = Δf_i(s^p)
end

while !terminated
   for each pair of (C_p, C_q)
      d(p, q) = |df(C_p) − df(C_q)|
   end

   C_new = C_p' ∪ C_q' where d(p, q) is the minimum
   let df(C_new) the average fitness difference of all strings in C_new
end
```

Fig. 7. The clustering algorithm. Superscripts in the figure index strings.

```
for each cluster C_p
   W_1 = {1, 2, ··· , i − 1, i + 1, ··· l}
   W_2 = {i}

   while |W_2| < k /* k is problem complexity */
      for each j ∈ W_1
         E_j = E(W_2 ∪ {j})
      end

      h = argmin_{j∈W_1} E_j
      W_1 = W_1 − {h}
      W_2 = W_2 ∪ h
      W_p = W_2
      E_p = E(W_2)
   end
end

/* for the case of non-overlapping building blocks */
p' = argmin_p E_p
let the linkage set V_i = W_p'

/* for the case of overlapping building blocks
V_i = ⋃_p W_p */
```

Fig. 8. The algorithm to construct linkage set

of fitness differences, (2) clustering strings according to the differences into some clusters and (3) estimation of the distribution in the clusters.

First, fitness differences by perturbations at the i-th variable are calculated for all strings:

$$\Delta f_i(\boldsymbol{s}) = f(s_1 s_2 \cdots s_i \cdots s_l) - f(s_1 s_2 \cdots \overline{s}_i \cdots s_l), \tag{12}$$

where $\overline{s}_i = 1 - s_i$ in binary strings, i.e. $0 \to 1$ or $1 \to 0$.

Then, strings are clustered into sub-populations $C_1, C_2, \cdots$ according to the fitness differences. The clustering algorithm is shown in Fig. 7.

Finally, the distributions of sub-populations are estimated in order to find variables which depend on the i-th variable by finding the set of variables which give the smallest entropy in the sub-populations. This stage is described in Fig. 8 and the reasons for doing the clustering and finding the minimum entropy are discussed later.

Based on the similarity of fitness differences, strings are clustered into sub-populations. Fig. 7 shows the clustering algorithm we employed. Although we employ a centroid method, other approaches such as k-means can also be applied. The clustering starts by assigning each string to its own cluster, then repeatedly joins together the two clusters which are closest. The distance between clusters is defined as the distance of their values. The value of a cluster is defined by averaging $\Delta f_i(\boldsymbol{s})$ of all strings within the cluster. The clustering is stopped when the number of clusters is sufficiently small and the clusters are too far apart to be joined together.

Fig. 8 shows how to construct a linkage set for an i-th variable. For all sub-populations (clusters) obtained from the previous stage, the following processes are iterated: First, the linkage set W_2 is initialized as $\{i\}$. The variable which gives the smallest entropy when taking part in W_2 is merged repeatedly until the size of W_2 exceeds the pre-defined problem complexity k. The entropy is defined as follows:

$$E(W_2) = -\sum_{x=1}^{2^{|W_2|}} p_x \log_2 p_x \tag{13}$$

where p_x is the appearance ratio of each sub-string x and $2^{|W_2|}$ is the number of all possible sub-strings defined over W_2.

This procedure is applied for all sub-populations except those including a small numbers of strings, because we cannot obtain reliable results with a small number of samples.

After the estimation of all sub-populations, the linkage set giving the smallest entropy of the sets from the sub-populations, is selected as linkage set V_i of variable i. If sub-functions are considered to be ovelapping, the union of all sets from the all sub-populations is used.

How and why the D^5 works

Here, we show the mechanism of the D^5. First, we give an explanation using an example and then we demonstrate a formal exposition. To illustrate how D^5

$s_1s_2\cdots s_l$	f	$\Delta f_1(s)$
111 000 001 111	114	−30
001 010 111 001	104	−26
010 111 010 000	102	−22
010 000 001 010	98	−22
100 001 000 111	98	14
010 001 000 010	98	−22
010 111 010 100	88	−22
011 111 001 111	86	30
110 111 010 111	82	22
011 111 000 010	80	30
000 010 000 101	78	−14
110 000 100 001	68	22
010 100 010 101	58	−22
001 011 101 111	56	−26
101 000 110 001	54	26
010 111 011 110	52	−22
001 101 101 001	52	−26
101 000 011 010	50	26
011 011 000 010	50	30
011 010 001 110	48	30
110 110 111 100	44	22
011 111 101 100	44	30
101 101 101 000	28	26
101 000 011 110	28	26
010 110 011 101	22	−22
010 101 101 101	22	−22
010 110 110 110	22	−22
100 011 110 101	14	14
011 101 110 110	0	30
110 101 101 011	0	22

$s_1s_2\cdots s_l$	$\Delta f_1(s)$
011 111 001 111	30
011 111 000 010	30
011 011 000 010	30
011 010 001 110	30
011 111 101 100	30
011 101 110 110	30
101 000 110 001	26
101 101 101 000	26
101 000 011 110	26
101 000 011 010	26
110 000 100 001	22
110 110 111 100	22
110 111 010 111	22
110 101 101 011	22
100 001 000 111	14
100 011 110 101	14
000 010 000 101	−14
010 111 010 000	−22
010 111 011 110	−22
010 001 000 010	−22
010 110 011 101	−22
010 101 101 101	−22
010 110 110 110	−22
010 111 010 100	−22
010 000 001 010	−22
010 100 010 101	−22
001 011 101 111	−26
001 101 101 001	−26
001 010 111 001	−26
111 000 001 111	−30

Fig. 9. The left table shows the original population with 30 strings. Strings are in the left column, their fitnesses are in the middle one, and the fitness differences by perturbation on the first variable ($i = 1$) are in the right one. The right table shows strings clustered according to the fitness differences. In each string, the first variable is perturbed and the first three variables show a particular sub-string in each sub-population.

works, we employ a sum of order-3 deceptive functions which was used to test the messy GA [3]:

$$f(s) = f_{3dec}(s_1s_2s_3) + f_{3dec}(s_4s_5s_6) + f_{3dec}(s_7s_8s_9) + f_{3dec}(s_{10}s_{11}s_{12})$$

$$f_{3dec}(s_is_js_k) = 30 \text{ if } 111,\quad 28 \text{ if } 000,\quad 26 \text{ if } 001,$$
$$22 \text{ if } 010,\quad 14 \text{ if } 100,\quad 0 \quad \text{otherwise.}$$

In this function, the linkage sets which must be identified are $\{1, 2, 3\}$, $\{4, 5, 6\}$, $\{7, 8, 9\}$ and $\{10, 11, 12\}$.

Fig. 9 shows the dependency detection for $i = 1$. The left table shows initial strings which are sorted based on their fitness. The right table shows the fitness differences by the perturbations at s_1. The right table shows the strings clustered into sub-populations according to their fitness differences. In the $\Delta f_1 = 30$ sub-population, a linkage set $\{1,2,3\}$ has the only 011 and $E(\{1,2,3\})$ is zero. On the other hand, a linkage set $\{1,4,5\}$ has $010, 001, 011$ and $E(\{1,4,5\})$ is larger than zero. The estimation algorithm (Fig. 8) starts with a linkage set $W_2 = \{1\}$. Then it tries to calculate entropies $E(\{1,j\})$ for variables $j = 2,3,\cdots,12$. In this case, the entropies are

$$E(\{1,2\}) = 0,\ E(\{1,3\}) = 0,\ E(\{1,4\}) \approx 0.92,\ E(\{1,5\}) \approx 0.65,\ \cdots.$$

Therefore, $j = 2$ or 3 is added to the linkage set W_2. In this case, we assume that $j = 3$ is added. Then, it calculates entropies $E(1,3,j)$ for $j = 2,4,5,6$.

$$E(\{1,3,2\}) = 0,\ E(\{1,3,4\}) \approx 0.92,\ E(\{1,3,5\}) \approx 0.65,\ E(\{1,3,6\}) \approx 0.65.$$

Then, $j = 2$ is added to W_2 and the set $\{1,2,3\}$ is considered to be the set of variables which depend on $i = 1$.

We give a further explanation about the mechanism of the D^5, using some equations.

As shown in section 3, we assume that a function is a sum of sub-functions as follows:

$$f(\boldsymbol{s}) = \sum_{j=1}^{m} f_j(\boldsymbol{s}_{\boldsymbol{v}_j}). \tag{14}$$

We can decompose sub-functions in (14) into two groups, where the sub-functions involve a variable i and those do not involve i:

$$f(\boldsymbol{s}) = \sum_{i \in V_j} f_j(\boldsymbol{s}_{\boldsymbol{v}_j}) + \sum_{i \notin V_i} f_j(\boldsymbol{s}_{\boldsymbol{v}_j}). \tag{15}$$

If V_j does not include i,

$$f_j(s_1 \cdots s_i \cdots) = f_j(s_1 \cdots \overline{s}_i \cdots) \tag{16}$$

and

$$\sum_{i \notin V_i} f_j(s_1 \cdots s_i \cdots) = \sum_{i \notin V_i} f_j(s_1 \cdots \overline{s}_i \cdots). \tag{17}$$

From equations (15) and (17)

$$\begin{aligned}
\Delta f_i(\boldsymbol{s}) &= f(s_1 \cdots s_i \cdots s_l) - f(s_1 \cdots \overline{s}_i \cdots s_l) \\
&= \{\sum_{i \in V_j} f_j(s_1 \cdots s_i \cdots) + \sum_{i \notin V_i} f_j(s_1 \cdots s_i \cdots)\} \\
&\quad -\{\sum_{i \in V_j} f_j(s_1 \cdots \overline{s}_i \cdots) + \sum_{i \notin V_i} (s_1 \cdots \overline{s}_i \cdots)\} \\
&= \sum_{i \in V_j} f_j(s_1 \cdots s_i \cdots) - \sum_{i \in V_j} f_j(s_1 \cdots \overline{s}_i \cdots)
\end{aligned} \tag{18}$$

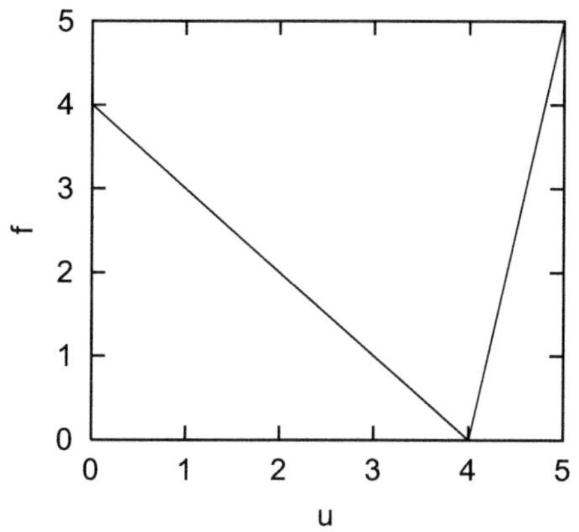

Fig. 10. Trap Function

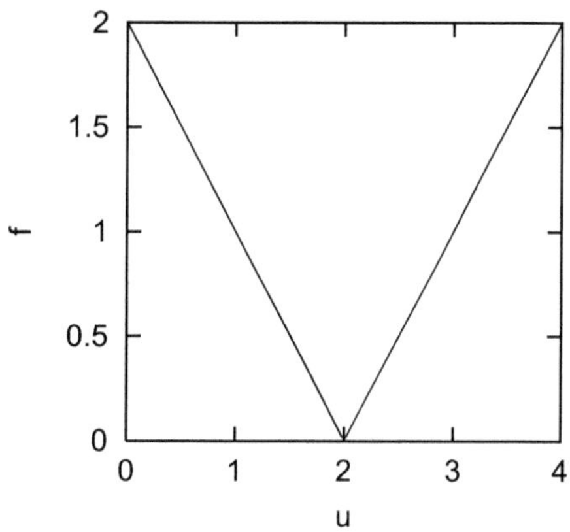

Fig. 11. Valley Function

In equation (18) the fitness difference $\Delta f_i(\boldsymbol{s})$ is independent of the variables in V_j such that $i \notin V_j$ and is dependent only on the variables in V_j such that $i \in V_j$. Therefore, after strings are clustered based on $\Delta f_i(\boldsymbol{s})$, the variables in the sets V_j such that $i \notin V_j$ take random values and their entropies become large. On the other hand, variables in V_j such that $i \in V_j$ take some particular sub-strings (sometimes a single sub-string) and their entropies should become small. Therefore, the set of variables which depend on i can be identified by finding the set of variables which minimizes the entropy.

If sub-functions do not overlap each other, i.e. $V_j \cap V_{j'} = \phi$ for all j', the linkage set V_j including i itself can be found. Otherwise, we must infer V_j by investigating all sets resulting perturbations at $i = 1, 2, \cdot, l$. This process will be described later.

Experiments

Here, we examine the linkage identification ability of PMs (the D^5 and LINC) for functions composed of non-overlapping sub-functions.

(1) Accuracy and Efficiency of Linkage Identification

We employ decomposable functions composed by non-overlapping sub-functions:

$$f(\boldsymbol{s}) = \sum_{j=1}^{m} f_j(\boldsymbol{s}_{\boldsymbol{v}_j}). \tag{19}$$

In this experiment, we employ two types of sub-functions as follows:

$$f_j(\boldsymbol{s}) = \mathrm{trap}(u) = \begin{cases} k & (u = k) \\ k - 1 - u & (\text{otherwise}) \end{cases} \tag{20}$$

$$f_i(\boldsymbol{s}) = \mathrm{valley}(u) = |u - \frac{k}{2}| \tag{21}$$

where k is the order of a sub-function — $k = 5$ for the trap and $k = 4$ for vally are used —, u is the number of ones in a sub-string.

Fig. 10 and Fig. 11 show examples of a trap function and valley function respectively. The trap function has deception and is known as a GA-difficult function. The valley function seems easier because there is no deception as in the trap function. However, there are only two sub-populations, the sub-populations with $\Delta f_i(\boldsymbol{s}) = 1$ and the sub-populations with $\Delta f_i(\boldsymbol{s}) = -1$. In both

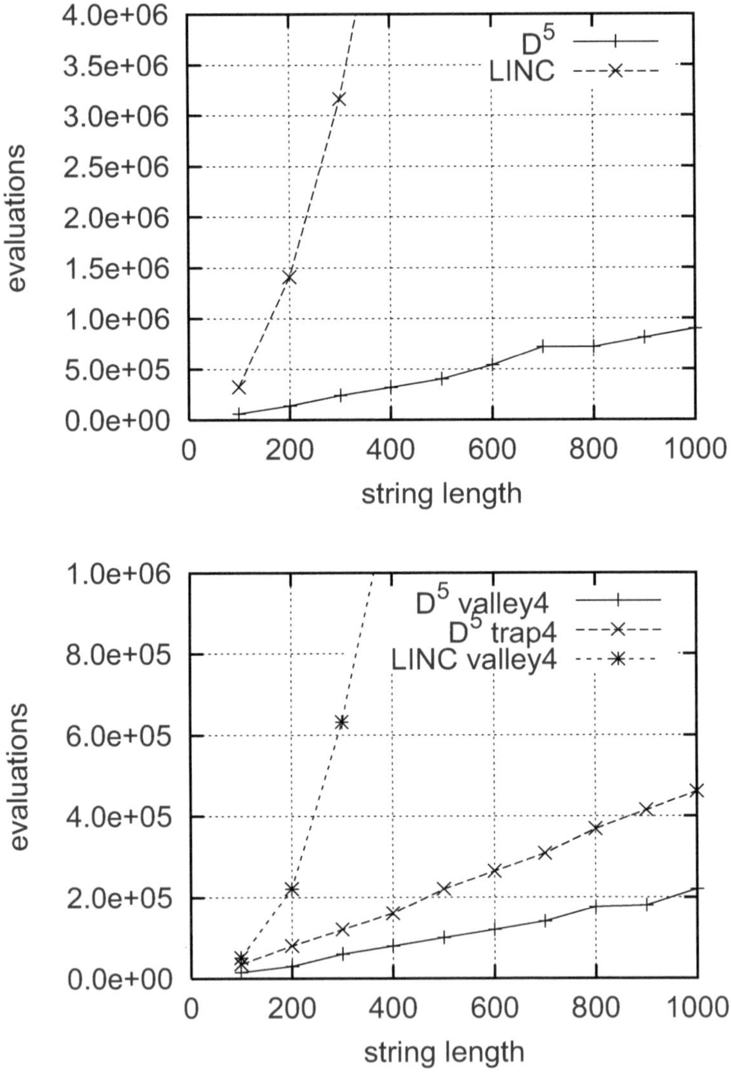

Fig. 12. D^5 and LINC for Trap Function (top) for Valley Function (bottom)

```
for each linkage set V_j
  if random(0, 1) < 0.5
    exchange sub-strings specified in V_j
  else
    do not exchange sub-strings specified in V_j
  end
end
```

Fig. 13. Linkage-wise uniform crossover for each pair of strings

sub-populations, there are several settings on V_j such that $i \in V_j$ and estimation of the distribution seems to be difficult.

For both test functions, string length l is varied by varying the number of sub-functions to examine the number of evaluations required to obtain correct linkage sets for each string length. Population size is gradually increased before obtaining the correct linkage sets for all the 10 independent runs.

Results for the 5-bit trap function and the 4-bit valley functions are shown in the top and bottom of Fig. 12 respectively. They show that the D^5 can detect linkages with smaller numbers of evaluations than the LINC for both functions. From the figure, the numbers of evaluations for the valley function are smaller than those for the trap function. As mentioned before, there are only two fitness differences 1 and -1 in the valley function and the entropy of a linkage set in each sub-population shold be larger, while the D^5 can exploit the half of the original population to search a linkage set in such relatively uniform distribution.

(2) Optimization Performance

In the previous experiments, we focused on linkage identification. Here, we investigate the whole optimization process including linkage identification and evolution. In this section, D^5-GA is compared with BOA [16]. Evan the way to combine overlapping overlapping building blocks will be described later, we employ linkage-wise uniform crossover shown in Fig. 13 here, for functions composed of non-overlapping sub-functions.

We employ test functions composed of uniformly-scaled and exponentially-scaled 5-bit trap functions as follows:

$$f(\boldsymbol{s}) = \sum_{j=1}^{m} w_j \times \text{trap}(u) \tag{22}$$

where $\text{trap}(u)$ is defined in (20) and w_i is the weight of each sub-function. We observe the number of evaluations for two cases, $w_j = 1$ and $w_j = 2^j$ $(j = 1, \cdots, m)$ which are called uniformly-scaled and exponentially-scaled trap functions respectively. In this experiment, we change the string length l and plot the numbers of evaluations required to converge to the optimal solution.

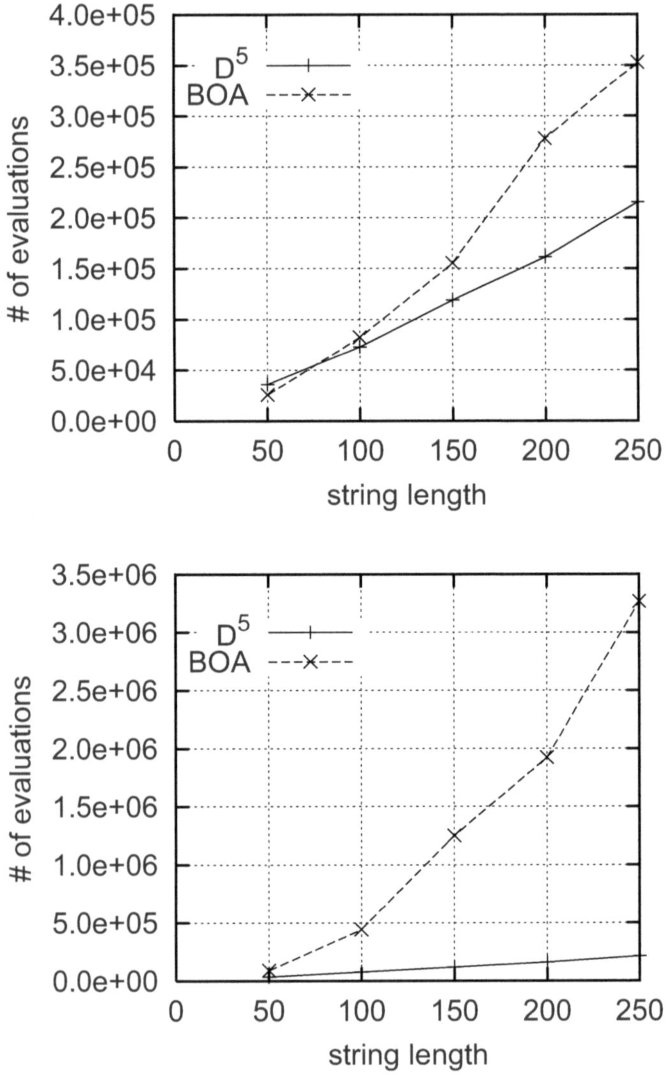

Fig. 14. D^5 and BOA for Uniformly-Scaled Trap Function (top) and Exponentially-Scaled Trap Function (bottom)

Results for the uniformly-scaled and the exponentially- scaled trap functions are shown in the top and bottom of Fig. 14 respectively. The numbers of evaluations required to converge to the global optimum in all the 10 independent runs are recorded. For the exponentially scaled functions, the numbers of evaluation in D^5 do not change from the uniformly scaled function and are always smaller than those in the BOA.

6 Context Dependent Crossover

So far, there had been several linkage identification techniques to identify variables that tightly linked to construct a building block. However, how to exploit the information they provide had not been studied enough. Yu et al. [20] proposed is the first known approach to exchange strongly overlapping building blocks. We extend their crossover method to combine complexly overlapping building blocks. It is called Context Dependent Crossover (CDC) and examines contexts of each pair of strings in addition to the linkage information to process building blocks.

Although it is not enough to divide variables into non-overlapping blocks for real world problems, it is difficult to combine overlapping building blocks. An example of such problem is shown in Fig. 15. In the figure, variables 4 and 5 belong to BB_1 and BB_2. If one of them is exchanged, the other one is disrupted. If both of them are exchanged, no information is exchanged.

Existing Method

Fig. 16 shows the crossover method proposed by Yu et al[20]. They investigated the relationship between an inaccurate linkage and the convergence time of GA to develop a graphical crossover method for problems with overlapping sub-functions and designed from the observation that the prevention of the detection failure error (the linkage model does not link those genes which are linked in reality) is critical to successful recombination. It constructs a graph $G = (N, E)$, where the nodes are linkage sets and the edges are overlapping relations between two nodes.

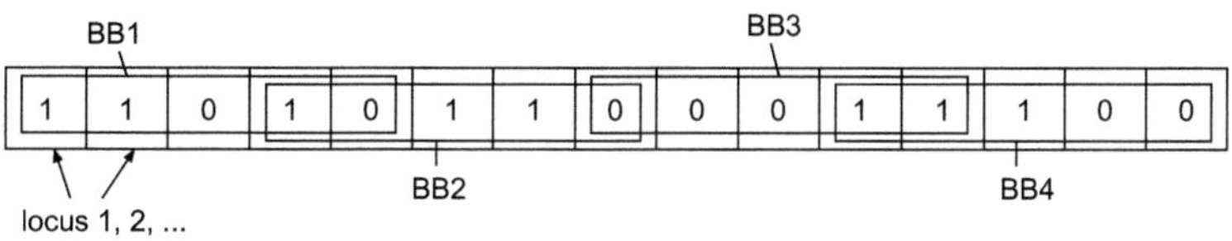

Fig. 15. An example of the overlapping sub-functions

1. Construct a graph $G = (N, E)$, where the nodes are linkage sets $\overline{V}_j$ and the edges are overlapping relations between two nodes.
2. For each crossover operator: Choose two nodes n_1, n_2 randomly. Then partition the graph G into two sub-graphs $G_1 = (N_1, E_1)$ and $G_2 = (N_2, E_2)$ which satisfy conditions: $n_1 \in N_1$, $n_2 \in N_2$ and $|E| - |E_1| - |E_2|$ is minimal.
3. Let $\mathcal{V} = \bigcup_{\overline{V}_j \in N_1} \overline{V}_j$ and exchange variables in $\mathcal{V}$.

Fig. 16. The crossover algorithm by Yu et al. [20]

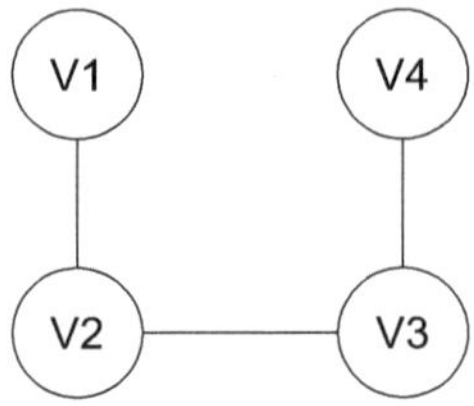

Fig. 17. A graph in the existing method. This represents the string in Fig. 15.

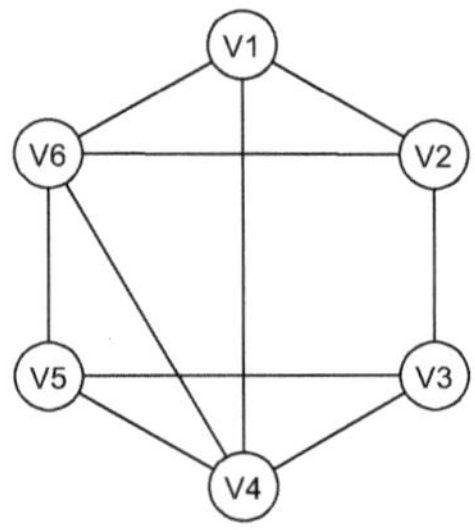

Fig. 18. An example of complex overlaps

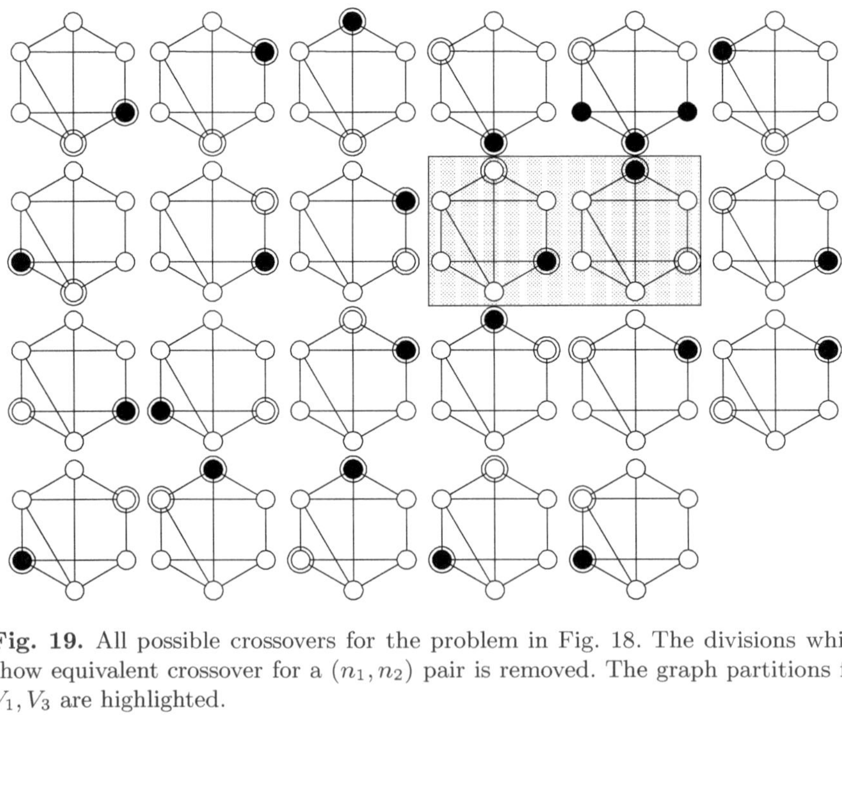

Fig. 19. All possible crossovers for the problem in Fig. 18. The divisions which show equivalent crossover for a (n_1, n_2) pair is removed. The graph partitions for V_1, V_3 are highlighted.

1. Construct a graph $G = (N, E)$, where the nodes are linkage sets $\overline{V}_j$ and the edges are overlapping relations between two nodes.
2. For each crossover operator for parent strings $\boldsymbol{s} = s_1 s_2 \cdots s_i \cdots s_l$ and $\boldsymbol{t} = t_1 t_2 \cdots t_i \cdots t_l$
   a) Remove nodes $\overline{V}_j$ where $\boldsymbol{s}_{\overline{\boldsymbol{v}}_j} = \boldsymbol{t}_{\overline{\boldsymbol{v}}_j}$.
   b) Remove edges between $\overline{V}_j$ and $\overline{V}_{j'}$ if the following conditions are hold :
      i. The exchange between BBs $\boldsymbol{s}_{\overline{\boldsymbol{v}}_j}$ and $\boldsymbol{t}_{\overline{\boldsymbol{v}}_j}$ does not disrupt BBs $\boldsymbol{s}_{\overline{\boldsymbol{v}}_{j'}}$ and $\boldsymbol{t}_{\overline{\boldsymbol{v}}_{j'}}$.
      ii. The exchange between BBs $\boldsymbol{s}_{\overline{\boldsymbol{v}}_{j'}}$ and $\boldsymbol{t}_{\overline{\boldsymbol{v}}_{j'}}$ does not disrupt BBs $\boldsymbol{s}_{\overline{\boldsymbol{v}}_j}$ and $\boldsymbol{t}_{\overline{\boldsymbol{v}}_j}$.
   c) Choose two nodes n_1, n_2 randomly. Then partition the graph G into two sub-graphs $G_1 = (N_1, E_1)$ and $G_2 = (N_2, E_2)$ which satisfy conditions: $n_1 \in N_1$, $n_2 \in N_2$ and $|E| - |E_1| - |E_2|$ is minimal.
   d) Let $\mathcal{V} = \bigcup_{\overline{V}_j \in N_1} \overline{V}_j$ and exchange variables in $\mathcal{V}$.

Fig. 20. The algorithm of Context Dependent Crossover

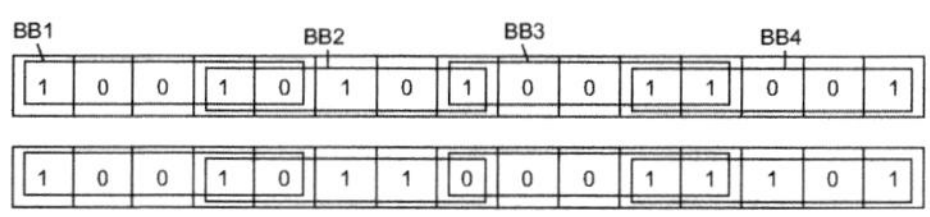

Fig. 21. Example of a pair of parental strings

However, when overlap becomes more complex, graph partition with minimum BB disruptions is restricted and the partition tends to divide a graph into a small graph and the remaining large one. For example, Let us consider a problem with complex overlaps shown in Fig. 18. There are ${}_6C_2 = 15$ choices for (n_1, n_2). For most pairs of nodes, the number of minimum disruption crossovers is small. All possible crossovers for the problem in Fig. 18 are shown in Fig. 19. Again, the double circles show (n_1, n_2) and the divisions which show equivalent crossover are removed. In the figure, there are a few crossover sets for each (n_1, n_2) pair and all except the one example exchange only one BB. For example, for a pair (V_1, V_3), only the two graph partitions highlighted in the figure 19 divide the pair with minimum disruptions for the complex problem.

Crossover for complex interactions and overlaps

Context Dependent Crossover (CDC) [19] has been proposed to modify the existing method to enrich a variety of crossovers without increasing BB disruptions. Fig. 20 shows the algorithm of the CDC. While the existing method searches the best division over a single graph G for all pairs of parents, the CDC

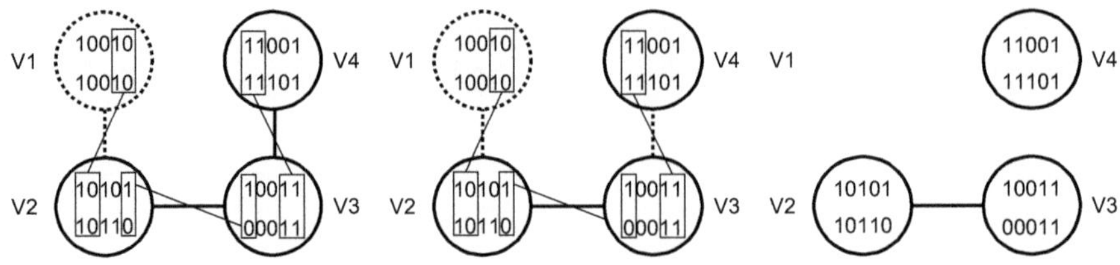

Fig. 22. Example of the CDC. left: remove same BBs, middle: remove edges where BB disruption does not occur, right: resulted graph.

reconstructs the graph for each pair of parent strings $\boldsymbol{s} = s_1 s_2 \cdots s_i \cdots s_l$ and $\boldsymbol{t} = t_1 t_2 \cdots t_i \cdots t_l$.

Fig. 22 shows an example of the CDC for parent strings shown in Fig. 21. First, the nodes on which sub-solutions (BBs) are identical are removed (Fig. 22, left), because whether such BBs are exchanged or not has no effect on the offspring. This operator ensures the obtained offsprings always differ from their parent strings. In our example, because BBs on the V_1 are same, the node for V_1 is removed.

Then, the edges where no BB disruption occurs practically are removed (Fig. 22, middle). In our example, the edge between V_3 and V_4 are removed because for parent sub-strings

10011001 with BBs 10011 and 11001
00011101 with BBs 00011 and 11101,

obtained sub-strings are

00011001 with BBs 00011 and 11011
10011101 with BBs 10011 and 11101,

or

10011101 with BBs 10011 and 11101
00011001 with BBs 00011 and 11001,

when V_3 or V_4 is exchanged respectively.

After that, the resulted graph (Fig. 22, right) is divided into two sub-graphs to minimize the number of cut edges. The right graph of Fig. 22 obtained by the CDC suggests that there is a graph partitioning without BB disruption. On the other hand, the existing method simply chooses a partitioning from

$\{V_1, V_2, V_3 | V_4\}$,
$\{V_1, V_2 | V_3, V_4\}$ which cuts one edge, and
$\{V_1 | V_2, V_3, V_4\}$ which creates no new string.

Moreover, the CDC can give various crossover sets even for the complex problems like the one shown in Fig. 18. The reason is as follows: For every crossover, the CDC removes nodes and edges to simplify the graph G. The reconstruction of the G depends on the values of parental strings. Therefore, different pairs should give different graphs. The different graphs result in various sub-graphs.

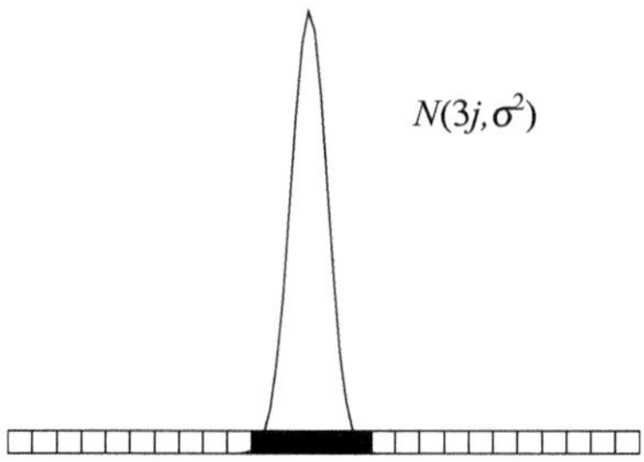

Fig. 23. A sub-function with small σ^2

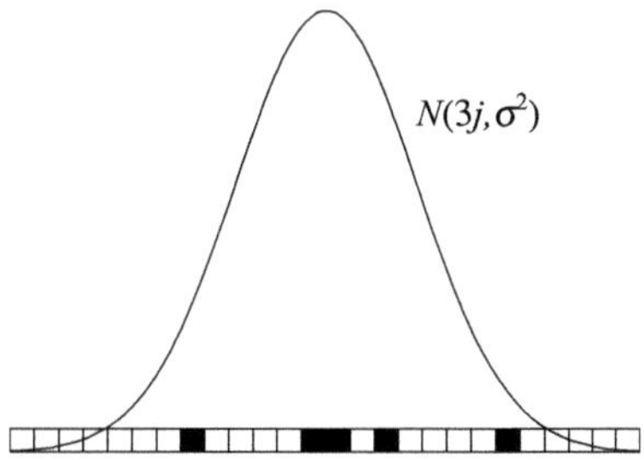

Fig. 24. A sub-function with large σ^2

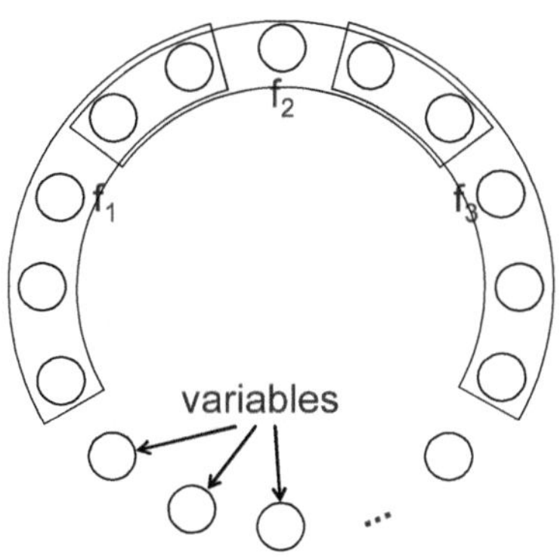

Fig. 25. A test function whose sub-functions overlap circulate

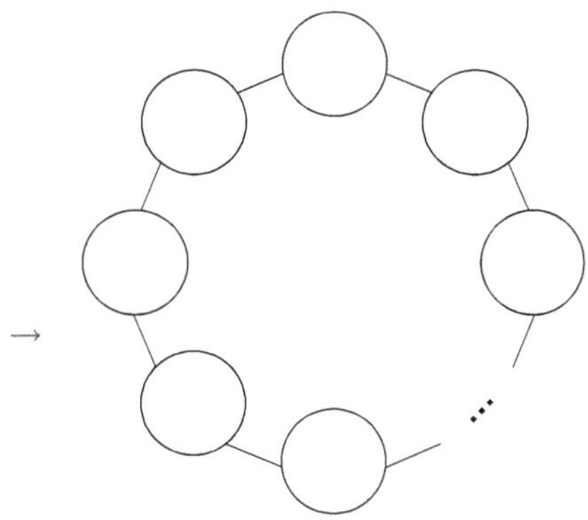

Fig. 26. Overlapping relationship in the test function shown in 25

Niching

If there are complex overlaps, it is difficult to find all V_j perfectly since the number of variables which depend on a variable becomes large. To overcome the failure in linkage identification, population diversity must be maintained. Therefore, we introduce niching to GAs for problems with overlapping sub-functions. Niching is also useful when optimal sub-solutions for overlapping sub-functions are contradictory to each other.

We employ the restricted tournament selection (RTR)[4] which compares offsprings to similar strings. For each offspring, the RTR selects a sub-population from the original population randomly. The size of the sub-population is called *window size*. Then, the offspring competes with the most similar string in the sub-population. If the fitness of the offspring is better than that of the string with the minimum Hamming distance, it replaces the string by the offspring; otherwise, the offspring is deleted.

Experiments

Test functions. A test function whose sub-functions overlap circulate and each overlapping length is 2 is employed in [20]. Fig. 25 shows the function and Fig. 26

shows the graph whose nodes and edges indicate the problem structure of the function. Each sub-function is a 5-bit trap function defined in (20) and decision variables of sub-function j are defined as:

$$\begin{aligned}\boldsymbol{v}_j = (\ & (3(j-1)+1) \mod l,\\ & (3(j-1)+2) \mod l,\\ & \cdots\\ & (3(j-1)+5) \mod l\ ),\end{aligned}$$

where $j = 1, 2, \cdots$. Then, the whole function is

$$\begin{aligned} f(\boldsymbol{s}) = & \operatorname{trap5}(s_1 s_2 s_3 s_4 s_5) + \operatorname{trap5}(s_4 s_5 s_6 s_7 s_8) \\ & + \cdots + \operatorname{trap5}(s_{l-2} s_{l-1} s_l s_1 s_2). \end{aligned} \tag{23}$$

Although the existing method could solve the function effectively, real world functions do not always have the regular structure as seen in the function.

We design a function with stochastic circularly overlapping sub-functions based on the function with circularly overlapping sub-functions. It enables us to control the complexity of overlap from systematic (circulate) one to random one. The variables which belong j-th sub-function are defined as follows:

$$\begin{aligned}\boldsymbol{v}_j = (\ & N(3j, \sigma^2) \mod l,\\ & N(3j, \sigma^2) \mod l,\\ & \cdots\\ & N(3j, \sigma^2) \mod l\ ),\end{aligned}$$

where $N(\mu, \sigma^2)$ is a normal distribution with mean μ and variance σ^2. If a sampled variable is already in $\boldsymbol{v}_j$, a new variable should be re-sampled from $N(3j, \sigma^2)$. The overlapping length can also be controlled to change the interval of μ. Fig. 23 and 24 show a sub-function with small σ^2 and one with large σ^2 respectively. The boxes in the figure show variables and painted boxes show the variables which belong the sub-function. The graphs in Fig. 27 are for the

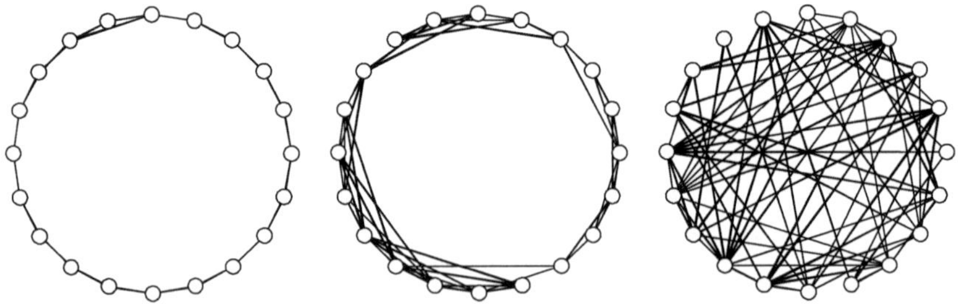

Fig. 27. Graphs representing functions with stochastic circularly overlapping sub-functions. $l = 60, \sigma^2 = 1$ (left), $l = 60, \sigma^2 = 25$ (middle) and $l = 60, \sigma^2 = 100^2$ (right). Nodes are linkage sets and edges are overlapping relations between nodes.

Table 1. The % of runs which obtain optimal solutions and the average number (#) of generations required to obtain the optimal solutions in GAs with CDC, existing method, and 1point-, 2point-, uniform-crossovers

Context Dependent Crossover								
	$\sigma^2 = 1$		$\sigma^2 = 4$		$\sigma^2 = 25$		$\sigma^2 = 100$	
l	%	#	%	#	%	#	%	#
60	100	16.2	100	16.7	100	17.9	100	21.6
90	93.3	24.6	90.0	24.3	100	25.5	96.7	45.4
120	86.6	37.9	86.7	32.2	90.0	33.3	73.3	56.0
Crossover by Yu et al.								
	$\sigma^2 = 1$		$\sigma^2 = 4$		$\sigma^2 = 25$		$\sigma^2 = 100$	
l	%	#	%	#	%	#	%	#
60	100	24.4	90.0	34.3	93.3	62.1	100	53.5
90	86.7	37.5	46.7	55.6	36.7	117.9	46.7	149.7
120	43.3	53.6	6.67	74.0	13.3	146	0.0	nan
2 Point Crossover								
	$\sigma^2 = 1$		$\sigma^2 = 4$		$\sigma^2 = 25$		$\sigma^2 = 100$	
l	%	#	%	#	%	#	%	#
60	100	39.7	100	46.7	53.3	108	6.67	168
90	100	58.4	90.0	76.6	23.3	126	0.0	nan
120	70.0	73.8	53.3	85.2	0.0	nan	0.0	nan
1 Point Crossover								
	$\sigma^2 = 1$		$\sigma^2 = 4$		$\sigma^2 = 25$		$\sigma^2 = 100$	
l	%	#	%	#	%	#	%	#
60	96.7	42.3	90.0	76.6	0.17	129	0	nan
90	46.7	68.6	43.3	76.9	0.0	nan	0.0	nan
120	36.7	84.5	3.33	91.0	0.0	nan	0.0	nan
Uniform Crossover								
	$\sigma^2 = 1$		$\sigma^2 = 4$		$\sigma^2 = 25$		$\sigma^2 = 100$	
l	%	#	%	#	%	#	%	#
60	0.0	nan	0.0	nan	0.0	nan	0.0	nan
90	0.0	nan	0.0	nan	0.0	nan	0.0	nan
120	0.0	nan	0.0	nan	0.0	nan	0.0	nan

functions with $l = 60$ and $\sigma^2 = 1, 25, 100^2$. The function with small σ^2 is similar to the original function with circularly overlapping sub-functions. Larger σ^2 gives more complex overlap relationship.

Experiments with known linkage sets. First, we perform experiments assuming that linkage sets are known to compare the CDC with the existing crossover method in terms of only crossover performance.

We employ functions with stochastic circularly overlapping sub-functions with $\sigma^2 = 1, 4, 25, 100$ and $l = 60, 90, 120$. For each test function and each crossover method, we perform 30 independent runs and record the percentage (%) of the runs which achieve the global optimal solution, and the number (#) of

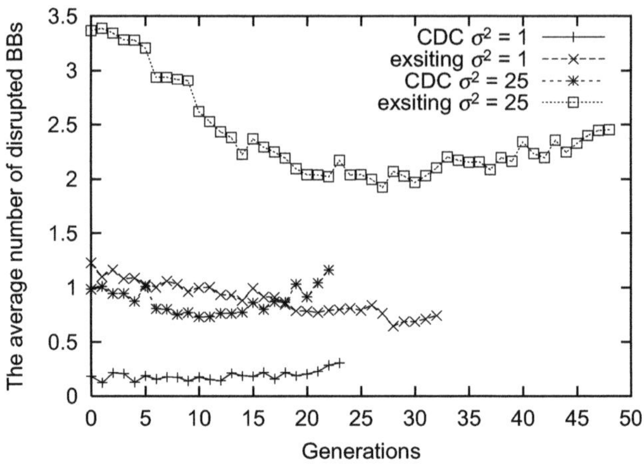

Fig. 28. The number of disrupted BBs in each generation by the CDC and existing method

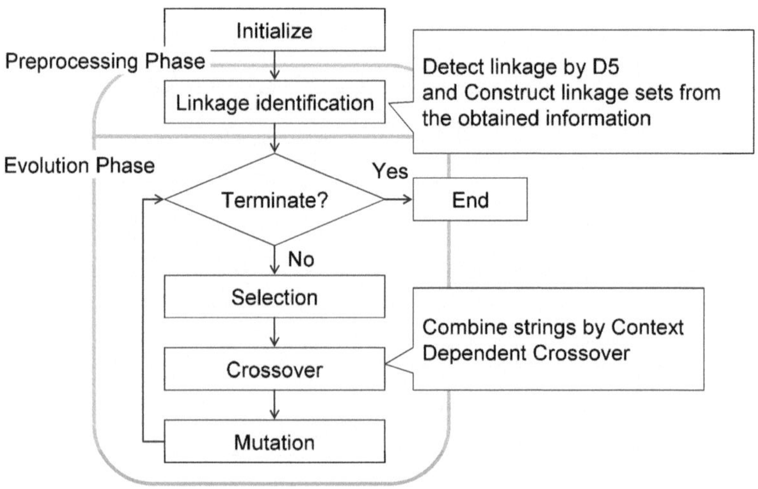

Fig. 29. The overall procedure of the D^5-GA

generations required to achieve the global optimal solution if it can be achieved. Population sizes are fixed for each σ^2 even when l is changed. Specifically, they are set to $n = 500, 600, 1200, 2000$ for $\sigma^2 = 1, 4, 25, 100$ respectively. The maximum number of generations set to 200. If there is no optimal solution at that time, the run is considered to be a failure. No mutation is employed to investigate performance of crossover. In addtion to the CDC and Yu's crossover, we perform GAs with 1-point, 2-point, and uniform crossover.

Table 1 shows the percentages of runs which obtain optimal solutions and the averages number of generations. Even for small σ^2, the CDC obtains the optimal solutions with higher probability than the existing method.

In a string of the stochastic circularly overlapping functions with relatively small σ^2, related variables are located tightly. Therefore, GAs with 1-point crossover and 2-point crossover can also obtain the optimal solutions. Especially, the GA with 2-point crossover works in a similar way to that the existing crossover for the stochastic circularly overlapping function with $\sigma^2 = 1$ whose graph constructs a ring topology, because the min-cut of its graph is approximate equivalent to the 2-point crossover for a sequence of variables. The

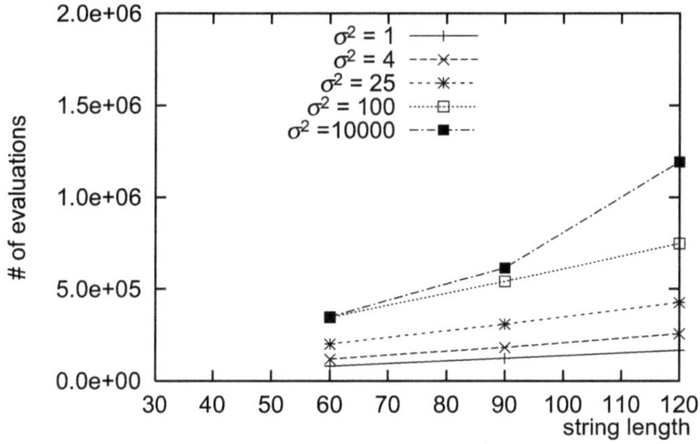

(a) D^5-GA with CDC

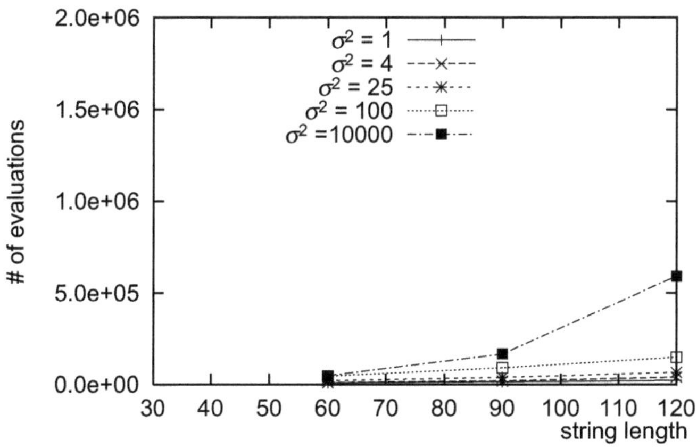

(b) D^5-GA with CDC (Evolution Phase Only)

Fig. 30. The numbers of evaluations required to obtain optimal solutions in all 30 runs

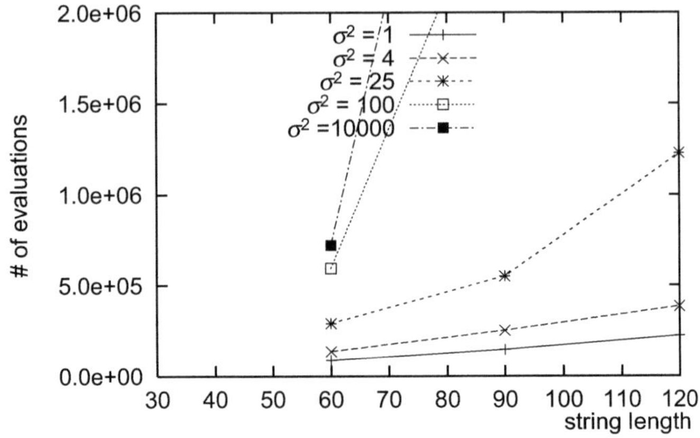

(c) D^5-GA with Existing Crossover Method

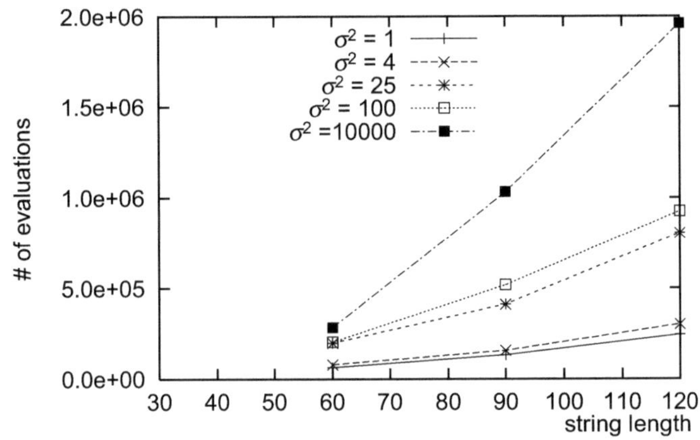

(d) BOA

Fig. 30. (*continued*)

performance of the GA with 1-point crossover is less than that of the GA with the 2-point crossover, because while the 1-point crossover exchanges $s_1 s_2 \cdots s_i$ variables where i is an arbitrary crossover site and the starting position s_1 is fixed, the 2-point crossover exchanges an arbitrary part of the ring.

For functions with larger σ^2, the GAs with 1-point and 2-point crossover cannot perform well because related variables are not allocated tightly in the functions. The GA with uniform crossover cannot find the optimal solutions for all cases, because it often disrupts building blocks due to the disregard of linkages. The GA with the CDC can find the optimal solution faster and more frequently than the other. Moreover, the difference becomes larger for larger σ^2.

Fig. 28 shows the number of disrupted BBs in each generation by the CDC and existing method for our test functions of $l = 90, \sigma^2 = 1$ and $l = 90, \sigma^2 = 25$. The interruption of the lines means that their evolutions finish. From the figure, it is clear that the existing method disrupts building blocks more frequently than the CDC.

Experiments with unknown linkage sets. The overall algorithm of the D^5-GA for problems with overlapping sub-functions is shown in Fig. 29. It can be divided into preprocessing phase and evolution phase. The preprocessing phase obtains linkage sets. The evolution phase evolves strings to find optimal solution(s) applying the CDC. The both phases use strings but the number of them could differ in each phase.

We investigate the numbers of evaluations required to obtain optimal solutions all 30 runs in the functions with stochastic circularly overlapping sub-functions with $\sigma^2 = 1, 4, 25, 100$, 10000 and $l = 60, 90, 120$. No mutation is used again. For comparison, not only the D^5-GA with the existing crossover but also BOA [16] with niching is performed. The BOA does not use crossover but builds a Bayesian network from promising strings and generates new strings from the network. The Bayesian network can also represent overlapping relations.

Fig. 30 (a)–(d) show the results. Fig. 30 (a) shows the number of evaluations in the whole phase of the D^5-GA with CDC and figure 30 (b) shows the number of evaluations in the evolution phase of the D^5-GA with CDC. From these two figures, it is shown that when BBs overlap complexly, the computation cost for the evolution phase becomes large. Comparing the (a) (c) and (d) in the figure, it is clear that the D^5-GA with CDC gives the best result for large l and large σ^2. Fig. 30 (c) shows that the D^5-GA with existing crossover requires enormous numbers of evaluations for problems with large σ^2 and l.

7 Conclusion

In this chapter, we have presented some techniques developed to realize competent GAs. Linkage identification methods using pertrubations such as LINC, LIMD, LIEM, $LIEM^2$ and D^5, which were developed to identify the linkages, have been reviewed. The crossover methods which can combine overlapping building blocks using the linkage information have also been presented.

Traditional GAs sometimes failed to process BBs due to the lack of interest in tight linkages between variables. In contrast, GAs with linkage identification presented in this chapter can optimize real-world problems even if the prior information of the target problems is not available.

References

1. Etxeberria, R., Larrañaga, P.: Global optimization with bayesian networks. In: Proceedings of the II Symposium on Artificial Intelligence CIMAF 1999, Special Session on Distributions and Evolutionary Optimization, pp. 332–339 (1999)

2. Goldberg, D.E., Deb, K., Kargupta, H., Harik, G.: Rapid accurate optimization of difficult problems using fast messy genetic algorithms. In: Forrest, S. (ed.) Proceedings of the Fifth Internatinal Conference on Genetic Algorithms, pp. 56–64. Morgan Kaufmann, San Mateo (1993)
3. Goldberg, D.E., Korb, B., Deb, K.: Messy genetic algorithms: Motivation, analysis, and first results. Complex Systems 3(5), 415–444 (1989)
4. Harik, G.R.: Finding multimodal solutions using restricted tournament selection. In: Eshelman, L. (ed.) Proceedings of the Sixth International Conference on Genetic Algorithms, pp. 24–31. Morgan Kaufmann, San Francisco (1995)
5. Heckendorn, R.B., Wright, A.H.: Efficient linkage discovery by limited probing. In: Cantú-Paz, E., Foster, J.A., Deb, K., Davis, L., Roy, R., O'Reilly, U.-M., Beyer, H.-G., Kendall, G., Wilson, S.W., Harman, M., Wegener, J., Dasgupta, D., Potter, M.A., Schultz, A., Dowsland, K.A., Jonoska, N., Miller, J., Standish, R.K. (eds.) GECCO 2003. LNCS, vol. 2723, pp. 1003–1014. Springer, Heidelberg (2003)
6. Holland, J.H.: Adaptation in Natural and Artificial Systems. University of Michigan Press (1975)
7. Kargupta, H.: The gene expression messy genetic algorithm. In: Proceedings of the IEEE International Conference on Evolutionary Computation (CEC), vol. 9, pp. 631–636 (1996)
8. Kargupta, H.: SEARCH, evolution, and the gene expression messy genetic algorithm. Technical Report LA-UR 96-60, Los Alamos National Laboratory, Los Alamos, NM (1996)
9. Kargupta, H., Park, B.-H.: Gene expression and fast construction of distributed evolutionary representation. Evolutionary Computation 9(1), 43–69 (2001)
10. Larrañaga, P., Lozano, J.A.: Estimation of Distribution Algorithms: A New Tool for Evolutionary Computation. Kluwer Academic Publishers, Dordrecht (2001)
11. Lobo, F.G., Goldberg, D.E., Pelikan, M.: Time complexity of genetic algorithms on exponentially scaled problems. In: Proceedings of the 2000 Genetic and Evolutionary Computation Conference, July 10-12, 2000, pp. 151–158. Morgan Kaufmann Publishers, San Francisco (2000)
12. Mühlenbein, H., Mahnig, T.: FDA – a scalable evolutionary algorithm for the optimization of additively decomposed functions. Evolutionary Computation 7(4), 353–376 (1999)
13. Munetomo, M.: Linkage identification based on epistasis measures to realize efficient genetic algorithms. In: Proceedings of the Congress on Evolutionary Computation - CEC 2002, pp. 445–452 (2002)
14. Munetomo, M.: Linkage identification with epistasis measure considering monotonicity conditions. In: Proceedings of the 4th Asia-Pacific Conference on Simulated Evolution and Learning (2002)
15. Munetomo, M., Goldberg, D.E.: Identifying linkage groups by nonlinearity/non-monotonicity detection. In: Proceedings of the Genetic and Evolutionary Computation Conference - GECCO 1999, vol. 7, pp. 433–440. Morgan Kaufmann Publishers, San Francisco (1999)
16. Pelikan, M., Goldberg, D.E., Cantú-Paz, E.: BOA: The Bayesian optimization algorithm. In: Proceedings of the Genetic and Evolutionary Computation Conference - GECCO 1999, pp. 525–532. Morgan Kaufmann Publishers, San Francisco (1999)
17. Pelikan, M., Goldberg, D.E., Lobo, F.G.: A survey of optimization by building and using probabilistic models. Computational Optimization and Applications 21(1), 5–20 (2002)

18. Thierens, D., Goldberg, D.E., Pereira, Â.G.: Domino convergence, drift and the temporalsalience structure of problems. In: Proceedings of the IEEE International Conference of Evolutionary Computation, pp. 535–540 (1998)
19. Tsuji, M., Munetomo, M., Akama, K.: A crossover for complex building blocks overlapping. In: Proceedings of the Genetic and Evolutionary Computation - GECCO 2006, pp. 1337–1344. ACM Press, New York (2006)
20. Yu, T.-L., Sastry, K., Goldberg, D.E.: Linkage learning, overlapping building blocks, and systematic strategy for scalable recombination. In: Proceedings of the Genetic and evolutionary computation conference, vol. 6, pp. 1217–1224 (2005)

Author Index

Araújo, Rui 219

Frigui, Hichem 51

Guterman, Hugo 77

Höppner, Frank 167

Jain, Lakhmi C. 1
Jain, Ravi 181

Klawonn, Frank 167
Koronios, Andy 181

Lerner, Boaz 77
Lim, Chee Peng 1

Mizuta, Masahiro 149
Munetomo, Masaharu 251

Nunes, Urbano 219

Oliveira, Luciano 219

Pedrycz, Witold 25
Peixoto, Paulo 219
Perfetti, Renzo 109

Ricci, Elisa 109

Sato-Ilic, Mika 195
Shimizu, Nobuo 149
Sousa, Pedro 219

Tan, Shing Chiang 1
Tsuji, Miwako 251

Windeatt, Terry 133

MIX
Papier aus verantwortungsvollen Quellen
Paper from responsible sources
FSC® C105338

If you have any concerns about our products,
you can contact us on
ProductSafety@springernature.com

In case Publisher is established outside the EU,
the EU authorized representative is:
Springer Nature Customer Service Center GmbH
Europaplatz 3, 69115 Heidelberg, Germany

Printed by Libri Plureos GmbH
in Hamburg, Germany